UNIVERSITY OF MAINE

RAYMOND H. FOGLER LIBRARY

Chemical Elements in the Environment

Springer

*Berlin
Heidelberg
New York
Barcelona
Budapest
Hong Kong
London
Milan
Paris
Santa Clara
Singapore
Tokyo*

Clemens Reimann · Patrice de Caritat

Chemical Elements in the Environment

Factsheets for the Geochemist
and Environmental Scientist

CLEMENS REIMANN
Section for Geochemistry
Hydrogeology
Geological Survey of Norway
PO Box 3006-Lade
N-7002 Trondheim
Norway

PATRICE DE CARITAT
Cooperative Research Centre
for Landscape Evolution
and Mineral Exploration
c/o Australian Geological
Survey Organisation
GPO Box 378, Canberra, ACT 2601
Australia

e-mail: chemical.elements@ngu.no

ISBN 3-540-63670-6 Springer-Verlag Berlin Heidelberg New York

Library of Congress Cataloging-in-Publication Data
Reimann, Clemens, 1952-
Chemical elements in the environment : factsheets for the geochemist and environmental scientist / Clemens Reimann, Patrice de Caritat.
p. cm.
ISBN 3-540-63670-6 (hardcover : alk. paper)
1. Environmental geochemistry. I. Caritat, Patrice de. II. Title .
QE516.4.R45 1998   97-39319
551.9--dc21   CIP

This work is subject to copyright. All rights are reserved, whether the whole or part of the material is concerned, specifically the rights of translation, reprinting, reuse of illustrations, recitation, broadcasting, reproduction on microfilms or in other way, and storage in data banks. Duplication of this publication or parts thereof is permitted only under the provisions of the German Copyright Law of September 9, 1965, in its current version, and permission for use must always be obtained from Springer-Verlag. Violations are liable for prosecution act under German Copyright Law.

© Springer-Verlag Berlin Heidelberg 1998
Printed in Germany

The use of general descriptive names, registered names, trademarks, etc. in this publication does not imply, even in the absence of a specific statement, that such names are exempt from the relevant protective laws and regulations and therefore free for general use.

Typesetting: Camera ready by author
Cover layout: de'blik, Berlin

SPIN: 1063670-6   32/3020 - 5 4 3 2 1 0 - Printed of acid-free paper

PREFACE

CHEMICAL ELEMENTS IN THE ENVIRONMENT

The idea of compiling geochemical and environmental data from existing literature probably has occurred to many of our readers. We can hardly claim any originality here! In our case, the constant frustration of searching, sometimes vainly, sometimes repeatedly, for data with which we could compare the results from a major multimedia, multielement ecogeochemistry project on the Kola peninsula, was the impetus for commencing this task. Thus, we set out to collect real data from regional, multielement studies available in existing literature. It was quite an eye-opener to realise how few such reliable studies exist, how limited in scope most are (either spatially or in terms of analytical breadth), and how haphazardly organised some previous attempts at data collation appear.

We have thus tried to collect reliable information on the 92 elements reported to occur in nature. The resulting factsheets, which form the core of this book, consist of four pages of data for each element. For each of these elements, we have started the factsheet by giving essential physico-chemical information, followed by the abundances in typical rocks and crustal materials, and an indication of which important minerals are likely to contain the element in question. Subsequently, we have listed the concentrations of each element in various geochemical or environmental sample media, ranging from rocks, to soils, plants, waters, and even to human body fluids. As explained below, we have selected carefully the sources from which we extracted the data. Finally, we close each factsheet with a discussion of the biological significance, uses, environmental pathways, mobility, action levels, production and prices of each element. We have done our utmost to reproduce these values faithfully, and we hope that our readers will inform us of any mistakes or omissions, so that possible later editions might be improved.

We would also like to mention that while preparing the manuscript for this book we met several scientists who were reluctant to contribute their data. Their unwillingness, rational as it may first appear, was generally fuelled by the suspicion that they would not be credited for their work, whereas we would. It is important to stress here that the purpose of this book is to prepare a collection of easily comparable data sets. We urge readers wanting to use these data in their own work to go back to, and cite, the original sources, or at least the contributor's name (e.g., "Manning as cited in Reimann and Caritat"). In any case, it is always strongly advisable to acquire and consult the original reference to find out more about sampling strategy, exact area covered, number of samples collected and analysed for each element, analytical details, and quality assurance and quality control procedures. The tendency to "hide" original data presents a real danger to the further development of science. In recent years, more and more papers have been published without data tables or even without data statistics. In other cases, only data for the three or four "most important" elements are given, even though very many more have obviously been analysed for. A good and easily accessible collection of existing data is fundamental to advance the understanding of the behaviour of chemical elements in the environment, to identify gaps in our knowledge and, last but not least, to generate new research ideas.

Chemical Elements in the Environment is intended as a workbook for the professional geochemist and environmental scientist, to which, we hope, they will often refer when interpreting their own data sets. While working on the manuscript of this book, we also bore in mind students in natural sciences at the university or college level who would be curious about the distribution of chemical elements on our planet - e.g. where are the main reservoirs of beryllium and rubidium? What do we know about tin in the environment? By how many orders of magnitude can the concentration of uranium in groundwater vary? The answers to these and many other questions can be found within these pages!

In addition to the libraries of our fellow scientists, we hope that this book will find a place on the shelf, or preferably in the hands, of legislators in environmental protection or management around the world. We believe that they, perhaps more than anyone else, need a book like this one to keep in touch with real data and hard facts about the environment.

We hope all readers will find this work useful. However, we cannot accept responsibility for any consequences resulting from the use of information in this book.

ACKNOWLEDGEMENTS

A book like this cannot be prepared without the help of many individuals. We would particularly like to thank the following colleagues for providing us with their data sets, for other scientific input and for helpful suggestions: David Banks (Norway), Charles Butt (Australia), Ulrich Berner (Germany), Matthias Cornelius (Australia), Rudolf Dutter (Austria), Wolfgang Engel (Germany), Colin Dunn (Canada), Bob Garrett (Canada), Phil Green (Great Britain), Jo Halleraker (Norway), Gwendy Hall (Canada), Chris Johnson (Great Britain), Kurt Konhauser (Great Britain), Bernt Markert (Germany), Joerg Matschullat (Germany), Gerry Moss (Great Britain), Heikki Niskavaara (Finland), Colin Pain (Australia), Jane Plant (Great Britain), Reijo Salminen (Finland), Tone Smith-Sivertsen (Norway), Ulrich Siewers (Germany), Brit Lisa Skjelkvåle (Norway), Ray Smith (Australia), Timo Tarvainen (Finland), and the Environmental Geochemistry students at Leoben Mining University (Austria).

In addition, at the Geological Survey of Norway, we have benefited from the library expertise of Anne Gaare and Grete Henriksen, from the data proficiency of Jan-Erik Kofoed, and from the secretarial know-how of Åse Minde.

Many others have contributed with hints, ideas or references. We express our sincere gratitude to everybody who has helped us, even if not explicitly named here.

Clemens Reimann and Patrice de Caritat
Trondheim, February 1998

CONTENTS

INTRODUCTION 1

FACTSHEET DESCRIPTION 5

GETTING MORE OUT OF THE FACTSHEETS 11

INFORMATION SOURCES 17

REFERENCES 23

FACTSHEETS 27

Ac	Actinium	28
Ag	Silver	30
Al	Aluminium	34
Ar	Argon	38
As	Arsenic	42
At	Astatine	46
Au	Gold	48
B	Boron	52
Ba	Barium	56
Be	Beryllium	60
Bi	Bismuth	64
Br	Bromine	68
C	Carbon	72
Ca	Calcium	76
Cd	Cadmium	80
Ce	Cerium	84
Cl	Chlorine	88
Co	Cobalt	92
Cr	Chromium	96
Cs	Cesium	100
Cu	Copper	104
Dy	Dysprosium	108
Er	Erbium	112
Eu	Europium	116
F	Fluorine	120
Fe	Iron	124
Fr	Francium	128
Ga	Gallium	130
Gd	Gadolinium	134
Ge	Germanium	138
H	Hydrogen	142
He	Helium	146

Hf	Hafnium	150
Hg	Mercury	154
Ho	Holmium	158
I	Iodine	162
In	Indium	166
Ir	Iridium	170
K	Potassium	174
Kr	Krypton	178
La	Lanthanum	182
Li	Lithium	186
Lu	Lutetium	190
Mg	Magnesium	194
Mn	Manganese	198
Mo	Molybdenum	202
N	Nitrogen	206
Na	Sodium	210
Nb	Niobium	214
Nd	Neodymium	218
Ne	Neon	222
Ni	Nickel	226
Np	Neptunium	230
O	Oxygen	232
Os	Osmium	236
P	Phosphorus	240
Pa	Protactinium	244
Pb	Lead	248
Pd	Palladium	252
Po	Polonium	256
Pr	Praseodymium	260
Pt	Platinum	264
Pu	Plutonium	268
Ra	Radium	270
Rb	Rubidium	274
Re	Rhenium	278
Rh	Rhodium	282
Rn	Radon	286
Ru	Ruthenium	290
S	Sulphur	294
Sb	Antimony	298
Sc	Scandium	302
Se	Selenium	306
Si	Silicon	310
Sm	Samarium	314
Sn	Tin	318
Sr	Strontium	322
Ta	Tantalum	326
Tb	Terbium	330
Te	Tellurium	334
Th	Thorium	338
Ti	Titanium	342

Tl	Thallium	346
Tm	Thulium	350
U	Uranium	354
V	Vanadium	358
W	Tungsten	362
Xe	Xenon	366
Y	Yttrium	370
Yb	Ytterbium	374
Zn	Zinc	378
Zr	Zirconium	382

APPENDIX 387

Table A1. Conversion between element and oxide masses 388
Table A2. Conversion between mg, mmol and meq 392
Table A3. Conversion between selected units 396
Table A4. Conversion between common concentration units 398

INTRODUCTION

Chemical Elements in the Environment: is a new book on this topic really needed? At a first glance one would believe that by now, after more than half a century of research into the distribution and behaviour of chemical elements in the environment and 20 years after Bowen's (1979) pioneering book focusing precisely on this topic ("Environmental Chemistry of the Elements"), we ought to know everything there is to know on the subject. After all, there are only about 90 chemical elements occurring naturally; it should not be such a major task to document their occurrence, abundance and distribution within the different compartments of the ecosystem.

When working with data from the "Kola Ecogeochemistry Project," an international multimedia, multielement geochemical mapping programme in the European Arctic (Reimann *et al.*, 1998; see also Internet site *http://www.ngu.no/Kola*), we were often in need of data with which to compare the Kola results. Many different media were collected in the different stages of the project, ranging from precipitation (rain and snow), to stream, lake and ground water, terrestrial moss and other vegetation, and soils and rocks. It was thus necessary to search beyond the borders of the geosciences when trying to interpret the results. As a general rule, we were only able to obtain data with relative ease for some ten elements. All media collected during the Kola Project were, however, analysed for more than 35 elements. The point was rapidly reached where good, relatively new, multielement data sets suited for comparison could not be located (though we are certain that excellent data sets exist but have just not reached the mainstream western publications yet). The only reference values remaining were then "world averages" for different media as originally given by Bowen (1979).

It is rarely realised that the values given in these "world average" tables do not originate from samples systematically collected and analysed by one project group over a sizeable area, and thus that they do not represent statistically relevant data sets. Usually they are based on single values, often from very few samples that may again have been taken at different places somewhere on earth. These results were then collated in retrospect and purported to represent concentrations for all the elements in the periodic table. A good example of this practice is the new North Pacific ocean water data set given in the factsheets. The author quoted as the source here (Nozaki, 1997) assembled these data from no less than 66 different original publications to be able to calculate the concentrations of 95 elements in "North Pacific ocean water." These original papers, it turns out, often give only one single value for one element, based on one or a few water samples taken somewhere in the ocean. The oldest source dates back to 1958, and not even all samples are actually from the North Pacific. The accompanying depth profiles presented in Nozaki (1997) give a good idea of how meaningful "world average" ocean water element concentrations are. In conclusion, many of the values presently used as "world average stream water, ocean water, rocks" etc. are at best a learned guess as to the range within which an element may occur in these different media. It is actually quite interesting to compare these "world average" values with median and concentration ranges derived from "real" data sets in the factsheets. Such a comparison shows that the "world averages" are reasonably reliable for some elements, while they may be up to several orders of magnitude (!) off-target for others. Awareness of these discrepancies is rather important, as many researchers tend to use these "world average" values as "natural background" levels in their work.

Another example is the very often cited "mean average crust" composition (also known as the "Clarke values"). Three different values are presented here for the bulk continental crust (upper and lower parts combined) and two different values for the upper continental crust alone. Discrepancies reflect conceptual differences among the various working groups as to the average composition of the continental crust in terms of proportions of the various rock types present. The resulting values are very interesting to the earth scientist, and the different conceptual approaches may as such be quite acceptable. These data are, however, often misinterpreted as representing the world-wide "natural background values" in environmental studies because non earth-scientists are not aware of the purely artificial nature of these data and the reasons why they were calculated.

Recent work by Reimann *et al.* (1996) has shown that the natural concentration of many elements in 150 bedrock ground-water samples taken in southern Norway varies by up to six orders of magnitude. There is every good reason to believe that similar variations in natural element concentrations occur in many other media as well. This clearly shows that data from different samples, taken at different sites by different working groups should preferably not be compiled to fill the existing gaps in our knowledge of chemical elements in the environment. What we really need are large regional data sets where as many elements as possible are analysed from the same set of samples.

We are thus confronted with the question of where one can find representative data sets to compare one's own, new results with. Science is founded upon such comparisons. As stated above, it is easy to find data sets for the major elements, or the main 10-12 heavy metals, but it becomes increasingly difficult the more elements are needed for comparison. It is surprising to realise that we do, in fact, not know very much about most "chemical elements in the environment," based on hard data and facts. Even geologists, used to working on a regional scale, do not have a clear idea of the regional distribution and variation of the elements in the different rock types. Similar gaps of knowledge exist when dealing with water, soil and plant resources. The cycling and fluxes of elements in the environment, which allow and sustain our very existence on the earth, are similarly incompletely understood. We consider the collection of regional databases on the chemistry of the various compartments of the geosphere to be a fundamental task of any Geological Survey organisation, a task that has been neglected in recent years. Plant *et al.* (1997) give an overview of existing regional data sets in western Europe, and IGCP Projects 259 and 360 (Darnley *et al.*, 1995) aim to fill existing gaps for some selected media on a world-wide scale in the years to come. We feel that there is a need for similar data bases to be collected in the field of biology to really achieve a better knowledge of "chemical elements in the environment."

Until recently, the collection of extensive, truly comparable, multielement data sets was hindered by analytical complications and costs. In the last decade, however, we have witnessed a revolution in analytical chemistry. The commercial availability of the inductively coupled plasma-mass spectrometer (ICP-MS) instrument now permits the precise and, to some extent, simultaneous determination of more than 60 elements with extremely low detection limits in environmental samples ranging from waters to soils, plants and rocks. We would like to think that this book could be used as a signpost to gaps in current data sets and as a guide to interesting research avenues in geochemistry.

Thus, the answer to the question, "Do we really need a new book on chemical elements in the environment?" can only be an emphatic "Yes, we probably need several!". At this stage, we think that it is of utmost importance to take stock of what (little) is known about element distributions in the environment, so that appropriate strategies can be adopted to address the identified weaknesses. In the future, we trust that the collection of new, regional, multielement data sets will result in a significantly improved understanding of what drives global elemental cycles and the interaction between the (mere) 92 naturally occurring elements at the surface of our planet.

FACTSHEET DESCRIPTION

In the factsheets, which form the core of the present book, we have tried to focus on the presentation of data that are of immediate interest to geochemists and environmental scientists, when trying to interpret their own analytical results. The final selection of presented data is largely based on personal experience (and frustrations) in trying to locate basic information on various elements. Much of this information is needed repeatedly during a geochemist's working day. It is, however, often necessary to scan through scores of pages in several different books and numerous publications only to discover that the information needed is in another source, useless, or non existent. Questions raised by our students, when teaching environmental geochemistry, also guided us in the selection of data. As a scientist, one is often caught up in one's own field of research, taking for granted what may not be obvious at all. For many years, students at Leoben Mining University have been collecting and presenting such data for a few selected elements as a practical exercise in their Environmental Geochemistry course. This has led to many interesting discussions on what is important information, what is not, and why the most interesting facts are so difficult to obtain.

All sources used for the data in the factsheets are detailed in the Information Sources section. The factsheet of every element presently consists of four pages. However, for those elements for which we had no actual data at all (e.g., actinium), we have left out the two empty middle pages. The following text tries to summarise some of our experiences in collecting these data and explains why certain sources were preferred over others.

Physico-chemical properties

One would suppose that at least such chemical constants as atomic mass and ionic radius are well established for all elements and that values quoted are the same everywhere. When comparing these data as cited by different sources, our surprise was considerable. Although discrepancies generally start at the second or third decimal place, there are cases where it seemed questionable whether the different data sources could really be referring to the same element!

The metallic character, element groups(s) and affinity have been included here because we have noted many misunderstandings in the literature. For example, one would imagine that the much used term "heavy metals" is well defined. This is not so. In the literature there exist different definitions, mostly based on a certain density (e.g., >4 or >4.5 g/cm^3) and the metallic character of the element. Other sources just give a list of elements. In the environmental literature, one can find examples where aluminium is referred to as one of the heavy metals although aluminium is by no means "heavy." Even more often, arsenic occurs on this list of heavy metals, although arsenic, in a chemical sense, is not a metal! We have thus characterised each element as "light metal," "heavy metal," "light non-metal," or "heavy non-metal," depending on its density (with the threshold being 4.5 g/cm^3) and its (predominantly) metallic character.

Readers interested in further information on the physico-chemical properties of all elements are referred to Emsley (1989), Lide (1996) or Winter (1997), as excellent starting points.

Concentrations in crust and rocks

The data cited as abundances in the earth's crust illustrate the disagreement between different authors on the average composition of the crust, depending on the assumed proportions of various crustal rock types. The final data are thus averages of hypothetical masses of equally hypothetical compositions of the rocks constituting the crust, and should be treated with care for reasons discussed above. Data given for the various rocks are mostly taken from a table in Koljonen (1992), which was the most complete source of which we were aware, but which is again based on data from a multitude of different sources and goes at least in parts back to Bowen (1979). These data are thus "expert guesses" rather than "real" values, but are sufficient to give a first impression of the differences in chemical composition of the main rock types. Other sources occasionally cite values for additional rock types; unfortunately, usually only for a very limited suite of elements. We did not consider these data as suited for inclusion here. Ongoing regional lithogeochemical projects, as for instance in Finland, Slovakia and Norway, will, in the years to come, produce more reliable data on median values and variation of element concentrations within and between the different rock types on a regional scale.

Minerals

The most widespread minerals of each element are cited in this book. For some elements, there are no widespread typical minerals, and the minerals quoted may be rather rare. In the case of elements that do not form their own minerals, we have indicated minerals in which the element plays an important role in the formula and/or structure. Possible host minerals for many minor elements are usually widespread rock-forming minerals (occasionally ore minerals) that can contain the element in traces (several 10s to 100s of mg/kg) or in rather large amounts (up to several wt.%). The presence of these minerals can thus give rise to naturally elevated element concentrations in environmental samples.

Mass in crust, ocean and plants

Cited masses in crust, ocean and plants are best estimates (they are again based on the "world average concentrations"). They give a relatively clear indication of where most of an element is concentrated on earth and of the size of the mass differences between the various compartments.

Results from the Kola Project

In this section, we have collated results from all media sampled during the regional phase of the Kola Ecogeochemistry Project (Reimann *et al.*, 1998). The presentation of these data within a single table allows easy comparison of concentrations in the different media, and illustrates the potential of regional, multimedia, multielement geochemical investigations in terms of interpreting sources and fate of elements. Data collected during the catchment phase of the Kola Ecogeochemistry Project (e.g., rain, snow, deposition) are given in other tables together with appropriate data from other sources.

Comparison of real data sets

Selection of data sets for presenting actual data from different media was preferentially based on the following criteria:
- samples collected from (and thus representing) a sizeable area (mostly >100,000 km^2),
- large population (several hundred samples collected),
- multielement analysis of the same sample set (more than 25 elements analysed for),
- data quality well documented,
- where applicable (e.g. soils and stream sediments), information given on digestion techniques and grain size fractions used.

Experienced readers will note that some of the data sets presented were produced using very different and often actually "incomparable" methods. As stated above, we have tried to cover the most widely used techniques and the most important information (grain size fraction and digestion technique for sediments/soils, unfiltered/filtered (filter size) for waters) is given in the table headers. While, for example, soil scientists widely use the <2 mm fraction of soil samples and an aqua regia extraction for analyses, other geoscientists will prefer finer grain size fractions (e.g., <0.18 mm or <0.063 mm) and a "total" extraction or analytical methods genuinely giving total concentrations (e.g., XRF, INAA). It should be noted that all values given for the crust and the different rock types are such total concentrations. Chemical attack or dissolution of samples is nearly always incomplete, though some extractions (e.g., ammonium acetate, aqua regia, 7N HNO_3) are typically weaker and more selective than others (e.g., HF-$HClO_4$-HCl-HNO_3). Most extraction methods will, for example, tend to significantly underestimate the concentration of elements locked in the matrix of silicate minerals. The choice of the extraction and analytical methods used often is based more on tradition than on logic. We felt that presenting results from these different techniques in an easily comparable form could stimulate reflection on the most appropriate techniques for future work and for forthcoming standardisation of methods.

When comparing analytical results from soils, it must be kept in mind that "soil" is a very heterogeneous material and that different soils (soil types or land uses) can display highly varying element levels. The term "soil" itself means different things to different professionals, some understanding it to mean organic soil, other taking it to refer to any unconsolidated sediment or weathered rock. The data presented here give an opportunity to assess the influence of all these factors on the median concentrations and range of any element. Differences in what is traditionally sampled for chemical analysis in the different sciences (the Ap horizon in soil sciences, mineral soils (B or C horizon) in the earth sciences, the O horizon in environmental studies) may be one of the reasons why different techniques have evolved and why so much confusion exists on, for instance, the question of contaminant action levels among decision-makers.

Especially for soils, there exists a multitude of other interesting data sets. One of these is the recent Chinese atlas of soil environmental background values (NEPA PRC, 1994). These data could, unfortunately, not be used here due to the report's unusual statistical treatment and presentation.

Most documented data for water are presently based on either unfiltered samples or on water samples filtered at <0.45 µm. Recent work has shown that, for trace element analysis of waters, the brand of filters used and the sampling protocol can influence the actual results (Horowitz *et al.*, 1996). Presently, the trend for collecting waters samples (especially stream and ground water) is tending toward the use of very fine filters (e.g., <0.2 or even <0.1 µm). We have again tried to include some data sets using a range of sampling protocols to facilitate comparison of possible effects in relation to analytical results.

In some cases (especially precipitation, air, yearly deposition and plants) we had to accept exceptions from the selection criteria named above as we could not locate good regional multielement data sets for certain materials. The most complete overview for some media is still given by Bowen (1979) and almost all newer sources just repeat these values. We have tried to avoid reproducing these data here, and feel that it is high time to produce some newer data sets.

The same is true of element concentrations in the human body or animal tissues. Much of the literature data are quite old and we opted for not including them here - see for example the discussion in Versieck and Cornelis (1980) on reported levels of elements in blood and serum. We have instead included one multielement data set on element concentrations in blood, serum and urine from Italy (Minoia et al., 1990).

Environmental geochemistry

The environmental geochemistry section is a compilation from a variety of sources, where we have attempted to organise and categorise the available information in a systematic, user-friendly way. Cited action levels for monitoring or remediation of contaminated water or soil are subject to change on short notice; readers are urged to consult the newest government regulations. For more detailed information and "the sources behind the sources," readers are referred to the books and papers cited in the Information Sources section. We have found yet again that much of the information presented in the different books can be traced back to Bowen (1979) and may be a little outdated. Many of the "world yearly production data" cited in literature can actually be traced back to Saager (1984) and are outdated for several elements. We have tried to include the newest statistical data wherever available and have quoted the year for which each production figure is valid. We have also indicated whether the data quoted refer to the element or an ore concentrate (and its grade).

Prices can vary widely according to type, source, quantity purchased, quality and application. Much confusion exists in the literature regarding market prices: occasionally prices are cited for a very pure form, a concentrate or even a pre-product (e.g., zirconium sand). Here, we have consistently cited two 1997 prices wherever available: one price for the form in which this element is traded in large quantities on the world market (form given), and another price for the purest form available on the market (purity given). When using the "purest form" price it should be noted that rare elements may appear cheap compared to some common elements because they may not be available in forms of comparable purity (compare zinc and zirconium). More information about prices can be found in the most recent issues of The Financial Times, The Wall Street Journal, or in magazines like Industrial Minerals.

Some readers may have desired data on element fluxes and a first attempt towards a differentiation between "geogenic" and "anthropogenic" sources. We too have followed with great interest the current literature on this topic, but consider that the data available thus far are too speculative (and often plainly wrong) to be included here. The same applies to data on the world economic reserves of the different elements - those who have argued over the Club of Rome's release on the "Limits of economic growth" in 1972 will understand why we have refrained from including such data.

Median values

We have chosen to quote, wherever available, the median values of existing data sets rather than the mean. This is because the chemical composition of natural materials seldom displays a normal distribution, at least not when a sizeable population representing a large region is considered. The median value (50th percentile) is a much more robust descriptor of non-normally distributed, strongly skewed populations, than the mean. It should be used systematically when dealing with (geo)chemical data from natural systems (Rock, 1988).

Units

There has been active debate between ourselves and with a number of colleagues as to which units we should use to present the data in this book. For rock analyses, major elements are usually quoted in weight percents (wt.%) of the respective oxide, while traces are given in parts per million (ppm = mg/kg) or, for some, in parts per billion (ppb = µg/kg) of the element itself. For environmental analyses of soils and plants, results usually are presented in mg/kg dry weight. For waters, major elements are often quoted in mg/l (assuming fresh water with a density of 1 g/cm^3, 1 mg/l = 1 ppm), while traces are presented in µg/l (= ppb) or even ng/l (= ppt). We finally decided to use only two directly comparable units for element concentrations in the different media throughout the book: mg/kg or mg/l (air and deposition being the only exceptions). This approach can lead to quite a number of "zeros" in cases of very low concentrations, or to unrealistically "precise" looking numbers (e.g., 0.000,015,5 mg/l) when the actual concentrations were reported in different units (e.g., in ng/l). Values have not been rounded off; we have trusted and preserved the original author's assessment of significant figures.

The choice of uniform units in this book (mostly mg/kg and mg/l) allows direct comparability without need for unit conversion. Despite the apparent ease of converting units, we have noted in the literature some blatant mistakes of three or more orders of magnitude, evidently caused by failures in proper conversion of units.

GETTING MORE OUT OF THE FACTSHEETS

This book can be used as a mere source of reference values for a multitude of purposes. We feel, however, that the unique combination of information and real data on the most important properties of all naturally occurring elements gives the book a much wider scope. It can, in fact, be used as a text- and working-book in teaching applied, environmental or exploration geochemistry when one tries to answer real questions from the data within the factsheets. From our own student days and from our teaching experience, we believe that it is much more effective to learn a subject by doing than by listening, and this book subscribes to that philosophy! As a starter, we have compiled a list of possible questions that can be answered from the factsheets (leading, in many cases, to further questions in the process).

How large is the natural variation in concentration of the different elements in various media?

We are used to thinking in terms of a single characteristic value, e.g., the average concentration of uranium in ground water. We prefer, often because we are too lazy to make use of the simplest statistics, to compare a measurement to such single values than to variances. But what does this single value actually mean? In fact, the natural difference between minimum and maximum concentrations can be five or more orders of magnitude for many elements and media (including uranium in ground water). This realisation should lead to greater care in applying "world average" values or citing natural background concentrations.

Which rock type contains the highest concentration of any given naturally occurring element?

This question has rather fundamental implications because local lithological highs or lows will have an influence on the element concentrations measured in all other media that interact with the rock substrate (soil, surface and ground water, plants, windblown dust). It is regrettable that there does not yet exist any really statistically reliable data set on the regional variation of bedrock geochemistry. Many resources in geosciences are allocated to the study of the deep earth, or the exploration of other planets, and too few geochemists are involved in documenting the "dirt" on which we base our existence. Background values cannot be determined without taking the local bedrock geology into consideration; a differentiation between anthropogenic and geogenic element sources is not possible without these fundamental data.

What is the order of magnitude of the mean concentration of a given element in various sample media?

From the answer to this question, we can begin to calculate the total reservoir of each element within each compartment of the ecosystem. The answer can also guide us in the choice of appropriate analytical techniques and can serve as a first check of our own analytical results. But beware: this book carries absolutely no guarantee that the oft-cited "world averages" are correct. Check them against real values from real data sets!

How does the often quoted "world average" concentration for different media compare with actual analytical data?

If the "world average" data are correct, they should usually be of approximately the same order of magnitude as real analytical data from the same medium. This is very often not the case, as many of the "world averages" are, in fact, very old, educated (but potentially wrong) guesses. Beware of one more factor: major differences can be introduced into data sets by varying choices of sample preparation method!

How about a concrete example? Is the cited "world average" concentration for scandium in stream waters realistic?

No, probably not. Most actually measured scandium concentrations in stream waters are all at least two orders of magnitude higher than the given "world average." However, if one really wishes to investigate this discrepancy, one avenue of enquiry might be to ask whether this rather reflects an analytical interference of scandium with silicon in modern ICP-MS techniques...

How do data sets on element concentrations from different countries, regions or geological settings compare?

Such discrepancies may be real or they may be ascribed to different sampling, preparation or analytical techniques. Some elements show surprisingly similar median concentrations in soils (for example), wherever in the world the samples are derived from, while others show large regional differences. We cannot provide a definitive, general answer to why this should be so; but the data in this book can be used to generate new ideas.

In which compartments of the ecosystem is an element particularly enriched or depleted?

By comparing the integrated mass for a compartment, it is possible to conclude which compartments of the ecosystem are most important for a given element. If the median concentration of an element in plants or humus is higher than in mineral soil, there is presumably some form of bioaccumulation of the element occurring. We can then start to ask what processes drive the enrichment of an element in the vegetation? If median concentrations of an element in plants or humus are generally lower than in mineral soil, but with maximum concentrations which are significantly higher, it is tempting to suggest that these maxima may be due to pollution, or maybe natural atmospheric fallout (soil dust). Comparison of data sets can thus yield information on atmospheric processes. For example, comparing the data on ocean water with surface waters, soils or mosses, it soon becomes very clear which elements will, for example, be enriched via input of sea spray at locations near the coast.

The effects of sample preparation (e.g., grain size fraction, extraction technique) on analytical results were mentioned above. How large is this influence and how do we cope with it when comparing analytical results from different laboratories?

Different branches of science, and even different laboratories within the same discipline, follow divergent "rituals" for sample preparation. Some may use the <2 mm fraction of a sediment, others may use <0.063 mm. Some may use an ammonium acetate extraction, others an aqua regia extraction, while still others may attempt to measure "total" contents *with a hydrofluoric acid digestion. It turns out that the measured concentration of different elements depends to differing degrees on preparation methods. For some elements, different sample preparation techniques do not seem to play an important role, while for others a difference in grain size fraction or extraction can lead to at least one order of magnitude discrepancy in the concentrations reported. For example, cadmium will be extracted almost fully by most acidic digestion techniques. In contrast, silicon may be locked up in mineral matrices and only appear in "total" digestion techniques or analytical methods that yield real total concentrations, such as X-ray fluorescence.*

How big is the difference between "total" and partial (e.g., aqua regia) extractions of a soil sample likely to be?

Especially for soil samples, very different extraction techniques are used. It is rarely realised by many environmental scientists that an aqua regia extraction (on which most soil action levels are based) is not a total extraction. The analytical results may be difficult to reproduce in differing laboratories as they depend on time and temperature of extraction, as well as acid concentrations and relative volumes. Additionally, the results depend strongly on the mineralogy of the sample. Looking through the real data sets given in the factsheets, it is possible to identify the elements that are very sensitive to extraction technique and for which difficulties in reproduction of analytical results may be expected. For some elements, partial and total concentrations can be similar, while for others, the partial concentration can be 10% or less of the total concentration.

How low a detection limit does one need to achieve to analyse for a given element in soils, sediments, waters or plants?

As a rule of thumb, geochemists assume that, in order to get reliable results, the detection limit of an analytical method should be about one order of magnitude below the expected concentration in the medium analysed. Generally, one should strive to reach a detection limit where as few as possible of the samples are below the detection limit in order to obtain a meaningful data set. At the same time, a very low detection limit is likely to be expensive. With some very sensitive techniques, problems with an upper limit of quantification may become apparent. The factsheets will help to determine the limit of detection for which one should aim. They can be a guide for the choice of analytical technique as well.

How can one compare one's own analytical results with those of other working groups?

The factsheets give the possibility for a rapid first check of whether new results seem to be of a correct order of magnitude. But beware: other working groups' values can be wrong as well! And be aware that discrepancies between new results and the results within these pages may not be due to analytical mistakes. They may be due to natural variations, or to pollution. In short, discrepancies may be telling something highly interesting. The exception proves the rule. Additionally, the factsheets may provide one with more suitable data to compare one's results to than just the "world average."

What is the world production of each element?

This question addresses not only the importance of the various elements to mankind, but also has some relevance to calculations of how much may be released to the environment by human activities. We have found that comparing data on human usage of elements with published data on total element fluxes can be a real eye-opener. In today's literature, the importance of geogenic fluxes for many elements is severely underestimated and the importance of anthropogenic fluxes overstated - often to the extreme, as can be seen when comparing with production data and keeping in mind that the elements are produced to do other things with them than just to contaminate the environment!

Can natural fluxes of toxic elements such as lead or mercury really be larger than anthropogenic fluxes?

Many readers may consider car exhausts to be the main source of lead to the environment, but have you ever considered the effect of forest fires? Lead fluctuations in time series from ice or sediment cores are a controversial topic. Where does the lead come from and what causes the variations that we see in the cores? Human activities? Mining and smelting during the last 2000 years? Take a look at the data on lead in the ash of spruce bark and try to imagine what kind of influence a climate and vegetation change, or a major forest fire, might have on the regional distribution of lead, including that of lead in the atmosphere. It is also interesting to note that large natural fluxes, e.g., exhalations from volcanoes, are easily accepted for very valuable elements like gold but not for many heavy metals!

How toxic are all these elements? Can they get into our bodies? Can we use them for anything?

These types of question raise the issue of whether or not a detected high element concentration is really of concern, either because it poses a risk to human health or because of its potential economic value. The existence of a risk presupposes the existence of a source for a toxic species, an environmental pathway and a receptor to be poisoned. It is surprising to note how little we generally know about toxicology, plant uptake (what species, especially what vegetables, take up, or are even enriched, in what element, and in which of their parts?), and speciation of these mere 92 elements! Descriptions of possible sources and pathways for each element are found in the Environmental Geochemistry section of each factsheet, together with their industrial uses and values.

Which is the most valuable of the elements?

You may think there is a straightforward answer to at least this query. It's probably one of the platinum group elements! Have a look at our data tables, there are many surprises waiting! The value of any element is an obscure concept, its price being determined by several things: rarity, whether it can be used for anything, quantity bought, whether it forms commercial deposits and its purity. In fact, the platinum group elements and gold are a real bargain: the highest elemental price cited in these pages is for protactinium, where the production cost of a mere 125 g in 1959-61 was US$ 500,000 (giving a kg production price of US$ 4,000,000!) followed by neptunium at US$ 660,000 per kg and then, surprisingly, boron (if bought as crystalline powder of high purity in g quantities), scandium (due to the fact that no commercial mineral deposits are known), and lutetium! There are even a few elements that are so rare that it would literally "cost you the earth" to provide even small quantities of them (e.g., astatine with <30 g in crust, and francium with <50 g in crust). The data can also be used to estimate the total value of any given element in the crust or in one of the compartments of the ecosystem.

For which elements have action levels been set by environmental authorities, and for which media?

The term action level is a somewhat ambiguous one. Some organisations set different action levels corresponding to (for example) the need for further investigation and the need for remediation or treatment. Although the term may be ambiguous, we have identified, in each individual case, to which regulatory limit the numbers are referring. Action levels in any form have, however, been set for only very few elements. Often these levels are selected more for reasons of ease of analysis of a particular element than its actual health hazard. For several elements (for example thorium), no action levels are yet defined, despite their potential toxicity.

And presumably the same applies to drinking water limits in different countries?

Of course. Drinking water limits are a type of action level. Substantial differences occur, however, from country to country. In fact, it is rather interesting to find out which countries have set action levels for which elements and how these values compare with real data reported in the factsheets. The results can be unexpected: Russia operates with a drinking water limit of 1.7 mg/l for uranium while Canada uses 0.02 mg/l. The maximum uranium concentration in drinking water reported in the factsheets is 2 mg/l, well above both action levels.

Beryllium is another interesting example: This element is rather toxic and yet there is no drinking water standard set for it either in the European Union or in Norway. Limits do exist elsewhere, but their toxicological basis is not easy to understand. In fact, a limit of 0.004 mg/l applies for beryllium in the USA, compared with the much more stringent (in contrast to uranium, see above) value of 0.0002 mg/l in Russia (how do they analyse to such low levels?). In these pages it is also possible to find out which industries might be emitting high beryllium concentrations, which geological settings could cause high natural beryllium concentrations in ground water and whether beryllium concentrations above the US drinking water limit might actually occur in "real life" drinking water samples (yes, see maximum concentration in Norwegian groundwater).

Sometimes, discrepancies in drinking water limits may reflect differences in water sources, management and use. Often, the setting of action levels is more guided by pragmatism than toxicology! For example, what does one do in a country where plumbing consists largely of lead (or copper) pipes? Can the solution be to set the drinking water limit high to ensure compliance (but note the US EPA's suggestion on lead in drinking water)?

Do we now know everything there is to know about the elements? Are there obvious gaps in our knowledge of element concentrations and behaviour in different media?

It is interesting to note how few elements are generally analysed for in environmental samples. We have had a lot of troubles locating data sets with more than 20 elements analysed in a satisfactorily large sample set from a large enough area to give at least some kind of statistical significance. For many elements there are obvious gaps in our knowledge that become clearly visible when data is collated systematically. We feel that the factsheets in this book can be directly used to identify many important research needs and possibilities in environmental sciences!

But maybe the reason that often only those 10-20 elements are analysed is that these are the only elements having importance for human health/environmental questions?

Unfortunately, it often seems to be ease and low cost of analysis (or available equipment) rather than human health concerns that govern the choice of elements analysed - see factsheets! Why, for example, is iodine so seldom analysed? Or tin? Or beryllium? Or thallium? Iodine determination requires a single-element analytical technique, and is therefore expensive. The same applies to tin, which requires a special extraction. Beryllium and thallium need rather low detection limits to get useful data.

We feel that the real value of this book does not lie only in providing a fast and user-friendly source of reference values, but especially in tempting the readers to ask (and try to answer) their own questions by actually using the factsheets. By doing so, the *user* will learn a lot about environmental geochemistry. In fact, we ourselves feel that we have learned much just by compiling these data, examining and comparing them. We don't want to hide the fact we had a lot of frustrations. But we also had a lot of fun discussing crazy geochemical ideas.

The book can thus be used for many different purposes: researchers may consult it to rapidly locate a reference value with which to compare their own results. They may want to identify exiting research possibilities by finding obvious gaps in our present knowledge. Regulators in environmental agencies may want to get a feel for some real values before setting action levels which have expensive economic implications. University and high school teachers may want to use the book in teaching environmental geochemistry or as a source of ideas for practical work with their students.

INFORMATION SOURCES

In preparing the elemental factsheets, a variety of information sources were used. The following sources were used for the various entries (in cases of several sources, the main source is cited first; if no main source can be identified, the sources are given alphabetically):

Physico-Chemical Properties
Atomic number: Lide (1996)
Atomic mass: IUPAC (1996)
Atomic radius: James and Lord (1992)
Main oxidation state(s): Lide (1996), Periodic Table of the Elements, Hollemann and Wiberg (1995), Streit (1994)
Ionic radius: James and Lord (1992), Streit (1994)
Electronegativity: Periodic Table of the Elements
Density: Lide (1996)
Melting/boiling point: Lide (1996)
Isotopes and isomers: Lide (1996)
Acid/base of oxide: Periodic Table of the Elements
State (at 300 K, 1 atm.): Periodic Table of the Elements
Metallic character: Periodic Table of the Elements; pred. = predominantly
Element group(s): Periodic Table of the Elements; Definition of heavy metal as (predominantly) metal with density >4.5 g/cm^3;
  Abbreviations: PGE = platinum group element(s), REE = rare earth element(s)
Affinity: Brownlow (1979), Mason and Moore (1982)
Data on isotopes: Lide (1996)

Concentrations in Crust/Rocks
Bulk continental crust: Wedepohl (1995) / Lide (1996) / Taylor and McLennan (1995)
Upper continental crust: Wedepohl (1995) / Taylor and McLennan (1995)
Ultramafic rock: Koljonen (1992)
Ocean ridge basalt: Koljonen (1992)
Gabbro, basalt: Koljonen (1992)
Granite, granodiorite: Koljonen (1992)
Sandstone: Koljonen (1992)
Greywacke: Wedepohl (1995)
Shale, schist: Koljonen (1992)
Limestone: Koljonen (1992)
Coal: Tauber (1988)

Typical Minerals: Wedepohl (1978), Hollemann and Wiberg (1995)
Possible Host Minerals: Wedepohl (1978), Hollemann and Wiberg (1995)

Mass in Continental Crust: calculation based on concentration given above for the bulk continental crust (from Wedepohl, 1995, or if missing, Lide, 1996) and mass of crust (2.36×10^{22} kg, Lide, 1996)
Mass in Oceans: The Open University (1989)
Mass in Plants: Markert (1992)

Concentrations in Media Sampled for the Kola Project
Moss: Reimann *et al.* (in prep.). N = 598. Area: 188,000 km^2
Humus: Reimann *et al.* (in prep.). N = 617. Area: 188,000 km^2
Topsoil: Reimann *et al.* (in prep.). N = 607. Area: 188,000 km^2
B horizon: Reimann *et al.* (in prep.). N = 609. Area: 188,000 km^2
C horizon: Reimann *et al.* (in prep.). N = 605. Area: 188,000 km^2
Lake water: Unpublished data. N = 120. Area: ca. 90,000 km^2

Concentrations in Soils and Sediments
Soil, World: Koljonen (1992). Compiled data
Agricultural soil, Canada: Garrett (1994, 1997). N = 1273. Area: 850,000 km^2
Agricultural soil, top (0-25 cm), Finland: Tarvainen (1997). N = 95. Area: ca. 250,000 km^2
Agricultural soil, bottom (50-75 cm), Finland: Tarvainen (1997). N = 95. Area: ca. 250,000 km^2
Topsoil (0-15 cm), England and Wales: McGrath and Loveland (1992). N = 5692 (differs slightly from element to element). Area: ca. 35,000 km^2
Urban soil, Trondheim: Ottesen *et al.* (1995). N = 314. Area: ca. 80 km^2
Forest soil, humus, Norway: Njåstad *et al.* (1994). N = 580. Area: 323,886 km^2
Forest soil, B horizon, Norway: Njåstad *et al.* (1994). N = 507. Area: 323,886 km^2
Forest soil, C horizon, Norway: Njåstad *et al.* (1994). N = 512. Area: 323,886 km^2
Till, fine, aqua regia, Finland: Koljonen (1992). N = 1057. Area: 337,032 km^2
Till, fine, total, Finland: Koljonen (1992). N = 1057. Area: 337,032 km^2
Laterite, Australia: Smith *et al.* (1992). N = 1072 to 2434 (depending on element). Area: ca. 500,000 km^2
Stream sediments, Austria: Thalmann *et al.* (1988). N = 29,717. Area: ca. 40,000 km^2
Stream sediments, southern Scotland: BGS (1993). N = 19,133 (differs slightly from element to element). Area: ca. 30,000 km^2
Stream sediment, Harz, Germany: Roostai (1997). N = 382. Area: 500 km^2
Overbank sediments, total, Norway: Ottesen *et al.* (in press). N = 688. Area: 323,886 km^2
Overbank sediments, conc. HNO$_3$, Norway: Ottesen *et al.* (in press). N = 688. Area: 323,886 km^2
Organic stream sediments, Finland: Lahermo *et al.* (1996). N = 1050 to 1166 (depending on element). Area: 337,032 km^2

Concentrations in Waters
Ocean water, world (1): Lide (1996). Compiled data
Ocean water, world (2): The Open University (1989). Compiled data
Ocean water, North Pacific: Nozaki (1997). Compiled data
Stream water, world: Koljonen (1992). Compiled data

Stream water, Nova Scotia, Canada: Hall et al. (1994), Hall and Pelchat (1995), Hall (1997). N = 513. Area: 5300 km^2

Stream water, Finland: Lahermo et al. (1996). N = 1122 to 1168 (depending on element; for Se: N = 207). Area: 337,032 km^2

Stream water, Romania: Siewers (1997). N = 113. Area: 237,500 km^2

Stream water, eastern India: Konhauser et al. (1997). N = 31. Catchment area: ca. 200,000 km^2

Stream water, Harz, Germany: Roostai (1997). N = 169 to 1023 (depending on element). Area: 500 km^2

Lake water, Norway: Skjelkvåle et al. (1996). N = 475. Area: 323,886 km^2

Ground water, Norway: Reimann et al. (1996). N = 150. Area: ca. 10,000 km^2 around Bergen, plus ca. 1500 km^2 around Oslo

Concentrations in Precipitation

Rain water, Kola, remote: Reimann et al. (1997). N = 15. Composite samples from 5 collectors placed at three stations distributed over an area of 20 km^2 and collected monthly during summer 1994

Rain water, Kola, polluted: Reimann et al. (1997). N = 17. Composite samples from 5 collectors placed at three stations distributed over an area of 24 km^2 and collected monthly during summer 1994

Snow meltwater, Kola, remote: Caritat et al. (1998). N = 7. Composite samples from snow cores collected at 7 stations distributed over an area of 20 km^2 and collected in spring 1994

Snow meltwater, Kola, polluted: Caritat et al. (1998). N = 10. Composite samples from snow cores collected at 10 stations distributed over an area of 24 km^2 and collected in spring 1994

Snow filter residue, Kola, remote: Caritat et al. (1998). N = 7. Composite samples from snow cores collected at 7 stations distributed over an area of 20 km^2 and collected in spring 1994

Snow filter residue, Kola, polluted: Caritat et al. (1998). N = 10. Composite samples from snow cores collected at 10 stations distributed over an area of 24 km^2 and collected in spring 1994

Rain water, Coastal, Norway: Berg et al. (1994). N = 98. Station Kårvatn, collected weekly 1989-1990

Rain water, Inland, Norway: Berg et al. (1994). N = 93. Station Osen, collected weekly 1989-1990

Concentrations in Air and Yearly Deposition

Air, world, remote: (Matschullat, 1995 - data often based on other sources, e.g., Bowen, 1979). Compiled data

Air, world, polluted: (Matschullat, 1995 - data often based on other sources, e.g., Bowen, 1979). Compiled data

Bulk deposition, West Germany: (Matschullat, 1995, compiled from Führer et al., 1988). Samples collected for a minimum of 1 year from over 100 stations. Area: 248,000 km^2

Throughfall deposition, West Germany: (Matschullat, 1995, compiled from Führer et al., 1988). Samples collected for a minimum of 1 year from over 100 stations. Area: 248,000 km^2

Bulk deposition, Kola, remote: Chekushin et al. (in prep.). Calculated from 1 year monitoring of precipitation chemistry in a 20 km^2 area

Bulk deposition, Kola, polluted: Chekushin et al. (in prep.). Calculated from 1 year monitoring of precipitation chemistry in a 24 km^2 area

Concentrations in Plants

Moss, Norway, 1990: Steinnes et al. (1992, 1993). N = 514. Area: 323,886 km^2

Moss, Germany, 1995: Siewers and Herpin (in prep.). N = 976. Area: 356,845 km^2

Lichen, Northwest Territories, Canada: Puckett and Finegan (1980). N = 20 to 36 (depending on element). *Cetraria cucullata* sampled from 36 sites spread all over the Northwest Territories (Area: 3,400,000 km^2)

Crustose lichen, Germany: Matschullat *et al.* (in prep.). N = 42. *Lecanora muralis* sampled from 1 east-west and 2 north-south transects across Germany (Area: 356,854 km^2)

Dandelion, Europe: Djingova and Kuleff (1993). Very few samples collected from 5 different background regions in Europe

Spruce bark, Canada: Dunn *et al.* (1991, 1992). N = 244. Area: 4000 km^2

Concentrations in Body Fluids

Human blood, Italy: Minoia *et al.*, 1990. N = 5 (Ga) to 959 (Pb). Lombardy region (Area: 24,000 km^2)

Human urine, Italy: Minoia *et al.*, 1990. N = 6 (Yb) to 879 (Cr). Lombardy region (Area: 24,000 km^2)

Human serum, Italy: Minoia *et al.*, 1990. N = 8 (Ir) to 916 (Al). Lombardy region (Area: 24,000 km^2)

Environmental Geochemistry

Biological impacts: Adriano (1986), Hirano and Suzuki (1996), Merian (1991), Streit (1994), US EPA Factsheets (1997)

Uses: Adriano (1986), Hirano and Suzuki (1996), Hollemann and Wiberg (1995), Markert (1992), Merian (1991), Skillen and Griffiths (1993), Streit (1994), US EPA Factsheets (1997)

Environmental pathways: own estimation, partly based on above sources on Uses, US EPA Factsheets (1997)

Environmental mobility: modified after Levinson (1980), US EPA factsheets (1997)

Geochemical barriers: modified after Levinson (1980)

Natural association: modified after Levinson (1980)

Action levels: Drinking water, Norway: Sosial- og helsedepartementet (Shd) (1995); Drinking water, US Environmental Protection Agency (US EPA): Fetter (1994); Drinking water, Canada: Canadian Water Quality Guidelines (CWQG) in Leeden *et al.* (1990);
Drinking water, Russia: Ministry of Health (MoH) in Kirjuhin *et al.* (1993);
Drinking water, world: World Health Organisation (WHO) in WHO (1996);
Water guidelines, The Netherlands: VROM (1994)
Soil guidelines, The Netherlands: VROM (1994); for a "standard" soil (10% organic material and 25% fine fraction)
Soil, Germany: Rosenkranz *et al.* (1988)
Abbreviations: MAC = Maximum Allowable Concentration

Remarks: Adriano (1986), Beissler *et al.* (1997), Bowen (1979), Hirano and Suzuki (1996), Hollemann and Wiberg (1995), Klöppel *et al.* (1997), Levinson (1980), Merian (1991), Skillen and Griffiths (1993), Streit (1994), US EPA Factsheets (1997), Wedepohl (1978)

Suggested analytical method(s): own experience, techniques selected after: a) ability to reach the necessary detection limit to obtain meaningful data (note that for some elements, ICP-AES is given for geological materials and ICP-MS for waters), b) multielement capability, and c) lowest market price of analysis;
Abbreviations: AAS = atomic absorption spectroscopy, CV-AAS = cold vapour-atomic

absorption spectroscopy, GF-AAS = graphite furnace-atomic absorption spectroscopy, HPGe detector = high-purity germanium detector, IC = ion chromatography, ICP-AES = inductively coupled plasma-atomic emission spectrometry, ICP-MS = inductively coupled plasma-mass spectrometry, INAA = instrumental neutron activation analysis, IR = infrared, XRF = X-ray fluorescence

World Yearly Production: 1995-data from Berner (1997); all other data from Lide (1996), Markert (1992), or Skillen and Griffiths (1993)

Price of Purest Form, in Small Quantities: Alfa (1997-1998)

Market Price: Mining Journal (1997), Industrial Minerals (1997)

REFERENCES

Adriano, D.C., 1986. *Trace Elements in the Terrestrial Environment*. Springer-Verlag, New York, USA, 533 pp.

Alfa, 1997-1998. *Research Chemical and Metals*. Johnson Matthey GmbH, Karlsruhe, Germany, 744 pp.

Beissler, H., Bächmann, K., Raes, F., Petrucci, G.A. and Omenetto, N., 1997. Applicability of gold as an atmospheric aerosol tracer. *Atmospheric Environment*, **31**: 2329-2336.

Berg, T., Røyset, O. and Steinnes, E., 1994. Trace elements in atmospheric precipitation at Norwegian background stations (1989-1990) measured by ICP-MS. *Atmospheric Environment*, **28**: 3519-3536.

Berner, U., 1997. Pers. comm. *Mineral Resources Database*. Bundesanstalt für Geowissenschaften und Rohstoffe, Hannover, Germany.

BGS, 1993. *Regional Geochemistry of Southern Scotland and Part of Northern England*. Keyworth, Nottingham, UK, 96 pp.

Bowen, H.J.M., 1979. *Environmental Chemistry of the Elements*. Academic Press, London, UK, 333 pp.

Brownlow, A.H., 1979. *Geochemistry*. Prentice Hall, Inc., Englewood Cliffs, USA, 498 pp.

Caritat, P. de, Äyräs, M., Niskavaara, H., Chekushin, V., Bogatyrev, I. and Reimann, C., 1998. Snow composition in eight catchments in the central Barents Euro-Arctic region. *Atmospheric Environment*, **32**: in press.

Chekushin, V.A., Bogatyrev, I.V., Caritat, P. de, Niskavaara, H. and Reimann, C., in prep. Annual atmospheric depositon of 16 elements in eight catchments of the central Barents region. Submitted.

Darnley, A., Björklund, A., Bølviken, B., Gustavson, N., Koval, P.V., Plant, J.A., Steenfelt, A., Tauchid, M. and Xuejing, X., 1995. *A Global Geochemical Database for Environmental and Resource Management. Recommendations for International Geochemical Mapping*. Final Report of IGCP Project 259, Unesco Publishing, Paris, France, 122 pp.

Djingova, R. and Kuleff, I., 1993. Monitoring of heavy metal pollution by *Taraxum officinale*. Chapter 20 in: Markert, B. (ed.): *Plants as Biomonitors*. Springer-Verlag, New York, USA, p. 435-459.

Dunn, C.E., Adcock, S.W. and Spirito, W.A., 1992. Reconnaissance biogeochemical survey, southeastern Cape Breton Island, Nova Scotia: Part 1-Black Spruce Bark. Geological Survey of Canada Open File 2558, Ottawa, Canada.

Dunn, C.E., Coker, W.B. and Rogers, P.J., 1991. Reconnaissance and detailed geochemical surveys for gold in eastern Nova Scotia using plants, lake sediment, soil and till. *Journal of Geochemical Exploration*, **40**: 143-163.

Emsley, J., 1989. *The Elements*. Second edition 1991, reprinted 1995. Oxford University Press, Oxford, UK, 250 pp.

Fetter, C.W., 1994. *Applied Hydrogeology*. Third edition. Macmillan, New York, USA, 691 pp.

Führer, H.W., Brechtel, H.M., Ernstberger, H. and Erpenbeck, C., 1988. *Ergebnisse von neuen Depositionsmessungen in der Bundesrepublik Deutschland und im benachbarten Ausland*. DVWK Mitteilungen 14, 122 pp. (in German).

Garrett, R.G., 1994. The distribution of cadmium in A horizon soils in the prairies of Canada and adjoining United States. In: *Current Research 1994-B*, Geological Survey of Canada, Ottawa, Canada, p. 73-82.

Garrett, R.G., 1997. Statistics for the prairies soil data set. Pers. comm.

Hall, G.E.M. and Pelchat, P., 1995. Hydrogeochemical survey in soutwest Cape Breton, 1995. In: *Mines and Energy Branches Report 95-2*, Nova Scotia Dept. of Natural Resources, Halifax, Canada, 50 pp.

Hall, G.E.M., 1997. Unpubl. results from stream water survey of Nova Scotia. Pers. comm.

Hall, G.E.M., Pelchat, P. and Balma, R.G., 1994. Hydrogeochemical exploration in central Nova Scotia. In: *Mines and Energy Branches Report 94-2*, Nova Scotia Dept. of Natural Resources, Halifax, Canada, 40 pp.

Hirano, S. and Suzuki, K.T., 1996. Exposure, metabolism, and toxicity of rare earths and related compounds. *Environmental Health Perspectives*, **104, Suppl. 1:** 85-95.

Hollemann, A.F. and Wiberg, E., 1995. *Hollemann-Wiberg Lehrbuch der anorganischen Chemie*. 101. Auflage. Walter de Gruyter and Co., Berlin, Germany, 2036 pp. (in German).

Horowitz, A.J., Lum, K.R., Garbarino, J.R., Hall, G.E.M., Lemieux, C. and Demas, C.R., 1996. Problems associated with using filtration to define dissolved trace element concentrations in natural water samples. *Environmental Science and Technology*, 30: 954-963.

Industrial Minerals, 1997. Metal Bulletin plc, London, July, 1997, 358: 70-71.

IUPAC, 1996. Atomic weights of the elements 1995. International Union of Pure and Applied Chemistry (IUPAC). *Pure and Applied Chemistry*, **68:** 2339-2359. See also *http://www.chem.qmw.ac.uk/iupac2/AtWt*.

James, A.M. and Lord, M.P., 1992. *Macmillan's Chemical and Physical Data*. Macmillan, London, UK, 565 pp.

Kirjuhin, V.A., Korotkov, A.N. and Shvarksev, C.L., 1993. *Gidrogeohimija* (Hydrogeochemistry). Nedra Publications, Moscow, Russia, 383 pp. (in Russian).

Klöppel, H., Fliedner, A. and Kördel, W., 1997. Behaviour and ecotoxicology of aluminium in soil and water - Review of the scientific literature. *Chemosphere*, 35: 353-363.

Koljonen, T. (ed.), 1992. *Geochemical Atlas of Finland, Part 2: Till*. Geological Survey of Finland, Espoo, Finland, 218 pp.

Konhauser, K.O., Powell, M.A., Fyfe, W.S., Longstaffe, F.J. and Tripathy, S., 1997. Trace element chemistry of major rivers in Orissa State, India. *Environmental Geology*, **29:** 132-141.

Lahermo, P., Väänänen, P., Tarvainen, T. and Salminen, R., 1996. *Geochemical Atlas of Finland, Part 3: Environmental Geochemistry, Stream Waters and Sediments*. Geological Survey of Finland, Espoo, Finland, 150 pp.

Leeden, van der, F., Troise, F.L. and Todd, D.K., 1990. *The Water Encyclopedia*. Second edition. Lewis Publishers, Chelsea, USA, 808 pp.

Levinson, A.A., 1980. *Introduction to Exploration Geochemistry*. Second edition. Applied Publishing Ltd., Wilmette, USA, 924 pp.

Lide, D.R. (editor-in-chief), 1996. *CRC Handbook of Chemistry and Physics*. 77th edition, 1996-1997. CRC Press, Boca Raton, USA.

Markert, B., 1992. Presence and significance of naturally occurring chemical elements of the periodic system in the plant organism and consequences for future investigations on inorganic environmental chemistry in ecosystems. *Vegetatio*, **103:** 1-30.

Mason, B.H. and Moore, C.B., 1982. *Principles of Geochemistry*. Fourth edition, Wiley, New York, USA, 344 pp.

Matschullat, J., 1995. Geochemische Flüsse in anthropogen beeinflussten Mitelgebirgen. *Heidelberger Beiträge zur Umwelt-Geochemie*, Bd. 1, Ruprecht Karls Universität, Heidelberg, Germany. p. 1-174. (in German).

Matschullat, J., Scharnweber, T., Garbe-Schönberg, C.D., Walther, A. and Wirth, V., in prep. Crustose lichens-monitors of atmospheric deposition of trace elements and organohalogens?

McGrath, S.P. and Loveland, P.J.,1992. *The Soil Geochemical Atlas of England and Wales*. Blackie Academic and Professional, London, UK, 101 pp.

Merian, E. (ed.), 1991. *Metals and Their Compounds in the Environment. Occurence, Analysis, and Biological Relevance*. VCH Verlagsgesellschaft, Weinheim, Germany, 1438 pp.

Mining Journal, 1997. London, January 10, 1997.

Minoia, C., Sabbioni, E., Apostoli, P., Pietra, R., Pozzoli, L., Gallorini, M., Nicolaou, G., Alessio, L. and Capodaglio, E., 1990. Trace element reference values in tissues from inhabitants of the European Community I. A study of 46 elements in urine, blood and serum of italian subjects. *The Science of the Total Environment*, **95:** 89-105.

NEPA PRC, 1994. National Environmental Protection Agency of the People's Republic of China. *The Atlas of Soil Environmental Background Value in the People's Republic of China.* China Environmental Science Press, Beijing, 196 pp.

Njåstad, O., Steinnes, E., Bølviken, B. and Ødegård, M., 1994. Landsomfattende kartlegging av elementsammensetning i naturlig jord: Resultater fra prøver innsamlet i 1977 og 1985 oppnåd ved ICP emisjonsspektrometri. Norges geologiske undersøkelse rapport nr. 94.027, Trondheim, Norway. 114 pp. (in Norwegian).

Nozaki, Y., 1997. A fresh look at element distribution in the North Pacific Ocean. *EOS*, May 27, 1997, p. 221. See also *http://www.agu.org/eos_elecas97025e.html*.

Ottesen, R.T., Almklov, P.G. and Tijhuis, L., 1995. Innhold av tungmetaller og organiske miljøgifter i overflatejord fra Trondheim. Datarapport. Rapport nr. TM 95/06. Trondheim kommune, Miljøavdelingen, Trondheim, Norway, 130 pp. (in Norwegian).

Ottesen, R.T., Bogen, J., Bølviken, B. and Volden, T., in press. Geokjemisk atlas for Norge. Norges geologiske undersøkelse, Trondheim, Norway. (in Norwegian).

Periodic Table of the Elements, Sargent-Welch Scientific Company, 1979. Catalog No. S-18805-50. See also: Periodic Table of the Elements, Editions A. De Boeck, Brussels, 1982.

Plant, J.A., Klaver, G., Locutura, J., Salminen, R., Vrana, K. and Fordyce, F.M., 1997. The forum of European Geological Surveys geochemistry task group inventory 1994-1996. *Journal of Geochemical Exploration*, 59: 123-146.

Reimann, C., Äyräs, M., Chekushin, V.A., Bogatyrev, I., Boyd, R., Caritat, P. de, Dutter, R., Finne, T.E., Halleraker, J.H., Jæger, Ø., Kashulina, G., Niskavaara, H., Lehto, O., Pavlov, V., Räisänen, M. L., Strand, T. and Volden, T., 1998. *Environmental Geochemical Atlas of the Central Barents Region.* NGU-GTK-CKE special publication, Geological Survey of Norway, Trondheim, Norway, 731 pp.

Reimann, C., Caritat, P. de, Halleraker, J.H., Volden, T., Äyräs, M., Niskavaara, H., Chekushin, V.A. and Pavlov, V.A., 1997. Rainwater composition in eight arctic catchments in northern Europe (Finland, Norway and Russia). *Atmospheric Environment*, **31**: 159-170.

Reimann, C., Hall, G.E.M., Siewers, U., Bjorvatn, K., Morland, G., Skarphagen, H. and Strand, T., 1996. Radon, fluoride and 62 elements as determined by ICP-MS in 145 Norwegian hardrock groundwaters. *The Science of the Total Environment*, **192**: 1-19.

Rock, N.M.S., 1988. *Numerical Geology.* Lecture Notes in Earth Sciences, 18. Springer-Verlag, Berlin, 427 pp.

Roostai, A.H., 1997. *Zur Geochemie der Oberflächengewässer des Brockenmassivs/Harz unter Berücksichtigung der Gewässerversauerung.* Dissertation, Friedrich-Alexander-Universität Erlangen-Nürnberg, Deutschland. (in German).

Rosenkranz, D., Bachmann, G., Einsele, G. and Harress, H.M., 1988. *Rosenkranz/Einsele/Harress Bodenschutz. Ergänzbares Handbuch der Massnahmen und Empfehlungen für Schutz, Pflege und Sanierung von Böden, Landschaft und Grundwasser.* Erich Schmidt Verlag, Berlin, Germany. 2 Volumes, updated to 11/94. (in German).

Saager, R., 1984. *Metallische Rohstoffe von Antimon bis Zirkonium.* Bank Vontobel, Zürich, Switzerland, 176 pp. (in German).

Siewers, U. and Herpin, U., in prep. Moos-Monitoring in Deutschland. Bundesanstalt für Geowissenschaften, Hannover (in German).

Siewers, U., 1997. First results of the stream water survey of Romania-statistics. Pers. comm.

Skillen, A.D. and Griffiths, J.B. (eds.), 1993. *Raw Materials for the Glass and Ceramics Industries.* Second edition. Industrial minerals, glass and ceramics survey 1993. Metal Bulletin, London, UK, 122 pp.

Skjelkvåle, B.L., Henriksen, A., Vadset, M. and Røyset, O., 1996. Sporelementer i norske innsjøer-Foreløpig resultat for 473 sjøer. Norsk institutt for vannforskning, Rapport, Løpenr. 3457-96, Oslo, Norway, 18 pp. (in Norwegian).

Smith, R.E., Anand, R.R., Churchward, H.M., Robertson, I.D.M., Grunsky, E.C., Gray, D.J., Wildman, J.E. and Perdrix, J.L., 1992. Laterite geochemistry for detecting concealed mineral deposits. Yilgarn Craton, Western Australia. Summary report for the CSIRO-AMIRA Project P240, Exploration Geoscience Restricted Report 236R, CSIRO, Perth, Australia.

Sosial- og helsedepartementet, 1995. Forskrift om vannforsyning og drikkevann m.m., Nr. 68, I-9/95, Oslo 1.februar 1995. 38 pp. (in Norwegian).

Steinnes, E., Røyset, O., Vadset, M. and Johansen, O., 1993. Atmosfærisk nedfall av tungmetaller i Norge. Landsomfattende undersøkelse 1990. Statens forurensningstilsyn, Rapport 523/93, Oslo, Norway, 36 pp. (in Norwegian).

Steinnes, E., Rambæk, J.P. and Hanssen, J.E., 1992. Large scale multi-element survey of atmospheric deposition using naturally growing moss as biomonitor. *Chemosphere*, **25:** 735-752.

Streit, B., 1994. *Lexikon Ökotoxikologie. Zweite, aktualisierte und erweiterte Auflage*. VCH Verlagsgesellschaft, Weinheim, Germany, 901 pp. (in German).

Tarvainen, T., 1997. *Suomen peltomaiden geokemiallinen kartoitus: hankkeen 33 53 alustavat tulokset*. Geological Survey of Finland, Archived Report S/42/0000/1/1997, 63 pp. (in Finnish).

Tauber, C., 1988. *Spurenelemente in Flugaschen*. Verlag TÜV Rheinland GmbH, Köln, Germany, 469 pp. (in German).

Taylor, S.R. and McLennan, S.M., 1995. The geochemical evolution of the continental crust. *Reviews of Geophysics*, **33:** 241-265.

Thalmann, F., Schermann, O., Schroll, E., and Hausberger, G., 1988. *Geochemischer Atlas der Republik Oesterreich. Boehmische Masse und Zentralzone der Ostalpen-Bachsedimente*. Geologische Bundesanstalt, Wien, Austria (in German).

The Open University, 1989. *Seawater: Its Composition, Properties and Behaviour*. Pergamon Press, Oxford, UK, 165 pp.

US EPA Factsheets, 1997. United States Environmental Protection Agency. *National Primary Drinking Water Regulations: Contaminant Specific Fact Sheets, Inorganic Chemicals*. EPA 811 F95002. See also *http://www.epa.gov/OGWDW/dwh/t-ioc.html*.

Versieck, J. and Cornelis, R., 1980. Normal levels of trace elements in human blood and plasma or serum. *Analytica Chimica Acta*, **116:** 217-254.

VROM, 1994. Ministerie van Volkshuisvesting, Ruimtelijke Ordening en Milieubeheer. *Leidraad bodembescherming, aflevering 9, oktober 1994*. Sdu Uitgeverij Koninginnegracht, Den Haag, 83-94. (In Dutch).

Wedepohl, K.H. (executive editor), 1978. *Handbook of Geochemistry*. Springer-Verlag, Berlin. 5 Volumes.

Wedepohl, K.H., 1995. The composition of the continental crust. *Geochimica et Cosmochimica Acta*, **59:** 1217-1232.

WHO, 1996. *Guidelines for Drinking-Water Quality*. Second edition. Volume 2, Health criteria and other supporting information. World Health Organization, Geneva, Switzerland, 973 pp.

Winter, M., 1997. WebElements. University of Sheffield, Department of Chemistry. World Wide Web Site *http://www.shef.ac.uk/~chem/web-elements*.

FACTSHEETS

Ac   ACTINIUM

Physico-Chemical Properties

Atomic number	89	Melting/boiling point (K)	1324 / 3471
Atomic mass	[227]	Isotopes & isomers	0 stable + 32 unstable
Atomic radius (pm)	-	Acid/base of oxide	-
Main oxidation state(s)	+3	State (at 300 K, 1 atm.)	solid
Ionic radius (pm)	126	Metallic character	metal
Electronegativity (Pauling)	1.1	Element group(s)	heavy metal
Density (g/cm^3)	10.07	Affinity	-

Naturally occurring isotopes	Natural abundance (%)	Atomic mass	Half-life
^{227}Ac	-	227.027,75	21.77 y
^{228}Ac	-	228.031,01	6.15 h

Unstable isotopes	Longest half-life
^{207}Ac to ^{234}Ac	21.77 y

Concentrations in Crust/Rocks (mg/kg)

Bulk continental crust	- / 5.5x10^{-10} / -
Upper continental crust	- / -
Ultramafic rock	-
Ocean ridge basalt	-
Gabbro, basalt	-
Granite, granodiorite	-
Sandstone	-
Greywacke	-
Shale, schist	-
Limestone	-
Coal	-

Typical Minerals
-

Possible Host Minerals
uraninite/pitchblende, brannerite, carnotite

Mass (kg) in

Continental crust 1.30x10^{+7}	Oceans -	Plants -

Concentrations in Media Sampled for the Kola Project (mg/kg)

Medium	Moss	Humus	Humus	Topsoil (0-5 cm)	B horizon	B horizon
	<2 mm	<2 mm	<2 mm	<2 mm	<2 mm	<2 mm
	conc. HNO$_3$		amm. acet.	total (HPGe)	aqua regia	total (XRF)
Median	-	-	-	3.73x10^{-9} (10 Bq/kg)	-	-
Min	-	-	-	1.49x10^{-10} (0.4 Bq/kg)	-	-
Max	-	-	-	3.03x10^{-8} (81.2 Bq/kg)	-	-

Medium	C horizon	C horizon	C horizon		Lake water
	<2 mm	<2 mm	<2 mm		unfiltered
	aqua regia	total (XRF)	total (INAA)		(mg/l)
Median	-	-	-		-
Min	-	-	-		-
Max	-	-	-		-

ACTINIUM   Ac

Environmental Geochemistry

Biological impacts	Considered non-essential
Uses	Production of neutrons
Environmental pathways	Occurs naturally wherever Th is present, of which it is a decay product
Environmental mobility	Oxidising conditions: - Acid conditions: - Reducing conditions: - Neutral to alkaline conditions: - Comments: -
Geochemical barriers	-
Natural association	U-Th-Ra-Ac (U deposits)
Action levels	-
Remarks	^{228}Ac is a decay product of ^{235}U/^{232}Th. Some plants on Th-rich soils seem to accumulate Ac (up to 1000 Bq/kg)
Suggested analytical method(s)	Gamma spectrometry

World Yearly Production (t/y)   -
Price of Purest Form, in Small Quantities ($/kg)   -
Market Price ($/kg)   -

Ag — SILVER

Physico-Chemical Properties

Atomic number	47	Melting/boiling point (K)	1234.93 / 2435
Atomic mass	107.868,2(2)	Isotopes & isomers	2 stable + 49 unstable
Atomic radius (pm)	175	Acid/base of oxide	amphoteric
Main oxidation state(s)	+1 (0,+2,+3)	State (at 300 K, 1 atm.)	solid
Ionic radius (pm)	81-142	Metallic character	metal
Electronegativity (Pauling)	1.93	Element group(s)	heavy metal, noble metal
Density (g/cm^3)	10.5	Affinity	chalcophile

Naturally occurring isotopes	Natural abundance (%)	Atomic mass	Half-life
^{107}Ag	51.84	106.905	stable
^{109}Ag	48.16	108.904,8	stable

Unstable isotopes	Longest half-life
^{94}Ag to ^{124}Ag	927 y

Concentrations in Crust/Rocks (mg/kg)

Bulk continental crust	0.07 / 0.075 / 0.08
Upper continental crust	0.055 / 0.05
Ultramafic rock	0.03
Ocean ridge basalt	0.03
Gabbro, basalt	0.1
Granite, granodiorite	0.05
Sandstone	0.003
Greywacke	-
Shale, schist	0.08
Limestone	0.01
Coal	0.1

Typical Minerals
argentite (Ag$_2$S), native Ag, cerargerite (AgCl), silver arsenide (Ag$_3$As)

Possible Host Minerals
various silicates at very low levels, galena, sphalerite, chalcopyrite, arsenides, tetrahedrite

Mass (kg) in

Continental crust	1.65x10^{+15}	Oceans	2.64x10^{+9}	Plants	3.68x10^{+8}

Concentrations in Media Sampled for the Kola Project (mg/kg)

Medium	Moss conc. HNO$_3$	Humus <2 mm conc. HNO$_3$	Humus <2 mm amm. acet.	Topsoil (0-5 cm) <2 mm total (INAA)	B horizon <2 mm aqua regia	B horizon <2 mm total (XRF)
Median	0.033	0.2	-	<5	0.019	-
Min	<0.01	0.025	-	<5	<0.001	-
Max	0.824	4.79	-	<5	0.378	-

Medium	C horizon <2 mm aqua regia	C horizon <2 mm total (XRF)	C horizon <2 mm total (INAA)	Lake water unfiltered (mg/l)
Median	0.008	-	<5	<0.000,01
Min	<0.001	-	<5	<0.000,01
Max	0.119	-	<5	0.000,2

SILVER   Ag

Concentrations in Soils and Sediments (mg/kg)

Medium	Soil	Agricultural soil - Ap horizon	Agricultural soil - Top (0-25 cm)	Agricultural Soil - Bottom (50-75 cm)	Topsoil (0-15 cm)	Urban soil (0-2 cm)
	World	Canada	Finland	Finland	England & Wales	Trondheim
	<2 mm	<2 mm	<2 mm	<2 mm	<2 mm	<2 mm
	total	total (AAS)	aqua regia	aqua regia	aqua regia	aqua regia
Median	0.07*	0.2	<0.5	<0.5	-	<1
Min	-	<0.2	<0.5	<0.5	-	<1
Max	-	0.9	0.518	0.661	-	2

*Estimated mean

Medium	Forest soil - Humus	Forest soil - B horizon	Forest soil - C horizon	Till (C horizon)	Till (C horizon)	Laterite (25 ±15 cm)
	Norway	Norway	Norway	Finland	Finland	Australia
	<2 mm	<2 mm	<2 mm	<0.063 mm	<0.063 mm	0.45-2 mm
	7N HNO$_3$	7N HNO$_3$	7N HNO$_3$	aqua regia	total	total
Median	<0.3	0.6	1	-	-	<0.05
Min	<0.3	<0.5	<0.5	-	-	<0.05
Max	2.9	3.1	3.2	-	-	3.4

Medium	Stream sediment	Stream sediment	Stream sediment	Overbank sediment	Overbank sediment	Organic stream sediment
	Austria	Southern Scotland	Harz, Germany	Norway	Norway	Finland
	<0.18 mm	<0.10 mm	<0.063 mm	<0.063 mm	<0.063 mm	<2 mm
	"total" (ICP-AES)	total (DCES)	total	total (XRF)	7N HNO$_3$	Conc. HNO$_3$
Median	0.08	<0.7	-	-	-	0.09
Min	-	<0.7	-	-	-	0.022#
Max	13	42	-	-	-	0.31^

#2nd percentile   ^98th percentile

Concentrations in Waters (mg/l)

Medium	Ocean water	Ocean water	Ocean water	Stream water	Stream water	Stream water
	World (1)	World (2)	North Pacific	World	Nova Scotia, Canada	Finland
					<0.45 µm	<0.45 µm
					ICP-MS	ICP-MS
Median	0.000,04*	0.000,002*	0.000,002*	0.000,05*	<0.000,05	<0.000,01
Min	-	-	-	-	<0.000,05	<0.000,01#
Max	-	-	-	-	<0.000,05	<0.000,01^

*Estimated mean   #2nd percentile   ^98th percentile

Medium	Stream water	Stream water	Stream water	Lake water	Ground water
	Romania	Eastern India	Harz, Germany	Norway	Southern Norway
	unfiltered	<0.2 µm	unfiltered	unfiltered	unfiltered
	ICP-MS	ICP-MS	ICP-MS	ICP-MS	ICP-MS
Median	0.000,005	0.000,19	0.000,03	<0.000,01	<0.000,001
Min	0.000,001	0.000,09	<0.000,01	<0.000,01	<0.000,001
Max	0.052	0.000,6	0.005,6	0.000,436	0.000,074

Ag SILVER

Concentrations in Precipitation (mg/l)

Medium	Rain water	Rain water	Snow melt-water	Snow melt-water	Snow filter residue	Snow filter residue
	Kola, remote <0.45 μm ICP-MS	Kola, polluted <0.45 μm ICP-MS	Kola, remote <0.45 μm ICP-MS	Kola, polluted <0.45 μm ICP-MS	Kola, remote >0.45 μm	Kola, polluted >0.45 μm
Median	<0.000,01	0.000,03	<0.000,01	<0.000,01	-	-
Min	<0.000,01	<0.000,01	<0.000,01	<0.000,01	-	-
Max	<0.000,01	0.000,14	<0.000,01	0.000,04	-	-

Medium	Rain water	Rain water				
	Coastal, Norway unfiltered ICP-MS	Inland, Norway unfiltered ICP-MS				
Median	-	-				
Min	-	-				
Max	-	-				

Concentrations in Air (ng/m^3) and Yearly Deposition (kg/km^2/yr)

Medium	Air	Air	Bulk deposition	Throughfall deposition	Bulk deposition	Bulk deposition
	World, remote	World, polluted	West Germany	West Germany	Kola, remote	Kola, polluted
Median	0.044	1	-	-	-	-
Min	0.001	0.04	-	-	-	-
Max	0.14	7	-	-	-	-

*Estimated value

Concentrations in Plants (mg/kg)

Medium	Moss	Moss	Crustose lichen	Lichen	Dandelion	Spruce bark
	Norway conc. HNO_3 (1990)	Germany conc. HNO_3 + H_2O_2 (1995)	Germany conc. HNO_3	Northwest Territories oven-dried	Europe oven-dried	Canada ashed total (INAA)
Median	0.1*	0.071	3.4	-	-	<2
Min	<0.05	0.01	0.14	-	-	<2
Max	1	1.9	166	-	-	5

*Mean #Geometric mean

Concentrations in Human Fluids (mg/l)

Medium	Human blood	Human serum	Human urine			
	Lombardy, Italy	Lombardy, Italy	Lombardy, Italy			
Mean	0.000,37	0.000,18	0.000,46			
Min	0.000,05	0.000,06	0.000,06			
Max	0.000,78	0.000,46	0.002,5			

SILVER   Ag

Environmental Geochemistry

Biological impacts	Considered to be non-essential. Toxic for many micro-organisms, fish. Low toxicity for humans. Does not appear to be carcinogenic
Uses	Photographic industry, coins, jewellery, batteries, brazing alloys, electroplating, electrical controls and conductors
Environmental pathways	Pb, Zn, Cu, Sn smelters, sewage sludge. In USA, 2.6 tons/yr of AgI are used to seed clouds to trigger rainfall. Natural pathways not well documented. Annual Ag loss to the environment is estimated at >2500 tons
Environmental mobility	Oxidising conditions: medium Acid conditions: high Reducing conditions: very low Neutral to alkaline conditions: very low Comments: -
Geochemical barriers	Presence of sulphide, pH increase, adsorption to organic matter, Fe-Mn oxides
Natural association	Pb-Zn-Cd-Ag-Hg-As-Sb-Se (complex sulphides), Ag-Ni-Co-Fe-S-As-Sb-Bi-(U) (Cobalt-type deposits), U-V-Se-As-Mo-Pb-Cu-Ag (red bed sandstone deposits), U-Cu-Ag-Co-Ni-As-V-Se-Au-Mo (unconformity vein U deposit), Au-Ag-Te-Hg (veins), Cu-Mo-Ag-(Au) (porphyry copper)
Action levels	Drinking water, MAC: 0.01 mg/l (Norway Shd); suggested MAC: 0.1 mg/l (US EPA); MAC: 0.05 mg/l (Canada CWQG); health-based recommended value not deemed necessary (WHO)
Remarks	High Ca, high Cl in water decrease toxicity. Ag-compounds can prevent uptake of Se, Cu, vitamin E. Aquatic plants tend to accumulate Ag. Mostly produced as by-product of Pb, Cu, Zn, Sn ores
Suggested analytical method(s)	GF-AAS, ICP-MS

World Yearly Production (t/y)	13,500 (in 1995)
Price of Purest Form, in Small Quantities ($/kg)	11,644 (99.999%)
Market Price ($/kg)	150

Al   ALUMINIUM

Physico-Chemical Properties

Atomic number	13	Melting/boiling point (K)	933.47 / 2792
Atomic mass	26.981,538(2)	Isotopes & isomers	1 stable + 16 unstable
Atomic radius (pm)	143	Acid/base of oxide	amphoteric
Main oxidation state(s)	+3	State (at 300 K, 1 atm.)	solid
Ionic radius (pm)	53-67.5	Metallic character	pred. metal
Electronegativity (Pauling)	1.61	Element group(s)	light metal
Density (g/cm^3)	2.698,9	Affinity	lithophile

Naturally occurring isotopes	Natural abundance (%)	Atomic mass	Half-life	Unstable isotopes	Longest half-life
^{27}Al	100	26.981,5	stable	^{22}Al to ^{36}Al	7.1x10^{+5} y

Concentrations in Crust/Rocks (mg/kg)

Bulk continental crust	79,600 / 82,300 / 84,100
Upper continental crust	77,440 / 80,400
Ultramafic rock	20,000
Ocean ridge basalt	89,000
Gabbro, basalt	83,000
Granite, granodiorite	73,000
Sandstone	37,000
Greywacke	71,456
Shale, schist	91,000
Limestone	4000
Coal	21,000

Typical Minerals
gibbsite (AlOH$_3$), boehmite (AlO(OH)), diaspore (AlO(OH)), sillimanite (Al$_2$SiO$_5$), corundum (Al$_2$O$_3$), cryolite (Na$_3$AlF$_6$), kaolinite (Al$_2$Si$_2$O$_5$(OH)$_4$)

Possible Host Minerals
feldspars, micas, clay minerals

Mass (kg) in

Continental crust	1.88x10^{+21}	Oceans	2.64x10^{+12}	Plants	1.47x10^{+11}

Concentrations in Media Sampled for the Kola Project (mg/kg)

Medium	Moss conc. HNO$_3$	Humus <2 mm conc. HNO$_3$	Humus <2 mm amm. acet.	Topsoil (0-5 cm) <2 mm	B horizon <2 mm aqua regia	B horizon <2 mm total (XRF)
Median	193	1890	84.5	-	18,900	74,155
Min	33.9	372	6.4	-	3200	33,081
Max	4850	20,600	2490	-	83,300	124,015

Medium	C horizon <2 mm aqua regia	C horizon <2 mm total (XRF)	C horizon <2 mm total (INAA)			Lake water unfiltered (mg/l)
Median	9910	73,800	-			0.047,05
Min	1840	29,200	-			0.005,76
Max	85,900	120,800	-			1.1

ALUMINIUM   Al

Concentrations in Soils and Sediments (mg/kg)

Medium	Soil	Agricultural soil - Ap horizon	Agricultural soil - Top (0-25 cm)	Agricultural Soil - Bottom (50-75 cm)	Topsoil (0-15 cm)	Urban soil (0-2 cm)
	World	Canada	Finland	Finland	England & Wales	Trondheim
	<2 mm	<2 mm	<2 mm	<2 mm	<2 mm	<2 mm
	total		aqua regia	aqua regia	aqua regia	aqua regia
Median	80,000*	-	7500	7420	27,917	18,650
Min	-	-	314	370	491	1770
Max	-	-	32,900	46,900	79,355	44,700

*Estimated mean

Medium	Forest soil - Humus	Forest soil - B horizon	Forest soil - C horizon	Till (C horizon)	Till (C horizon)	Laterite (25 ±15 cm)
	Norway	Norway	Norway	Finland	Finland	Australia
	<2 mm	<2 mm	<2 mm	<0.063 mm	<0.063 mm	0.45-2 mm
	7N HNO$_3$	7N HNO$_3$	7N HNO$_3$	aqua regia	total	total
Median	1700	24,300	19,900	13,000	74,000	43,535
Min	240	4400	4000	-	-	7251
Max	27,700	92,500	80,100	-	-	91,040

Medium	Stream sediment	Stream sediment	Stream sediment	Overbank sediment	Overbank sediment	Organic stream sediment
	Austria	Southern Scotland	Harz, Germany	Norway	Norway	Finland
	<0.18 mm	<0.10 mm	<0.063 mm	<0.063 mm	<0.063 mm	<2 mm
	total (XRF)	total	total (XRF)	total (XRF)	7N HNO$_3$ ICP-AES	Conc. HNO$_3$
Median	88,000	-	52,189	69,000	6600	11,400
Min	-	-	7834	27,400	2000	4400#
Max	204,800	-	85,059	108,700	58,600	36,700^

#2nd percentile   ^98th percentile

Concentrations in Waters (mg/l)

Medium	Ocean water	Ocean water	Ocean water	Stream water	Stream water	Stream water
	World (1)	World (2)	North Pacific	World	Nova Scotia, Canada	Finland
					<0.45 µm	<0.45 µm
					ICP-MS	ICP-MS
Median	0.002*	0.002*	0.000,03*	0.3*	0.096	0.095
Min	-	-	-	-	<0.002	0.000,8#
Max	-	-	-	-	0.629	0.532^

*Estimated mean   #2nd percentile   ^98th percentile

Medium	Stream water	Stream water	Stream water	Lake water	Ground water
	Romania	Eastern India	Harz, Germany	Norway	Southern Norway
	unfiltered	<0.2 µm	unfiltered	unfiltered	unfiltered
	ICP-MS	ICP-MS	ICP-MS	ICP-MS	ICP-MS
Median	0.398	-	0.108	-	0.012
Min	0.013	-	0.005	-	<0.002
Max	30.9	-	2.11	-	2.537

Al — ALUMINIUM

Concentrations in Precipitation (mg/l)

Medium	Rain water	Rain water	Snow melt-water	Snow melt-water	Snow filter residue	Snow filter residue
	Kola, remote <0.45 µm ICP-MS	Kola, polluted <0.45 µm ICP-MS	Kola, remote <0.45 µm ICP-MS	Kola, polluted <0.45 µm ICP-MS	Kola, remote >0.45 µm ICP-AES	Kola, polluted >0.45 µm ICP-AES
Median	0.002,9	0.105	0.003,04	0.121,5	0.005	0.285
Min	0.000,9	0.013,4	0.002,24	0.088	0.004	0.121
Max	0.008,6	0.272	0.014	0.279	0.025	2.5

Medium	Rain water	Rain water
	Coastal, Norway unfiltered ICP-MS	Inland, Norway unfiltered ICP-MS
Median	-	-
Min	-	-
Max	-	-

Concentrations in Air (ng/m^3) and Yearly Deposition (kg/km^2/yr)

Medium	Air	Air	Bulk deposition	Throughfall deposition	Bulk deposition	Bulk deposition
	World, remote	World, polluted	West Germany	West Germany	Kola, remote	Kola, polluted
Median	46 / 70	1300	-	-	9.13*	664*
Min	0.33	150	8	17	-	-
Max	66 (1200)	3500	90	360	-	-

*Estimated value

Concentrations in Plants (mg/kg)

Medium	Moss	Moss	Crustose lichen	Lichen	Dandelion	Spruce bark
	Norway conc. HNO$_3$ (1990)	Germany conc. HNO$_3$ + H$_2$O$_2$ (1995)	Germany conc. HNO$_3$	Northwest Territories oven-dried total (INAA)	Europe oven-dried	Canada ashed aqua regia
Median	430	463	-	633.8*	-	11,945
Min	120	98	-	-	-	2120
Max	6000	16,200	-	-	-	30,620

*Mean   #Geometric mean

Concentrations in Human Fluids (mg/l)

Medium	Human blood	Human serum	Human urine			
	Lombardy, Italy	Lombardy, Italy	Lombardy, Italy			
Mean	-	0.000,6	0.010,9			
Min	-	0.000,1	0.001			
Max	-	0.010,9	0.031			

ALUMINIUM   Al

Environmental Geochemistry

Biological impacts	Considered to be essential for some organisms. Toxic to fish at low pH. Toxic to plants. Free Al ions toxic to humans
Uses	Packaging industry (e.g., beverages), transportation, electrical, consumer goods, outside building decoration, abrasives, tanning, textiles, sewage treatment, cement accelerator, and many others
Environmental pathways	Geogenic and anthropogenic dust (clay minerals), Al smelters (often + Be, F), cement plants, sewage sludge. Cryolite used as insecticide
Environmental mobility	Oxidising conditions: very low Acid conditions: very low Reducing conditions: very low Neutral to alkaline conditions: very low Comments: Solubility increases at pH<5.5, e.g., under acid deposition
Geochemical barriers	pH
Natural association	Al-Si (substitution in most silicates)
Action levels	Drinking water, MAC: 0.2 mg/l (Norway Shd); suggested range: 0.05-0.2 mg/l (US EPA)
Remarks	Easily soluble Al-compounds have high acute toxic effects (e.g., Al-chloride, Al-nitrate). The European Commission is considering classifying a number of Al-compounds and Al-powder as dangerous substances. In contact with the atmosphere and fresh water, metallic Al develops an oxide layer preventing corrosion. Al-sulphate used as flocculant in treatment of drinking and waste waters. Acid deposition yields Al-hydroxy-sulphates in soils, dissolving at pH<4.2. Several plants (e.g., mustard, sweet leaf, holly (*Ilex*), tea, moss) can accumulate Al. Available Al may be toxic to plants at pH<5. Growth of some plants positively affected by Al. Al can block uptake of phosphate. Relation between Al and Alzheimer's disease has been discussed. Concentrations of 0.2 mg/l in drinking water may give rise to consumer complaints due to deposition and discoloration. Al smelting is very energy consuming. Bauxite, an important ore of aluminium and the product of rock weathering, is a mixture of gibbsite, boehmite and diaspore
Suggested analytical method(s)	XRF, ICP-AES

World Yearly Production (t/y)	25,093,000 (in 1995)
Price of Purest Form, in Small Quantities ($/kg)	11,969 (99.999,9%)
Market Price ($/kg)	1.7

Ar ARGON

Physico-Chemical Properties

Atomic number	18	Melting/boiling point (K)	83.8 / 87.3
Atomic mass	39.948(1)	Isotopes & isomers	3 stable + 14 unstable
Atomic radius (pm)	88	Acid/base of oxide	-
Main oxidation state(s)	0	State (at 300 K, 1 atm.)	gas
Ionic radius (pm)	-	Metallic character	-
Electronegativity (Pauling)	-	Element group(s)	noble gas
Density (g/cm^3)	(0.001,784)	Affinity	atmophile

Naturally occurring isotopes	Natural abundance (%)	Atomic mass	Half-life
^{36}Ar	0.336,5	35.967,5	stable
^{38}Ar	0.063,2	37.962,7	stable
^{40}Ar	99.600,3	39.962,4	stable

Unstable isotopes	Longest half-life
^{31}Ar to ^{47}Ar	268 y

Concentrations in Crust/Rocks (mg/kg)

Bulk continental crust	- / 3.5 / -
Upper continental crust	- / -
Ultramafic rock	-
Ocean ridge basalt	-
Gabbro, basalt	-
Granite, granodiorite	-
Sandstone	-
Greywacke	-
Shale, schist	-
Limestone	-
Coal	-

Typical Minerals
-

Possible Host Minerals
-

Mass (kg) in

Continental crust	8.26x10^{+16}	Oceans	5.68x10^{+14}	Plants	-

Concentrations in Media Sampled for the Kola Project (mg/kg)

| Medium | Moss | Humus <2 mm | Humus <2 mm | Topsoil (0-5 cm) <2 mm | B horizon <2 mm | B horizon <2 mm |
	conc. HNO$_3$		amm. acet.	aqua regia	aqua regia	total (XRF)
Median	-	-	-	-	-	-
Min	-	-	-	-	-	-
Max	-	-	-	-	-	-

Medium	C horizon <2 mm aqua regia	C horizon <2 mm total (XRF)	C horizon <2 mm total (INAA)		Lake water unfiltered (mg/l)
Median	-	-	-		-
Min	-	-	-		-
Max	-	-	-		-

ARGON   Ar

Concentrations in Soils and Sediments (mg/kg)

Medium	Soil	Agricultural soil -	Agricultural soil - Top	Agricultural Soil - Bottom	Topsoil (0-15 cm)	Urban soil (0-2 cm)
	World	*Ap horizon*	*(0-25 cm)*	*(50-75 cm)*	*England & Wales*	*Trondheim*
		Canada	*Finland*	*Finland*		
	<2 mm	*<2 mm*	*<2 mm*	*<2 mm*	*<2 mm*	*<2 mm*
	total		*aqua regia*	*aqua regia*	*aqua regia*	*aqua regia*
Median	-	-	-	-	-	-
Min	-	-	-	-	-	-
Max	-	-	-	-	-	-

*Estimated mean

Medium	Forest soil - Humus	Forest soil - B horizon	Forest soil - C horizon	Till (C horizon)	Till (C horizon)	Laterite (25 ±15 cm)
	Norway	*Norway*	*Norway*	*Finland*	*Finland*	*Australia*
	<2 mm	*<2 mm*	*<2 mm*	*<0.063 mm*	*<0.063 mm*	*0.45-2 mm*
	7N HNO$_3$	*7N HNO$_3$*	*7N HNO$_3$*	*aqua regia*	*total*	*total*
Median	-	-	-	-	-	-
Min	-	-	-	-	-	-
Max	-	-	-	-	-	-

Medium	Stream sediment	Stream sediment	Stream sediment	Overbank sediment	Overbank sediment	Organic stream sediment
		Southern	*Harz, Germany*			*Finland*
	Austria	*Scotland*		*Norway*	*Norway*	
	<0.18 mm	*<0.10 mm*	*<0.063 mm*	*<0.063 mm*	*<0.063 mm*	*<2 mm*
		total	*total*	*total (XRF)*	*7N HNO$_3$*	*Conc. HNO$_3$*
Median	-	-	-	-	-	-
Min	-	-	-	-	-	-
Max	-	-	-	-	-	-

#2nd percentile ^98th percentile

Concentrations in Waters (mg/l)

Medium	Ocean water	Ocean water	Ocean water	Stream water	Stream water	Stream water
	World (1)	*World (2)*	*North Pacific*	*World*	*Nova Scotia, Canada*	*Finland*
					<0.45 μm	*<0.45 μm*
Median	0.45*	0.43*	0.62*	-	-	-
Min	-	-	-	-	-	-
Max	-	-	-	-	-	-

*Estimated mean   #2nd percentile   ^98th percentile

Medium	Stream water	Stream water	Stream water	Lake water	Ground water
	Romania	*Eastern India*	*Harz, Germany*	*Norway*	*Southern Norway*
	unfiltered	*<0.2 μm*	*unfiltered*	*unfiltered*	*unfiltered*
	ICP-MS	*ICP-MS*		*ICP-MS*	
Median	-	-	-	-	-
Min	-	-	-	-	-
Max	-	-	-	-	-

Ar ARGON

Concentrations in Precipitation (mg/l)

Medium	Rain water	Rain water	Snow melt-water	Snow melt-water	Snow filter residue	Snow filter residue
	Kola, remote <0.45 μm	Kola, polluted <0.45 μm	Kola, remote <0.45 μm	Kola, polluted <0.45 μm	Kola, remote >0.45 μm	Kola, polluted >0.45 μm
Median	-	-	-	-	-	-
Min	-	-	-	-	-	-
Max	-	-	-	-	-	-

Medium	Rain water	Rain water				
	Coastal, Norway unfiltered ICP-MS	Inland, Norway unfiltered ICP-MS				
Median	-	-				
Min	-	-				
Max	-	-				

Concentrations in Air (ng/m^3) and Yearly Deposition (kg/km^2/yr)

Medium	Air	Air	Bulk deposition	Throughfall deposition	Bulk deposition	Bulk deposition
	World, remote	World, polluted	West Germany	West Germany	Kola, remote	Kola, polluted
Median	-	-	-	-	-	-
Min	-	-	-	-	-	-
Max	-	-	-	-	-	-

*Estimated value

Concentrations in Plants (mg/kg)

Medium	Moss	Moss	Crustose lichen	Lichen	Dandelion	Spruce bark
	Norway conc. HNO$_3$ (1990)	Germany conc. HNO$_3$ + H$_2$O$_2$ (1995)	Germany conc. HNO$_3$	Northwest Territories oven-dried	Europe oven-dried	Canada ashed
Median	-	-	-	-	-	-
Min	-	-	-	-	-	-
Max	-	-	-	-	-	-

*Mean   #Geometric mean

Concentrations in Human Fluids (mg/l)

Medium	Human blood	Human serum	Human urine			
	Lombardy, Italy	Lombardy, Italy	Lombardy, Italy			
Mean	-	-	-			
Min	-	-	-			
Max	-	-	-			

ARGON   Ar

Environmental Geochemistry

Biological impacts	Considered non-essential
Uses	Welding, light bulbs, fluorescent tubes, glow tubes, electronic industry (protective atmosphere for growing Si and Ge crystals), carrier gas in analytical instruments
Environmental pathways	Natural trace component of atmosphere
Environmental mobility	Oxidising conditions: very high Acid conditions: very high Reducing conditions: very high Neutral to alkaline conditions: very high Comments: Inert gas
Geochemical barriers	-
Natural association	-
Action levels	-
Remarks	Occurs in atmosphere at low concentrations, together with the other noble gasses. Ar is decay product of ^{40}K
Suggested analytical method(s)	-

World Yearly Production (t/y)	-
Price of Purest Form, in Small Quantities ($/kg)	7.5 $/l (99.999,5%)
Market Price ($/kg)	8.2 $/m^3

As — ARSENIC

Physico-Chemical Properties

Atomic number	33
Atomic mass	74.921,60(2)
Atomic radius (pm)	133
Main oxidation state(s)	-3,+3 (+5)
Ionic radius (pm)	47.5-72
Electronegativity (Pauling)	2.18
Density (g/cm^3)	5.73 (grey) 1.97 (yellow)
Melting/boiling point (K)	1090 (triple point) / 887 (sublimes)
Isotopes & isomers	1 stable + 24 unstable
Acid/base of oxide	weak acid
State (at 300 K, 1 atm.)	solid
Metallic character	pred. non-metal
Element group(s)	heavy non-metal (grey)
Affinity	chalcophile

Naturally occurring isotopes	Natural abundance (%)	Atomic mass	Half-life
^{75}As	100	74.921,59	stable

Unstable isotopes	Longest half-life
^{65}As to ^{87}As	80.3 d

Concentrations in Crust/Rocks (mg/kg)

Bulk continental crust	1.7 / 1.8 / 1
Upper continental crust	2 / 1.5
Ultramafic rock	0.7
Ocean ridge basalt	1
Gabbro, basalt	0.7
Granite, granodiorite	3
Sandstone	0.5
Greywacke	-
Shale, schist	13
Limestone	1.5
Coal	10

Typical Minerals
arsenopyrite (FeSAs), realgar (AsS), orpiment (As$_2$S$_3$), arsenolite (As$_2$O$_3$)

Possible Host Minerals
feldspars, magnetite, ilmenite, pyrite, galena, sphalerite, apatite

Mass (kg) in

Continental crust	Oceans	Plants
4.01x10^{+16}	4.89x10^{+12}	1.84x10^{+8}

Concentrations in Media Sampled for the Kola Project (mg/kg)

Medium	Moss conc. HNO$_3$	Humus <2 mm conc. HNO$_3$	Humus <2 mm amm. acet.	Topsoil (0-5 cm) <2 mm total (INAA)	B horizon <2 mm aqua regia	B horizon <2 mm total (XRF)
Median	0.173	1.16	<0.05	<0.05	1.1	-
Min	<0.05	0.364	<0.05	<0.05	<0.1	-
Max	3.42	43.5	4.9	16	77.8	-

Medium	C horizon <2 mm aqua regia	C horizon <2 mm total (XRF)	C horizon <2 mm total (INAA)		Lake water unfiltered (mg/l)
Median	0.5	-	<0.5		0.000,12
Min	<0.1	-	<0.5		<0.000,05
Max	30.7	-	54		0.002,94

ARSENIC As

Concentrations in Soils and Sediments (mg/kg)

Medium	Soil	Agricultural soil - Ap horizon	Agricultural soil - Top (0-25 cm)	Agricultural Soil - Bottom (50-75 cm)	Topsoil (0-15 cm)	Urban soil (0-2 cm)
	World	Canada	Finland	Finland	England & Wales	Trondheim
	<2 mm	<2 mm	<2 mm	<2 mm	<2 mm	<2 mm
	total	total (INAA)	aqua regia	aqua regia	aqua regia	aqua regia
Median	5*	6.6	1.55	1.54	-	0.5
Min	-	0.2	<0.2	<0.2	-	83
Max	-	19	13.4	12	-	-

*Estimated mean

Medium	Forest soil - Humus	Forest soil - B horizon	Forest soil - C horizon	Till (C horizon)	Till (C horizon)	Laterite (25 ±15 cm)
	Norway	Norway	Norway	Finland	Finland	Australia
	<2 mm	<2 mm	<2 mm	<0.063 mm	<0.063 mm	0.45-2 mm
	7N HNO$_3$	7N HNO$_3$	7N HNO$_3$	aqua regia	total	total
Median	-	-	-	-	2.6	3
Min	-	-	-	-	-	<0.5
Max	-	-	-	-	-	170

Medium	Stream sediment	Stream sediment	Stream sediment	Overbank sediment	Overbank sediment	Organic stream sediment
	Austria	Southern Scotland	Harz, Germany	Norway	Norway	Finland
	<0.18 mm	<0.10 mm	<0.063 mm	<0.063 mm	<0.063 mm	<2 mm
	Gutzeit	total (AAS/XRF)	total (ICP-MS)	total (XRF)	7N HNO$_3$ AAS	Conc. HNO$_3$
Median	2	12	22	-	5.3	2.9
Min	<2	<4	<5.0	-	<0.05	0.53#
Max	1305	150	334	-	57.6	23.8^

#2nd percentile   ^98th percentile

Concentrations in Waters (mg/l)

Medium	Ocean water	Ocean water	Ocean water	Stream water	Stream water	Stream water
	World (1)	World (2)	North Pacific	World	Nova Scotia, Canada	Finland
					<0.45 µm	<0.45 µm
					ICP-MS	ICP-MS
Median	0.003,7*	0.003,7*	0.001,205*	0.004*	0.000,2	0.000,36
Min	-	-	-	-	<0.000,1	<0.000,05#
Max	-	-	-	-	0.003,3	0.002,36^

*Estimated mean   #2nd percentile   ^98th percentile

Medium	Stream water	Stream water	Stream water	Lake water	Ground water
	Romania	Eastern India	Harz, Germany	Norway	Southern Norway
	unfiltered	<0.2 µm	unfiltered	unfiltered	unfiltered
	ICP-MS	ICP-MS	ICP-MS	ICP-MS	ICP-MS
Median	0.001,4	0.001,1	0.000,6	-	0.000,2
Min	0.000,1	0.000,1	<0.000,01	-	<0.000,025
Max	0.023	0.013,2	0.025	-	0.011,1

As — ARSENIC

Concentrations in Precipitation (mg/l)

Medium	Rain water	Rain water	Snow melt-water	Snow melt-water	Snow filter residue	Snow filter residue
	Kola, remote	Kola, polluted	Kola, remote	Kola, polluted	Kola, remote	Kola, polluted
	<0.45 µm	<0.45 µm	<0.45 µm	<0.45 µm	>0.45 µm	>0.45 µm
	ICP-MS	ICP-MS	ICP-MS	ICP-MS	ICP-AES	ICP-AES
Median	0.000,08	0.012,3	<0.000,1	0.001,05	<0.000,3	0.004,65
Min	<0.000,05	0.003,6	<0.000,1	0.000,18	<0.000,3	0.002,5
Max	0.000,18	0.084,4	0.000,11	0.001,76	0.000,3	0.016,9

Medium	Rain water	Rain water				
	Coastal, Norway unfiltered ICP-MS	Inland, Norway unfiltered ICP-MS				
Median	0.000,3	0.000,37				
Min	<0.000,01	<0.000,01				
Max	0.003,2	0.005,1				

Concentrations in Air (ng/m^3) and Yearly Deposition (kg/km^2/yr)

Medium	Air	Air	Bulk deposition	Throughfall deposition	Bulk deposition	Bulk deposition
	World, remote	World, polluted	West Germany	West Germany	Kola, remote	Kola, polluted
Median	0.5 / 1	15	-	-	0.157*	2.818*
Min	0.007	1.5	-	-	-	-
Max	2.3	190	-	-	-	-

*Estimated value

Concentrations in Plants (mg/kg)

Medium	Moss	Moss	Crustose lichen	Lichen	Dandelion	Spruce bark
	Norway conc. HNO$_3$ (1990)	Germany conc. HNO$_3$ + H$_2$O$_2$ (1995)	Germany conc. HNO$_3$	Northwest Territories oven-dried total (INAA)	Europe oven-dried total (INAA)	Canada ashed total (INAA)
Median	0.27	0.246	-	0.26*	0.12#	10
Min	<0.03	0.001	-	-	-	2.4
Max	3.2	3.6	-	-	-	100

*Mean #Geometric mean

Concentrations in Human Fluids (mg/l)

Medium	Human blood	Human serum	Human urine			
	Lombardy, Italy	Lombardy, Italy	Lombardy, Italy			
Mean	0.007,9	-	0.016,7			
Min	0.000,4	-	0.001			
Max	0.070,5	-	0.065,4			

ARSENIC   As

Environmental Geochemistry

Biological impacts	Essential for some organisms (e.g., humans). Toxic. Teratogenic
Uses	Alloys, wood preservative, ammunition, semi-conductors, batteries, paints, textile, tanning
Environmental pathways	Coal combustion (esp. brown coal), sulphidic ore roasting and smelting. Pig and poultry sewage. Phosphate fertilisers. Insecticide, fungicide
Environmental mobility	Oxidising conditions: medium Acid conditions: medium Reducing conditions: very low Neutral to alkaline conditions: medium Comments: -
Geochemical barriers	In water bodies, As is removed relatively fast from the water and loosely bound into the sediment. Presence of sulphide, adsorption to Fe-hydroxides and clay
Natural association	Au-As or Ag-As (hydrothermal veins), Cu-Ni-Co-As-Ag-Fe (Ni-Cu massive sulphides), U-As (veins), Cu-As (Cu shales), Cu-V-U-Ag-As (Cu sandstone deposits), some phosphate deposits
Action levels	Drinking water, MAC: 0.01 mg/l (Norway Shd), 0.05 mg/l (US EPA), 0.05 mg/l (Canada CWQG); recommended: 0.7 (WHO); MAC: 0.05 mg/l (Russia MoH). Ground water, background: 0.01 mg/l; remediate: 0.06 mg/l (Netherlands VROM). Soil, background: 29 mg/kg; remediate: 55 mg/kg (Netherlands VROM). Agricultural soil, maximum tolerable concentration: 20 mg/kg (Germany)
Remarks	Toxicity dependent on valence: As^{5+} compounds less toxic than As^{3+} compounds. Toxicity inhibited when ingested with Se. Promotes growth in animals (used in pig and poultry farming). High natural As levels in ground water used as drinking water have been reported to cause widespread disease (e.g., India and Bangladesh). Some plants can highly concentrate As (radish). Cu-arsenates used as insecticide, fungicide (e.g., vineyards). As-compounds used extensively as herbicide, esp. along railroad tracks. High levels in some black shales
Suggested analytical method(s)	Hydride-AAS, ICP-MS, INAA

World Yearly Production (t/y)	30,453 (in 1995)
Price of Purest Form, in Small Quantities ($/kg)	6960 (99.999,99%)
Market Price ($/kg)	1.1

At   ASTATINE

Physico-Chemical Properties

Atomic number	85	Melting/boiling point (K)	575 / 610 (estimated)
Atomic mass	[210]	Isotopes & isomers	0 stable + 34 unstable
Atomic radius (pm)	143	Acid/base of oxide	-
Main oxidation state(s)	-1,+1 (+3,+5,+7)	State (at 300 K, 1 atm.)	solid
Ionic radius (pm)	76 (+7)	Metallic character	pred. metal
Electronegativity (Pauling)	2.2	Element group(s)	halogen
Density (g/cm^3)	-	Affinity	-

Naturally occurring isotopes	Natural abundance (%)	Atomic mass	Half-life
-	-	-	-

Unstable isotopes	Longest half-life
^{196}At to ^{223}At	8.1 h

Concentrations in Crust/Rocks (mg/kg)

Bulk continental crust	- / - / -
Upper continental crust	- / -
Ultramafic rock	-
Ocean ridge basalt	-
Gabbro, basalt	-
Granite, granodiorite	-
Sandstone	-
Greywacke	-
Shale, schist	-
Limestone	-
Coal	-

Typical Minerals
-

Possible Host Minerals
uraninite/pitchblende, carnotite

Mass (kg) in

Continental crust	-	Oceans	-	Plants	-

Concentrations in Media Sampled for the Kola Project (mg/kg)

Medium	Moss conc. HNO$_3$	Humus <2 mm	Humus <2 mm amm. acet.	Topsoil (0-5 cm) <2 mm	B horizon <2 mm aqua regia	B horizon <2 mm total (XRF)
Median	-	-	-	-	-	-
Min	-	-	-	-	-	-
Max	-	-	-	-	-	-

Medium	C horizon <2 mm aqua regia	C horizon <2 mm total (XRF)	C horizon <2 mm total (INAA)	Lake water unfiltered (mg/l)
Median	-	-	-	-
Min	-	-	-	-
Max	-	-	-	-

ASTATINE   At

Environmental Geochemistry

Biological impacts	Considered non-essential
Uses	Organic compounds for medical purposes
Environmental pathways	Poorly understood
Environmental mobility	Oxidising conditions: - Acid conditions: - Reducing conditions: - Neutral to alkaline conditions: - Comments: Probably very high under all conditions
Geochemical barriers	-
Natural association	-
Action levels	-
Remarks	Accumulates in thyroid. About 0.000,05 mg pure At produced world-wide so far. Minimum quantities of At exist in nature, where it is in equilibrium with U- and Th-isotopes. Total mass in the earth's crust is less than 30 g. At behaves chemically very much like iodine
Suggested analytical method(s)	-

World Yearly Production (t/y)	-
Price of Purest Form, in Small Quantities ($/kg)	-
Market Price ($/kg)	-

Au GOLD

Physico-Chemical Properties

Atomic number	79	Melting/boiling point (K)	1337.33 / 3129
Atomic mass	196.966,55(2)	Isotopes & isomers	1 stable + 45 unstable
Atomic radius (pm)	179	Acid/base of oxide	amphoteric
Main oxidation state(s)	0,+1,+3 (+2,+5)	State (at 300 K, 1 atm.)	solid
Ionic radius (pm)	71-151	Metallic character	pred. metal
Electronegativity (Pauling)	2.54	Element group(s)	heavy metal, noble metal
Density (g/cm^3)	19.3	Affinity	siderophile

Naturally occurring isotopes	Natural abundance (%)	Atomic mass	Half-life
^{197}Au	100	196.966,5	stable

Unstable isotopes	Longest half-life
^{172}Au to ^{205}Au	186.12 d

Concentrations in Crust/Rocks (mg/kg)

Bulk continental crust	0.002,5 / 0.004 / 0.003
Upper continental crust	- / 0.001,8
Ultramafic rock	0.000,5
Ocean ridge basalt	0.000,2
Gabbro, basalt	0.001,5
Granite, granodiorite	0.002
Sandstone	0.000,5
Greywacke	0.004,8
Shale, schist	0.002
Limestone	0.000,1
Coal	0.01

Typical Minerals
native Au, calaverite (Au$_2$Te), krennerite ((Au,Ag)Te$_2$), sylvanite ((Ag,Au)Te$_4$), petzite (Ag$_3$AuTe$_2$)

Possible Host Minerals
tellurides, silver, native Au in quartz

Mass (kg) in

Continental crust	5.90x10^{+13}	Oceans	5.29x10^{+9}	Plants	1.84x10^{+6}

Concentrations in Media Sampled for the Kola Project (mg/kg)

Medium	Moss	Humus	Humus	Topsoil (0-5 cm)	B horizon	B horizon
		<2 mm	<2 mm	<2 mm	<2 mm	<2 mm
	conc. HNO$_3$	amm. acet.	total (INAA)	aqua regia	total (XRF)	
Median	-	-	-	<0.002	-	-
Min	-	-	-	<0.002	-	-
Max	-	-	-	0.029	-	-

Medium	C horizon	C horizon	C horizon			Lake water
	<2 mm	<2 mm	<2 mm			unfiltered
	aqua regia	total (XRF)	total (INAA)			(mg/l)
Median	-	-	<0.002			-
Min	-	-	<0.002			-
Max	-	-	0.981			-

GOLD   Au

Concentrations in Soils and Sediments (mg/kg)

Medium	Soil	Agricultural soil - Ap horizon	Agricultural soil - Top (0-25 cm)	Agricultural Soil - Bottom (50-75 cm)	Topsoil (0-15 cm)	Urban soil (0-2 cm)
	World	Canada	Finland	Finland	England & Wales	Trondheim
	<2 mm	<2 mm	<2 mm	<2 mm	<2 mm	<2 mm
	total	total (INAA)	aqua regia	aqua regia	aqua regia	aqua regia
Median	0.001,3*	<0.002	-	-	-	-
Min	-	<0.002	-	-	-	-
Max	-	2.3	-	-	-	-

*Estimated mean

Medium	Forest soil - Humus	Forest soil - B horizon	Forest soil - C horizon	Till (C horizon)	Till (C horizon)	Laterite (25 ±15 cm)
	Norway	Norway	Norway	Finland	Finland	Australia
	<2 mm	<2 mm	<2 mm	<0.063 mm	<0.063 mm	0.45-2 mm
	7N HNO$_3$	7N HNO$_3$	7N HNO$_3$	aqua regia	total	total
Median	-	-	-	-	0.000,6	<0.001
Min	-	-	-	-	-	<0.001
Max	-	-	-	-	-	0.02

Medium	Stream sediment	Stream sediment	Stream sediment	Overbank sediment	Overbank sediment	Organic stream sediment
	Austria	Southern Scotland	Harz, Germany	Norway	Norway	Finland
	<0.18 mm	<0.10 mm	<0.063 mm	<0.063 mm	<0.063 mm	<2 mm
	total	total	total (XRF)	7N HNO$_3$	7N HNO$_3$	Conc. HNO$_3$
Median	-	-	-	-	-	-
Min	-	-	-	-	-	-
Max	-	-	-	-	-	-

#2nd percentile   ^98th percentile

Concentrations in Waters (mg/l)

Medium	Ocean water World (1)	Ocean water World (2)	Ocean water North Pacific	Stream water World	Stream water Nova Scotia, Canada	Stream water Finland
					<0.45 µm	<0.45 µm
Median	0.000,004*	0.000,004*	0.000,000,02*	-	-	-
Min	-	-	-	-	-	-
Max	-	-	-	-	-	-

*Estimated mean   #2nd percentile   ^98th percentile

Medium	Stream water	Stream water	Stream water	Lake water	Ground water
	Romania	Eastern India	Harz, Germany	Norway	Southern Norway
	unfiltered	<0.2 µm	unfiltered	unfiltered	unfiltered
	ICP-MS	ICP-MS		ICP-MS	
Median	-	<0.000,01	-	-	-
Min	-	<0.000,01	-	-	-
Max	-	0.000,02	-	-	-

Au GOLD

Concentrations in Precipitation (mg/l)

Medium	Rain water Kola, remote <0.45 µm	Rain water Kola, polluted <0.45 µm	Snow melt- water Kola, remote <0.45 µm	Snow melt- water Kola, polluted <0.45 µm	Snow filter residue Kola, remote >0.45 µm	Snow filter residue Kola, polluted >0.45 µm
Median	-	-	-	-	-	-
Min	-	-	-	-	-	-
Max	-	-	-	-	-	-

Medium	Rain water Coastal, Norway unfiltered ICP-MS	Rain water Inland, Norway unfiltered ICP-MS				
Median	-	-				
Min	-	-				
Max	-	-				

Concentrations in Air (ng/m^3) and Yearly Deposition (kg/km^2/yr)

Medium	Air World, remote	Air World, polluted	Bulk deposition West Germany	Throughfall deposition West Germany	Bulk deposition Kola, remote	Bulk deposition Kola, polluted
Median	- / 0.000,09	-	-	-	-	-
Min	0.000,08	0.001	-	-	-	-
Max	0.001,5	0.3	-	-	-	-

*Estimated value

Concentrations in Plants (mg/kg)

Medium	Moss Norway conc. HNO$_3$ (1990)	Moss Germany conc. HNO$_3$ + H$_2$O$_2$ (1995)	Crustose lichen Germany conc. HNO$_3$	Lichen Northwest Territories oven-dried	Dandelion Europe oven-dried total (INAA)	Spruce bark Canada ashed total (INAA)
Median	-	-	-	-	0.011#	0.005
Min	-	-	-	-	-	<0.005
Max	-	-	-	-	-	0.078

*Mean   #Geometric mean

Concentrations in Human Fluids (mg/l)

Medium	Human blood Lombardy, Italy	Human serum Lombardy, Italy	Human urine Lombardy, Italy			
Mean	0.000,045	0.000,012	0.000,07			
Min	0.000,002	0.000,001	0.000,003			
Max	0.000,06	0.000,09	0.000,85			

GOLD — Au

Environmental Geochemistry

Biological impacts	Considered non-essential. Some Au compounds, especially Au^{3+} complexes (similar effects as As), are toxic to plants and animals
Uses	Jewellery, galvanising, electronics, dentistry, electroplating, catalyst, anti-arthritic drugs
Environmental pathways	Gold mining and smelting, other metal smelters, dentistry, wear of catalysts
Environmental mobility	Oxidising conditions: medium Acid conditions: high Reducing conditions: very low Neutral to alkaline conditions: very low Comments: -
Geochemical barriers	Mechanical
Natural association	Au-Ag-Si-As-Sb-S-Fe (gold quartz veins), Fe-Zn-Pb-Cu-Mo-Au (sulphide deposits), Au-U (placers)
Action levels	-
Remarks	Due to its high price, emission of Au by human activities are minimal, but enhanced Au levels in urban hare hair and city dust are reported. For unknown reasons, Au is enriched in atmospheric particulate material compared to mean crust and sea water concentrations. Volcanoes seem to emit large amounts of Au in both gaseous and aerosols forms (e.g., Mt Etna, Italy: 880 kg/yr; Mt St Helens, USA: 1100 kg/yr). Some organisms can enrich Au considerably (cyanogenic plants, e.g., brown algae)
Suggested analytical method(s)	ICP-MS, INAA, GF-AAS

World Yearly Production (t/y)	2000 (in 1995)
Price of Purest Form, in Small Quantities ($/kg)	154,375 (99.999,9%)
Market Price ($/kg)	11,399

B   BORON

Physico-Chemical Properties

Atomic number	5	Melting/boiling point (K)	2348 / 4273
Atomic mass	10.811(7)	Isotopes & isomers	2 stable + 9 unstable
Atomic radius (pm)	117	Acid/base of oxide	weak acid
Main oxidation state(s)	+3	State (at 300 K, 1 atm.)	solid
Ionic radius (pm)	15-41	Metallic character	non-metal
Electronegativity (Pauling)	2.04	Element group(s)	light non-metal
Density (g/cm^3)	2.34 (crystalline)	Affinity	lithophile

Naturally occurring isotopes	Natural abundance (%)	Atomic mass	Half-life
^{10}B	19.9	10.012,937	stable
^{11}B	80.1	11.009,306	stable

Unstable isotopes	Longest half-life
^{7}B to ^{19}B	0.77 s

Concentrations in Crust/Rocks (mg/kg)

Bulk continental crust	11 / 10 / 10
Upper continental crust	17 / 15
Ultramafic rock	3
Ocean ridge basalt	4
Gabbro, basalt	8
Granite, granodiorite	15
Sandstone	35
Greywacke	37
Shale, schist	130
Limestone	20
Coal	50

Typical Minerals

borax (Na$_2$B$_4$O$_7$.10H$_2$O), tourmaline ((Na,Ca)(Mg, Fe^{2+},Fe^{3+},Al,Li)Al$_6$(BO$_3$)$_3$Si$_6$O$_{18}$OH$_4$), colemanite (Ca$_2$B$_6$O$_{11}$.5H$_2$O)

Possible Host Minerals

feldspars, micas

Mass (kg) in

Continental crust	2.60x10^{+17}	Oceans	5.82x10^{+15}	Plants	7.36x10^{+10}

Concentrations in Media Sampled for the Kola Project (mg/kg)

Medium	Moss conc. HNO$_3$	Humus <2 mm conc. HNO$_3$	Humus <2 mm amm. acet.	Topsoil (0-5 cm) <2 mm	B horizon <2 mm aqua regia	B horizon <2 mm total (XRF)
Median	1.76	2.15	<0.1	-	<3	-
Min	<0.5	<0.8	<0.1	-	<3	-
Max	21.6	13	0.4	-	10	-

Medium	C horizon <2 mm aqua regia	C horizon <2 mm total (XRF)	C horizon <2 mm total (INAA)	Lake water unfiltered (mg/l)
Median	<3	-	-	0.001,365
Min	<3	-	-	0.000,24
Max	11	-	-	0.022,3

BORON   B

Concentrations in Soils and Sediments (mg/kg)

Medium	Soil	Agricultural soil - Ap horizon	Agricultural soil - Top (0-25 cm)	Agricultural Soil - Bottom (50-75 cm)	Topsoil (0-15 cm)	Urban soil (0-2 cm)
	World	Canada	Finland	Finland	England & Wales	Trondheim
	<2 mm	<2 mm	<2 mm	<2 mm	<2 mm	<2 mm
	total		aqua regia	aqua regia	aqua regia	aqua regia
Median	30*	-	-	-	-	5
Min	-	-	-	-	-	<3
Max	-	-	-	-	-	28

*Estimated mean

Medium	Forest soil - Humus	Forest soil - B horizon	Forest soil - C horizon	Till (C horizon)	Till (C horizon)	Laterite (25 ±15 cm)
	Norway	Norway	Norway	Finland	Finland	Australia
	<2 mm	<2 mm	<2 mm	<0.063 mm	<0.063 mm	0.45-2 mm
	7N HNO_3	7N HNO_3	7N HNO_3	aqua regia	total	total
Median	2.5	<0.3	<0.3	-	-	-
Min	<0.2	<0.3	<0.3	-	-	-
Max	58.7	9.5	15.7	-	-	-

Medium	Stream sediment	Stream sediment	Stream sediment	Overbank sediment	Overbank sediment	Organic stream sediment
	Austria	Southern Scotland	Harz, Germany	Norway	Norway	Finland
	<0.18 mm	<0.10 mm	<0.063 mm	<0.063 mm	<0.063 mm	<2 mm
		total (DCES)	total (ICP-MS)	total (XRF)	7N HNO_3	Conc. HNO_3
Median	-	53	23	-	-	2.21
Min	-	<8	<1.0	-	-	<0.5#
Max	-	133	110	-	-	16.1^

#2nd percentile   ^98th percentile

Concentrations in Waters (mg/l)

Medium	Ocean water	Ocean water	Ocean water	Stream water	Stream water	Stream water
	World (1)	World (2)	North Pacific	World	Nova Scotia, Canada	Finland
					<0.45 µm	<0.45 µm
						ICP-MS
Median	4.44*	4.4*	4.5*	0.01*	-	0.002,78
Min	-	-	-	-	-	0.000,6#
Max	-	-	-	-	-	0.074^

*Estimated mean   #2nd percentile   ^98th percentile

Medium	Stream water	Stream water	Stream water	Lake water	Ground water
	Romania	Eastern India	Harz, Germany	Norway	Southern Norway
	unfiltered	<0.2 µm	unfiltered	unfiltered	unfiltered
	ICP-MS	ICP-MS	ICP-MS	ICP-MS	ICP-MS
Median	0.056	0.388	0.007,5	0.001,41	0.020,6
Min	0.001	0.324	0.001,3	<0.000,2	0.000,5
Max	2.97	0.466	0.086	2.513	0.854

B BORON

Concentrations in Precipitation (mg/l)

Medium	Rain water Kola, remote <0.45 µm ICP-MS	Rain water Kola, polluted <0.45 µm ICP-MS	Snow melt- water Kola, remote <0.45 µm ICP-MS	Snow melt- water Kola, polluted <0.45 µm ICP-MS	Snow filter residue Kola, remote >0.45 µm ICP-AES	Snow filter residue Kola, polluted >0.45 µm ICP-AES
Median	<0.000,5	0.000,73	<0.000,5	<0.000,5	<0.000,3	<0.000,3
Min	<0.000,5	<0.000,5	<0.000,5	<0.000,5	<0.000,3	<0.000,3
Max	0.001,17	0.002,36	<0.000,5	0.000,65	<0.000,3	0.000,4

Medium	Rain water Coastal, Norway unfiltered ICP-MS	Rain water Inland, Norway unfiltered ICP-MS
Median	-	-
Min	-	-
Max	-	-

Concentrations in Air (ng/m^3) and Yearly Deposition (kg/km^2/yr)

Medium	Air World, remote	Air World, polluted	Bulk deposition West Germany	Throughfall deposition West Germany	Bulk deposition Kola, remote	Bulk deposition Kola, polluted
Median	-	-	-	-	-	-
Min	-	-	-	-	-	-
Max	-	-	-	-	-	-

*Estimated value

Concentrations in Plants (mg/kg)

Medium	Moss Norway conc. HNO$_3$ (1990)	Moss Germany conc. HNO$_3$ + H$_2$O$_2$ (1995)	Crustose lichen Germany conc. HNO$_3$	Lichen Northwest Territories oven-dried	Dandelion Europe oven-dried	Spruce bark Canada ashed aqua regia
Median	3.1	5.785	-	-	-	96
Min	0.4	0.1	-	-	-	16
Max	21	73	-	-	-	284

*Mean #Geometric mean

Concentrations in Human Fluids (mg/l)

Medium	Human blood Lombardy, Italy	Human serum Lombardy, Italy	Human urine Lombardy, Italy
Mean	-	-	1.89
Min	-	-	0.47
Max	-	-	780

BORON   B

Environmental Geochemistry

Biological impacts	Essential for some organisms (probably not for humans). Elemental B and borates considered non-toxic to humans. Toxic to plants at high levels (soils in semi-arid, arid climates, soils on marine sediments)
Uses	Laundry products, alloys, abrasives, ceramics, glass, enamel, fertilisers, insulating fibreglass, flame retardant (textile industry), antiseptics, nuclear industry
Environmental pathways	Sea spray, coal combustion, sewage, fertilisers
Environmental mobility	Oxidising conditions: very high Acid conditions: very high Reducing conditions: very low Neutral to alkaline conditions: very high Comments: -
Geochemical barriers	Adsorption on clay minerals
Natural association	Li-Be-B-Nb-Sn (pegmatites), B-Be-Cu-Zn-Pb-Mo-W (skarns), B-Be-Sn-F-W (greisens), B-Na-K-Li-Mg-Ca (evaporites)
Action levels	Drinking water; recommended: 0.3 mg/l (Norway Shd), 0.3 mg/l (WHO); MAC: 5 mg/l (Canada CWQG), 0.5 mg/l (Russia MoH). Agricultural soil, maximum tolerable concentration: 25 mg/kg (Germany)
Remarks	Adsorption to soil is pH-dependent. When using B-rich fertilisers, over-fertilisation is possible on acid, organic soils. Pine has low B-tolerance, grass has high tolerance. Some plants are indicators of B deposits. Plant toxicity successfully used in mineral exploration. Water with more than 0.3 mg B/l should not be used for irrigation
Suggested analytical method(s)	ICP-MS

World Yearly Production (t/y)   494,000 (in 1995)
Price of Purest Form, in Small Quantities ($/kg)   586,813 (99.999,5%)
Market Price ($/kg)   0.3 (colemanite, ca. 40% B_2O_3)

Ba — BARIUM

Physico-Chemical Properties

Atomic number	56
Atomic mass	137.327(7)
Atomic radius (pm)	278
Main oxidation state(s)	+2
Ionic radius (pm)	149-175
Electronegativity (Pauling)	0.89
Density (g/cm^3)	3.5
Melting/boiling point (K)	1000 / 2170
Isotopes & isomers	7 stable + 33 unstable
Acid/base of oxide	strong base
State (at 300 K, 1 atm.)	solid
Metallic character	metal
Element group(s)	alkaline earth, light metal
Affinity	lithophile

Naturally occurring isotopes	Natural abundance (%)	Atomic mass	Half-life
^{130}Ba	0.106	129.906,3	stable
^{132}Ba	0.101	131.905,06	stable
^{134}Ba	2.417	133.904,5	stable
^{135}Ba	6.592	134.905,68	stable
^{136}Ba	7.854	135.904,57	stable
^{137}Ba	11.23	136.905,82	stable
^{138}Ba	71.71	137.905,24	stable

Unstable isotopes	Longest half-life
^{117}Ba to ^{149}Ba	10.53 y

Concentrations in Crust/Rocks (mg/kg)

Bulk continental crust	584 / 425 / 250
Upper continental crust	668 / 550
Ultramafic rock	5
Ocean ridge basalt	60
Gabbro, basalt	330
Granite, granodiorite	600
Sandstone	300
Greywacke	426
Shale, schist	550
Limestone	90
Coal	250

Typical Minerals
barite (BaSO$_4$), witherite (BaCO$_3$)

Possible Host Minerals
K-feldspar, micas, apatite, calcite

Mass (kg) in

Continental crust 1.38x10^{+19}	Oceans 2.64x10^{+13}	Plants 7.36x10^{+10}

Concentrations in Media Sampled for the Kola Project (mg/kg)

Medium	Moss	Humus	Humus	Topsoil (0-5 cm)	B horizon	B horizon
	<2 mm	<2 mm	<2 mm	<2 mm	<2 mm	<2 mm
	conc. HNO$_3$	conc. HNO$_3$	amm. acet.	total (INAA)	aqua regia	total (XRF)
Median	19	76.2	25.4	390	28.7	-
Min	6.71	13.9	1.77	<50	5.6	-
Max	175	290	92	3000	385	-

Medium	C horizon	C horizon	C horizon			Lake water
	<2 mm	<2 mm	<2 mm			unfiltered
	aqua regia	total (XRF)	total (INAA)			(mg/l)
Median	43.5	-	575			0.004,27
Min	4.7	-	210			0.000,69
Max	1300	-	3000			0.026,7

BARIUM   Ba

Concentrations in Soils and Sediments (mg/kg)

Medium	Soil	Agricultural soil - Ap horizon	Agricultural soil - Top (0-25 cm)	Agricultural Soil - Bottom (50-75 cm)	Topsoil (0-15 cm)	Urban soil (0-2 cm)
	World	Canada	Finland	Finland	England & Wales	Trondheim
	<2 mm	<2 mm	<2 mm	<2 mm	<2 mm	<2 mm
	total	total (INAA)	aqua regia	aqua regia	aqua regia	aqua regia
Median	500*	660	43.2	37.3	121	72.2
Min	-	120	1.86	4.79	11	18
Max	-	4300	209	272	2973	385

*Estimated mean

Medium	Forest soil - Humus	Forest soil - B horizon	Forest soil - C horizon	Till (C horizon)	Till (C horizon)	Laterite (25 ±15 cm)
	Norway	Norway	Norway	Finland	Finland	Australia
	<2 mm	<2 mm	<2 mm	<0.063 mm	<0.063 mm	0.45-2 mm
	7N HNO$_3$	7N HNO$_3$	7N HNO$_3$	aqua regia	total	total
Median	46	33.1	53.8	51.3	570	228
Min	5.3	5.5	5.9	-	-	13
Max	483.3	267	475.3	-	-	1257

Medium	Stream sediment	Stream sediment	Stream sediment	Overbank sediment	Overbank sediment	Organic stream sediment
	Austria	Southern Scotland	Harz, Germany	Norway	Norway	Finland
	<0.18 mm	<0.10 mm	<0.063 mm	<0.063 mm	<0.063 mm	<2 mm
	"total" (ICP-AES)	total (DCES)	total (XRF)	total (XRF)	7N HNO$_3$ ICP-AES	Conc. HNO$_3$
Median	490	607	376	554	64	90
Min	-	54	127	120	8.9	33#
Max	5930	134	5686	2988	672	309^

#2nd percentile   ^98th percentile

Concentrations in Waters (mg/l)

Medium	Ocean water	Ocean water	Ocean water	Stream water	Stream water	Stream water
	World (1)	World (2)	North Pacific	World	Nova Scotia, Canada	Finland
					<0.45 µm	<0.45 µm
					ICP-MS	ICP-MS
Median	0.013*	0.02*	0.015*	0.02*	0.006,9	0.01
Min	-	-	-	-	<0.000,2	0.001,95#
Max	-	-	-	-	0.127,4	0.039^

*Estimated mean   #2nd percentile   ^98th percentile

Medium	Stream water	Stream water	Stream water	Lake water	Ground water
	Romania	Eastern India	Harz, Germany	Norway	Southern Norway
	unfiltered	<0.2 µm	unfiltered	unfiltered	unfiltered
	ICP-MS	ICP-MS	ICP-MS	ICP-MS	ICP-MS
Median	0.056	0.042,7	0.032	0.003,08	0.016,65
Min	0.004	0.019,3	0.002	0.000,11	0.000,2
Max	1.39	0.176,4	3.37	0.147	0.237,1

Ba — BARIUM

Concentrations in Precipitation (mg/l)

Medium	Rain water	Rain water	Snow melt-water	Snow melt-water	Snow filter residue	Snow filter residue
	Kola, remote <0.45 µm ICP-MS	Kola, polluted <0.45 µm ICP-MS	Kola, remote <0.45 µm ICP-MS	Kola, polluted <0.45 µm ICP-MS	Kola, remote >0.45 µm ICP-AES	Kola, polluted >0.45 µm ICP-AES
Median	0.000,47	0.001,07	0.000,15	0.001,88	0.000,15	0.001,96
Min	0.000,4	0.000,6	0.000,13	0.001,45	<0.000,1	0.000,85
Max	0.001,1	0.001,09	0.014	0.002,71	0.000,67	0.005,38

Medium	Rain water	Rain water
	Coastal, Norway unfiltered ICP-MS	Inland, Norway unfiltered ICP-MS
Median	0.000,17	0.000,85
Min	<0.000,01	<0.000,01
Max	0.003,5	0.028

Concentrations in Air (ng/m^3) and Yearly Deposition (kg/km^2/yr)

Medium	Air	Air	Bulk deposition	Throughfall deposition	Bulk deposition	Bulk deposition
	World, remote	World, polluted	West Germany	West Germany	Kola, remote	Kola, polluted
Median	0.4 / 0.8	12	-	-	-	-
Min	0.016	0.2	-	-	-	-
Max	0.73 (10)	90	-	-	-	-

*Estimated value

Concentrations in Plants (mg/kg)

Medium	Moss	Moss	Crustose lichen	Lichen	Dandelion	Spruce bark
	Norway conc. HNO$_3$ (1990)	Germany conc. HNO$_3$ + H$_2$O$_2$ (1995)	Germany conc. HNO$_3$	Northwest Territories oven-dried	Europe oven-dried total (INAA)	Canada ashed total (INAA)
Median	24	19.5	75	-	49#	2000
Min	4.5	4	18	-	-	560
Max	108	306	650	-	-	8500

*Mean #Geometric mean

Concentrations in Human Fluids (mg/l)

Medium	Human blood	Human serum	Human urine
	Lombardy, Italy	Lombardy, Italy	Lombardy, Italy
Mean	0.001,2	-	0.002,7
Min	0.000,47	-	0.000,25
Max	0.002,9	-	0.010,1

BARIUM   Ba

Environmental Geochemistry

Biological impacts	May be essential for some organisms. Soluble compounds toxic to humans, animals and plants.
Uses	Drilling mud, glass hardener, television tubes (radiation absorbent), paint, rubber, paper industry, electronics, pyrotechnics, ceramics
Environmental pathways	Poorly understood. Probably mostly windblown dust, weathering of Ba-bearing minerals. Generally, geogenic sources more important than anthropogenic ones, which include Cu smelting, automobile industry and steel works
Environmental mobility	Oxidising conditions: low Acid conditions: low Reducing conditions: very low Neutral to alkaline conditions: low Comments: In water, the more toxic soluble Ba salts are likely to be converted to insoluble salts, which precipitate. Ba does not bind to most soils, and may migrate to ground water. Low tendency to accumulate in aquatic life
Geochemical barriers	Presence of sulphate, carbonate. Adsorption
Natural association	K-Rb-Ba (granitoids), Ba-Pb-Zn (sulphide ores, carbonatites)
Action levels	Drinking water, recommended: 0.1 mg/l (Norway Shd); MAC: 2 mg/l (US EPA), 1 mg/l (Canada CWQG), 0.1 mg/l (Russia MoH). Ground water, background: 0.05 mg/l; remediate: 0.625 mg/l (Netherlands VROM). Soil, background: 200 mg/kg; remediate: 625 mg/kg (Netherlands VROM)
Remarks	Liming and addition of sulphates immobilises Ba in soils. Ba is especially enriched in Mn-coatings in soils. Ba-carbonate used as rat poison. Technically used in sewage plants. About 90% of the world production of Ba goes to drilling mud production. The US EPA suggests that drinking water exceeding action levels can lead to gastrointestinal disturbances and muscular weakness in the short term, and to high blood pressure in the long term
Suggested analytical method(s)	XRF, ICP-AES, ICP-MS (water)

World Yearly Production (t/y)	2,900,000 (in 1995, from 4,900,000 $BaSO_4$)
Price of Purest Form, in Small Quantities ($/kg)	26,025 (99.9%)
Market Price ($/kg)	50 (barite, drilling mud), 300 (barite, filler)

Be — BERYLLIUM

Physico-Chemical Properties

Atomic number	4
Atomic mass	9.012,182(3)
Atomic radius (pm)	140
Main oxidation state(s)	+2
Ionic radius (pm)	30-49
Electronegativity (Pauling)	1.57
Density (g/cm^3)	1.848
Melting/boiling point (K)	1560 / 2744
Isotopes & isomers	1 stable + 7 unstable
Acid/base of oxide	amphoteric
State (at 300 K, 1 atm.)	solid
Metallic character	pred. metal
Element group(s)	alkaline earth, light metal
Affinity	lithophile

Naturally occurring isotopes	Natural abundance (%)	Atomic mass	Half-life
^{9}Be	100	9.012,182	stable

Unstable isotopes	Longest half-life
^{6}Be to ^{14}Be	1,520,000 y

Concentrations in Crust/Rocks (mg/kg)

Bulk continental crust	2.4 / 2.8 / 1.5
Upper continental crust	3.1 / 3
Ultramafic rock	0.3
Ocean ridge basalt	1
Gabbro, basalt	1
Granite, granodiorite	5
Sandstone	0.7
Greywacke	-
Shale, schist	3
Limestone	0.5
Coal	1

Typical Minerals
beryl ($Be_3Al_2Si_6O_{18}$), bertrandite ($Be_2SiO_4 \cdot H_2O$)

Possible Host Minerals
plagioclases, micas, pyroxenes, clay minerals

Mass (kg) in

Continental crust 5.66x10^{+16} Oceans 7.40x10^{+9} Plants 1.84x10^{+6}

Concentrations in Media Sampled for the Kola Project (mg/kg)

Medium	Moss	Humus <2 mm	Humus <2 mm	Topsoil (0-5 cm) <2 mm	B horizon <2 mm	B horizon <2 mm
	conc. HNO$_3$	conc. HNO$_3$	amm. acet.		aqua regia	total (XRF)
Median	<0.03	0.04	-	-	0.28	-
Min	<0.03	<0.02	-	-	0.08	-
Max	1.51	1.87	-	-	8.28	-

Medium	C horizon <2 mm	C horizon <2 mm	C horizon <2 mm	Lake water unfiltered
	aqua regia	total (XRF)	total (INAA)	(mg/l)
Median	0.23	-	-	<0.000,05
Min	0.06	-	-	<0.000,05
Max	14	-	-	0.000,2

BERYLLIUM — Be

Concentrations in Soils and Sediments (mg/kg)

Medium	Soil	Agricultural soil - Ap horizon	Agricultural soil - Top (0-25 cm)	Agricultural Soil - Bottom (50-75 cm)	Topsoil (0-15 cm)	Urban soil (0-2 cm)
	World	Canada	Finland	Finland	England & Wales	Trondheim
	<2 mm total	<2 mm	<2 mm aqua regia	<2 mm aqua regia	<2 mm aqua regia	<2 mm aqua regia
Median	3*	-	<0.5	<0.5	-	-
Min	-	-	<0.5	<0.5	-	-
Max	-	-	1.5	2.08	-	-

*Estimated mean

Medium	Forest soil - Humus	Forest soil - B horizon	Forest soil - C horizon	Till (C horizon)	Till (C horizon)	Laterite (25 ±15 cm)
	Norway	Norway	Norway	Finland	Finland	Australia
	<2 mm 7N HNO$_3$	<2 mm 7N HNO$_3$	<2 mm 7N HNO$_3$	<0.063 mm aqua regia	<0.063 mm total	0.45-2 mm total
Median	0.2	1.5	1.6	-	-	<0.2
Min	<0.05	0.1	0.3	-	-	<0.2
Max	3	4.8	8	-	-	5.3

Medium	Stream sediment	Stream sediment	Stream sediment	Overbank sediment	Overbank sediment	Organic stream sediment
	Austria	Southern Scotland	Harz, Germany	Norway	Norway	Finland
	<0.18 mm "total" (ICP-AES)	<0.10 mm total (DCES)	<0.063 mm total (ICP-MS)	<0.063 mm total (XRF)	<0.063 mm 7N HNO$_3$	<2 mm Conc. HNO$_3$
Median	4	2	4.1	-	-	0.33
Min	-	<0.3	0.8	-	-	<0.1#
Max	30	105	70	-	-	2.45^

#2nd percentile   ^98th percentile

Concentrations in Waters (mg/l)

Medium	Ocean water	Ocean water	Ocean water	Stream water	Stream water	Stream water
	World (1)	World (2)	North Pacific	World	Nova Scotia, Canada	Finland
					<0.45 µm ICP-MS	<0.45 µm ICP-MS
Median	0.000,005,6*	0.000,005,6*	0.000,000,21*	0.000,1*	0.000,015	<0.000,1
Min	-	-	-	-	<0.000,005	<0.000,1#
Max	-	-	-	-	0.000,135	0.000,228^

*Estimated mean   #2nd percentile   ^98th percentile

Medium	Stream water	Stream water	Stream water	Lake water	Ground water
	Romania	Eastern India	Harz, Germany	Norway	Southern Norway
	unfiltered ICP-MS	<0.2 µm ICP-MS	unfiltered ICP-MS	unfiltered ICP-MS	unfiltered ICP-MS
Median	0.000,021	<0.000,1	0.000,18	<0.000,01	0.000,04
Min	0.000,005	<0.000,1	<0.000,01	<0.000,01	<0.000,002
Max	0.008,3	<0.000,1	0.006,2	0.001,34	0.002,82

Be   BERYLLIUM

Concentrations in Precipitation (mg/l)

Medium	Rain water	Rain water	Snow melt-water	Snow melt-water	Snow filter residue	Snow filter residue
	Kola, remote <0.45 µm ICP-MS	Kola, polluted <0.45 µm ICP-MS	Kola, remote <0.45 µm ICP-MS	Kola, polluted <0.45 µm ICP-MS	Kola, remote >0.45 µm	Kola, polluted >0.45 µm
Median	<0.000,1	<0.000,1	<0.000,3	<0.000,3	-	-
Min	<0.000,1	<0.000,1	<0.000,3	<0.000,3	-	-
Max	<0.000,1	<0.000,1	<0.000,3	<0.000,3	-	-

Medium	Rain water	Rain water				
	Coastal, Norway unfiltered ICP-MS	Inland, Norway unfiltered ICP-MS				
Median	0.000,02	0.000,02				
Min	<0.000,01	<0.000,01				
Max	0.000,18	0.000,29				

Concentrations in Air (ng/m^3) and Yearly Deposition (kg/km^2/yr)

Medium	Air	Air	Bulk deposition	Throughfall deposition	Bulk deposition	Bulk deposition
	World, remote	World, polluted	West Germany	West Germany	Kola, remote	Kola, polluted
Median	-	-	-	-	-	-
Min	-	-	-	-	-	-
Max	-	-	-	-	-	-

*Estimated value

Concentrations in Plants (mg/kg)

Medium	Moss	Moss	Crustose lichen	Lichen	Dandelion	Spruce bark
	Norway conc. HNO$_3$ (1990)	Germany conc. HNO$_3$ + H$_2$O$_2$ (1995)	Germany conc. HNO$_3$	Northwest Territories oven-dried	Europe oven-dried	Canada ashed aqua regia
Median	<0.02	0.032	-	-	-	<0.2
Min	<0.02	<0.000,1	-	-	-	<0.2
Max	0.32	0.47	-	-	-	1.3

*Mean   #Geometric mean

Concentrations in Human Fluids (mg/l)

Medium	Human blood	Human serum	Human urine			
	Lombardy, Italy	Lombardy, Italy	Lombardy, Italy			
Mean	-	0.000,15	0.000,4			
Min	-	<0.000,08	<0.000,02			
Max	-	0.000,36	0.000,82			

BERYLLIUM   Be

Environmental Geochemistry

Biological impacts	Considered non-essential. Toxic. Carcinogenic for several animal classes. Metal and compounds are allergenic.
Uses	High power electronic circuitry, automotive ignition systems, telecommunications, laser bores, microwave radar devices, alloys (esp. with Cu), aircraft, spacecraft, missiles, satellites, nuclear industry, rocket fuel
Environmental pathways	Coal combustion, Cu rolling, non-ferrous metal smelting, Al smelting, rock dust
Environmental mobility	Oxidising conditions: low Acid conditions: low Reducing conditions: very low Neutral to alkaline conditions: low Comments: Very little is known about the fate of Be when released to the environment
Geochemical barriers	Adsorption on clay minerals, organic matter, Fe-Mn oxides
Natural association	Li-Be-B-Rb-Cs (pegmatites), B-Be-Cu-Zn-Pb-Mo-W (skarns), B-Be-Sn-F-W (greisens)
Action levels	Drinking water: listed in Norwegian regulations but no value given; MAC: 0.004 mg/l (US EPA); no adequate data to permit recommendation of heath-based guideline value (WHO); MAC: 0.000,2 mg/l (Russia MoH)
Remarks	Be is toxic if inhaled. Hickory known to accumulate Be. Be can replace Si in many silicates (similar ionic radii). Be-rich soils (up to 300 mg/kg) known to occur naturally. BeO ceramics are used for their high thermal conductivity and good electrical insulation properties. The US EPA suggests that drinking water exceeding action levels can lead to damage to bones and lungs, and to cancer in the long term
Suggested analytical method(s)	GF-AAS, ICP-MS

World Yearly Production (t/y)   327 (in 1995)
Price of Purest Form, in Small Quantities ($/kg)   12,075 (99.7%)
Market Price ($/kg)   0.08 (beryl, 10% BeO)

Bi — BISMUTH

Physico-Chemical Properties

Atomic number	83	Melting/boiling point (K)	544.55 / 1837
Atomic mass	208.980,38(2)	Isotopes & isomers	1 stable + 40 unstable
Atomic radius (pm)	163	Acid/base of oxide	weak acid
Main oxidation state(s)	+3 (+5)	State (at 300 K, 1 atm.)	solid
Ionic radius (pm)	110-131 (+3), 90 (+5)	Metallic character	pred. metal
Electronegativity (Pauling)	2.02	Element group(s)	heavy metal
Density (g/cm^3)	9.747	Affinity	chalcophile

Naturally occurring isotopes	Natural abundance (%)	Atomic mass	Half-life	Unstable isotopes	Longest half-life
^{209}Bi	100	208.980,38	stable	^{187}Bi to ^{216}Bi	3,000,000 y

Concentrations in Crust/Rocks (mg/kg)

Bulk continental crust	0.085 / 0.008,5 / 0.06
Upper continental crust	0.123 / 0.127
Ultramafic rock	0.01
Ocean ridge basalt	0.01
Gabbro, basalt	0.05
Granite, granodiorite	0.2
Sandstone	0.05
Greywacke	-
Shale, schist	0.25
Limestone	0.1
Coal	-

Typical Minerals
bismuthinite (Bi$_2$S$_3$), bismite (Bi$_2$O$_3$)

Possible Host Minerals
apatite, galena, sphalerite, chalcopyrite

Mass (kg) in

Continental crust	2.01x10^{+15}	Oceans	2.64x10^{+10}	Plants	1.84x10^{+7}

Concentrations in Media Sampled for the Kola Project (mg/kg)

Medium	Moss	Humus	Humus	Topsoil (0-5 cm)	B horizon	B horizon
	<2 mm	<2 mm	<2 mm	<2 mm	<2 mm	<2 mm
	conc. HNO$_3$	conc. HNO$_3$	amm. acet.		aqua regia	total (XRF)
Median	0.018	0.159	-	-	0.031	-
Min	<0.004	0.029	-	-	<0.005	-
Max	0.544	1.12	-	-	0.308	-

Medium	C horizon	C horizon	C horizon	Lake water
	<2 mm	<2 mm	<2 mm	unfiltered
	aqua regia	total (XRF)	total (INAA)	(mg/l)
Median	0.026	-	-	<0.000,01
Min	<0.005	-	-	<0.000,01
Max	3.89	-	-	0.000,02

BISMUTH   Bi

Concentrations in Soils and Sediments (mg/kg)

Medium	Soil	Agricultural soil - Ap horizon	Agricultural soil - Top (0-25 cm)	Agricultural Soil - Bottom (50-75 cm)	Topsoil (0-15 cm)	Urban soil (0-2 cm)
	World	Canada	Finland	Finland	England & Wales	Trondheim
	<2 mm	<2 mm	<2 mm	<2 mm	<2 mm	<2 mm
	total		aqua regia	aqua regia	aqua regia	aqua regia
Median	0.3*	-	<0.02	<0.02	-	-
Min	-	-	<0.02	<0.02	-	-
Max	-	-	0.136	0.183	-	-

*Estimated mean

Medium	Forest soil - Humus	Forest soil - B horizon	Forest soil - C horizon	Till (C horizon)	Till (C horizon)	Laterite (25 ±15 cm)
	Norway	Norway	Norway	Finland	Finland	Australia
	<2 mm	<2 mm	<2 mm	<0.063 mm	<0.063 mm	0.45-2 mm
	7N HNO$_3$	7N HNO$_3$	7N HNO$_3$	aqua regia	total	total
Median	-	-	-	-	-	0.2
Min	-	-	-	-	-	<0.1
Max	-	-	-	-	-	9.5

Medium	Stream sediment	Stream sediment	Stream sediment	Overbank sediment	Overbank sediment	Organic stream sediment
	Austria	Southern Scotland	Harz, Germany	Norway	Norway	Finland
	<0.18 mm	<0.10 mm	<0.063 mm	<0.063 mm	<0.063 mm	<2 mm
		total (DCES)	total (ICP-MS)	total (XRF)	7N HNO$_3$ AAS	Conc. HNO$_3$
Median	-	<4	0.53	-	0.12	0.08
Min	-	<4	0.2	-	<0.02	0.03#
Max	-	18	2.1	-	2.24	0.31^

#2nd percentile   ^98th percentile

Concentrations in Waters (mg/l)

Medium	Ocean water	Ocean water	Ocean water	Stream water	Stream water	Stream water
	World (1)	World (2)	North Pacific	World	Nova Scotia, Canada	Finland
					<0.45 µm	<0.45 µm
					ICP-MS-Hydride	ICP-MS
Median	0.000,02*	0.000,02*	0.000,000,03*	0.000,005*	<0.000,004	<0.000,03
Min	-	-	-	-	<0.000,004	<0.000,03#
Max	-	-	-	-	0.000,012	0.000,036^

*Estimated mean   #2nd percentile   ^98th percentile

Medium	Stream water	Stream water	Stream water	Lake water	Ground water	
	Romania	Eastern India	Harz, Germany	Norway	Southern Norway	
	unfiltered	<0.2 µm	unfiltered	unfiltered	unfiltered	
	ICP-MS	ICP-MS	ICP-MS	ICP-MS	ICP-MS	
Median	0.000,006	0.000,04	<0.000,02	<0.000,02	0.000,001	
Min	0.000,001	0.000,03	<0.000,02	<0.000,02	<0.000,001	
Max	0.000,28	0.001,15	0.001,3	0.003,62	0.000,16	

Bi — BISMUTH

Concentrations in Precipitation (mg/l)

Medium	Rain water Kola, remote <0.45 μm ICP-MS	Rain water Kola, polluted <0.45 μm ICP-MS	Snow melt-water Kola, remote <0.45 μm ICP-MS	Snow melt-water Kola, polluted <0.45 μm ICP-MS	Snow filter residue Kola, remote >0.45 μm	Snow filter residue Kola, polluted >0.45 μm
Median	<0.000,03	<0.000,03	<0.000,02	<0.000,02	-	-
Min	<0.000,03	<0.000,03	<0.000,02	<0.000,02	-	-
Max	<0.000,03	0.000,08	<0.000,02	<0.000,02	-	-

Medium	Rain water Coastal, Norway unfiltered ICP-MS	Rain water Inland, Norway unfiltered ICP-MS				
Median	<0.000,01	<0.000,01				
Min	<0.000,01	<0.000,01				
Max	0.000,06	0.000,34				

Concentrations in Air (ng/m^3) and Yearly Deposition (kg/km^2/yr)

Medium	Air World, remote	Air World, polluted	Bulk deposition West Germany	Throughfall deposition West Germany	Bulk deposition Kola, remote	Bulk deposition Kola, polluted
Median	-	-	-	-	-	-
Min	-	-	-	-	-	-
Max	-	-	-	-	-	-

*Estimated value

Concentrations in Plants (mg/kg)

Medium	Moss Norway conc. HNO$_3$ (1990)	Moss Germany conc. HNO$_3$ + H$_2$O$_2$ (1995)	Crustose lichen Germany conc. HNO$_3$	Lichen Northwest Territories oven-dried	Dandelion Europe oven-dried	Spruce bark Canada ashed
Median	0.03	0.03	0.2	-	-	-
Min	<0.01	0.007	0.08	-	-	-
Max	0.91	0.48	16	-	-	-

*Mean #Geometric mean

Concentrations in Human Fluids (mg/l)

Medium	Human blood Lombardy, Italy	Human serum Lombardy, Italy	Human urine Lombardy, Italy			
Mean	0.000,49	-	0.001,2			
Min	0.000,12	-	0.000,2			
Max	0.000,89	-	0.002,55			

BISMUTH   Bi

Environmental Geochemistry

Biological impacts	Considered non-essential. Rare toxic effects to humans reported (e.g., uptake from cosmetics). Interferes with growth of micro-organisms
Uses	Low melting point alloys, fire detection, fire extinguishing systems, catalysts (e.g., production of acrylic fibres), cosmetics, pharmaceutical industry, batteries, magnets, dentistry
Environmental pathways	Pb, Cu, Ag, Au smelting
Environmental mobility	Oxidising conditions: low — Acid conditions: low Reducing conditions: very low — Neutral to alkaline conditions: low Comments: -
Geochemical barriers	Fe-oxides
Natural association	Mo-Sn-W-Cu-Pb-Ag-Au (polymetallic deposits)
Action levels	Drinking water, MAC: 0.1 mg/l (Russia MoH)
Remarks	Usually forms insoluble compounds. No known environmental damage related to Bi. Most diamagnetic of all metals. Bi usually recovered from Pb, Cu, Sn, Ag, Au production processes
Suggested analytical method(s)	ICP-MS, hydride-AAS

World Yearly Production (t/y)	3100 (in 1995)
Price of Purest Form, in Small Quantities ($/kg)	5528 (99.999,9%)
Market Price ($/kg)	6.83

Br BROMINE

Physico-Chemical Properties

Atomic number	35	Melting/boiling point (K)	265.95 / 331.95
Atomic mass	79.904(1)	Isotopes & isomers	2 stable + 31 unstable
Atomic radius (pm)	112	Acid/base of oxide	strong acid
Main oxidation state(s)	-1,+1,+3,+5,+7	State (at 300 K, 1 atm.)	liquid
Ionic radius (pm)	182 (-1), 73 (+3), 45 (+5), 39-53 (+7)	Metallic character	non-metal
Electronegativity (Pauling)	2.96	Element group(s)	halogen, light non-metal
Density (g/cm^3)	3.12	Affinity	atmophile, lithophile

Naturally occurring isotopes	Natural abundance (%)	Atomic mass	Half-life
^{79}Br	50.69	78.918,34	stable
^{81}Br	49.31	80.916,29	stable

Unstable isotopes	Longest half-life
^{69}Br to ^{94}Br	2.376 d

Concentrations in Crust/Rocks (mg/kg)

Bulk continental crust	1 / 2.4 / -
Upper continental crust	1.6 / 2
Ultramafic rock	1
Ocean ridge basalt	-
Gabbro, basalt	1
Granite, granodiorite	2
Sandstone	0.3
Greywacke	-
Shale, schist	6
Limestone	6
Coal	20

Typical Minerals
bromargyrite (AgBr), carnallite (KMg(Cl,Br)$_3$.6H$_2$O)

Possible Host Minerals
amphiboles, biotite, apatite, eudialyte, sodalite

Mass (kg) in

Continental crust	2.36x10^{+16}	Oceans	8.86x10^{+16}	Plants	7.36x10^{+9}

Concentrations in Media Sampled for the Kola Project (mg/kg)

Medium	Moss <2 mm conc. HNO$_3$	Humus <2 mm	Humus <2 mm amm. acet.	Topsoil (0-5 cm) <2 mm total (INAA)	B horizon <2 mm aqua regia	B horizon <2 mm total (XRF)
Median	-	-	-	4.5	-	-
Min	-	-	-	<0.5	-	-
Max	-	-	-	240	-	-

Medium	C horizon <2 mm aqua regia	C horizon <2 mm total (XRF)	C horizon <2 mm total (INAA)		Lake water unfiltered (mg/l)
Median	-	-	3.7		<0.03
Min	-	-	<0.5		<0.03
Max	-	-	56		0.16

BROMINE   Br

Concentrations in Soils and Sediments (mg/kg)

Medium	Soil	Agricultural soil - Ap horizon	Agricultural soil - Top (0-25 cm)	Agricultural Soil - Bottom (50-75 cm)	Topsoil (0-15 cm)	Urban soil (0-2 cm)
	World	Canada	Finland	Finland	England & Wales	Trondheim
	<2 mm	<2 mm	<2 mm	<2 mm	<2 mm	<2 mm
	total	total (INAA)	aqua regia	aqua regia	aqua regia	aqua regia
Median	10*	4.7	-	-	-	-
Min	-	0.7	-	-	-	-
Max	-	163	-	-	-	-

*Estimated mean

Medium	Forest soil - Humus	Forest soil - B horizon	Forest soil - C horizon	Till (C horizon)	Till (C horizon)	Laterite (25 ±15 cm)
	Norway	Norway	Norway	Finland	Finland	Australia
	<2 mm	<2 mm	<2 mm	<0.063 mm	<0.063 mm	0.45-2 mm
	7N HNO₃	7N HNO₃	7N HNO₃	aqua regia	total	total
Median	-	-	-	-	4.3	-
Min	-	-	-	-	-	-
Max	-	-	-	-	-	-

Medium	Stream sediment	Stream sediment	Stream sediment	Overbank sediment	Overbank sediment	Organic stream sediment
	Austria	Southern Scotland	Harz, Germany	Norway	Norway	Finland
	<0.18 mm	<0.10 mm	<0.063 mm	<0.063 mm	<0.063 mm	<2 mm
	total	total	total	total (XRF)	7N HNO₃	Conc. HNO₃
Median	-	-	-	-	-	-
Min	-	-	-	-	-	-
Max	-	-	-	-	-	-

#2nd percentile ^98th percentile

Concentrations in Waters (mg/l)

Medium	Ocean water	Ocean water	Ocean water	Stream water	Stream water	Stream water
	World (1)	World (2)	North Pacific	World	Nova Scotia, Canada	Finland
					<0.45 μm	<0.45 μm
					Dionex	IC
Median	67.3*	67*	67*	0.03*	<0.05	<0.5
Min	-	-	-	-	<0.05	<0.5#
Max	-	-	-	-	<0.05	<0.5^

*Estimated mean   #2nd percentile   ^98th percentile

Medium	Stream water	Stream water	Stream water	Lake water	Ground water
	Romania	Eastern India	Harz, Germany	Norway	Southern Norway
	unfiltered	<0.2 μm	unfiltered	unfiltered	unfiltered
	ICP-MS	ICP-MS	ICP-MS	ICP-MS	ICP-MS
Median	0.07	0.033	-	-	0.035,1
Min	0.002	0.022	-	-	0.003,32
Max	4.47	0.321	-	-	1.43

69

Br BROMINE

Concentrations in Precipitation (mg/l)

Medium	Rain water	Rain water	Snow melt-water	Snow melt-water	Snow filter residue	Snow filter residue
	Kola, remote <0.45 µm IC	Kola, polluted <0.45 µm IC	Kola, remote <0.45 µm IC	Kola, polluted <0.45 µm IC	Kola, remote >0.45 µm	Kola, polluted >0.45 µm
Median	<0.2	<0.2	<0.2	<0.2	-	-
Min	<0.2	<0.2	<0.2	<0.2	-	-
Max	<0.2	<0.2	<0.2	<0.2	-	-

Medium	Rain water	Rain water				
	Coastal, Norway unfiltered ICP-MS	Inland, Norway unfiltered ICP-MS				
Median	-	-				
Min	-	-				
Max	-	-				

Concentrations in Air (ng/m^3) and Yearly Deposition (kg/km^2/yr)

Medium	Air	Air	Bulk deposition	Throughfall deposition	Bulk deposition	Bulk deposition
	World, remote	World, polluted	West Germany	West Germany	Kola, remote	Kola, polluted
Median	- / 3.5	350	-	-	-	-
Min	0.29	8	-	-	-	-
Max	33	2500	-	-	-	-

*Estimated value

Concentrations in Plants (mg/kg)

Medium	Moss	Moss	Crustose lichen	Lichen	Dandelion	Spruce bark
	Norway conc. HNO$_3$ (1985)	Germany conc. HNO$_3$ + H$_2$O$_2$ (1995)	Germany conc. HNO$_3$	Northwest Territories oven-dried	Europe oven-dried total (INAA)	Canada ashed total (INAA)
Median	4.8	-	-	-	13#	50.5
Min	1.4	-	-	-	-	16
Max	34.1	-	-	-	-	230

*Mean #Geometric mean

Concentrations in Human Fluids (mg/l)

Medium	Human blood	Human serum	Human urine			
	Lombardy, Italy	Lombardy, Italy	Lombardy, Italy			
Mean	-	-	-			
Min	-	-	-			
Max	-	-	-			

BROMINE — Br

Environmental Geochemistry

Biological impacts	Essential for some organisms (humans?). Toxic, presents a serious health hazard
Uses	Antiknock agent, flame proofing, insecticide, herbicide, rodenticide, dyes, photographic and pharmaceutical industry
Environmental pathways	Sea spray, traffic, agriculture
Environmental mobility	Oxidising conditions: very high Acid conditions: very high Reducing conditions: very high Neutral to alkaline conditions: very high Comments: -
Geochemical barriers	Evaporation
Natural association	Cl-Na-Mg-Br-S (sea spray and brines)
Action levels	Drinking water, MAC: 0.2 mg/l (Russia MoH). Agricultural soil, maximum tolerable concentration: 10 mg/kg (Germany)
Remarks	Can occur in gaseous, liquid and solid forms in the environment. Commercially obtained from natural brines. Only liquid, non-metallic element. Many harmful organic compounds (e.g., bromacil, bromadiolon, bromchlordifluomethane, brommethane) widely used. Bromacil and brommethane have been found in ground water, and bromchlordifluomethane in troposphere
Suggested analytical method(s)	IC, INAA

World Yearly Production (t/y)	432,250 (in 1995)
Price of Purest Form, in Small Quantities ($/kg)	1070 (99.998%)
Market Price ($/kg)	1.3 (bulk, 99.95%, ex-producer)

C — CARBON

Physico-Chemical Properties

Atomic number	6
Atomic mass	12.010,7(8)
Atomic radius (pm)	91
Main oxidation state(s)	+4 (-4,-2,0,+2,+3)
Ionic radius (pm)	6-30
Electronegativity (Pauling)	2.55
Density (g/cm^3)	1.8-3.5
Melting/boiling point (K)	4765 (triple point) / 3915 (sublimes)
Isotopes & isomers	2 stable + 11 unstable
Acid/base of oxide	weak acid
State (at 300 K, 1 atm.)	solid
Metallic character	non-metal
Element group(s)	light non-metal
Affinity	atmophile, biophile, lithophile, siderophile

Naturally occurring isotopes	Natural abundance (%)	Atomic mass	Half-life
^{12}C	98.89	12.000,0	stable
^{13}C	1.11	13.003,355	stable

Unstable isotopes	Longest half-life
^{8}C to ^{20}C	5715 y

Concentrations in Crust/Rocks (mg/kg)

Bulk continental crust	1990 / 200 / -
Upper continental crust	3240 / -
Ultramafic rock	-
Ocean ridge basalt	-
Gabbro, basalt	-
Granite, granodiorite	-
Sandstone	-
Greywacke	-
Shale, schist	-
Limestone	-
Coal	698,000

Typical Minerals
graphite (C), diamond (C)

Possible Host Minerals
carbonates

Mass (kg) in

Continental crust	4.70x10^{+19}	Oceans	3.70x10^{+16}	Plants	8.19x10^{+14}

Concentrations in Media Sampled for the Kola Project (mg/kg)

Medium	Moss conc. HNO$_3$	Humus <2 mm CHN-anal.	Humus <2 mm amm. acet.	Topsoil (0-5 cm) <2 mm	B horizon <2 mm aqua regia	B horizon <2 mm total (XRF)
Median	-	450,000	-	-	-	-
Min	-	153,000	-	-	-	-
Max	-	508,000	-	-	-	-

Medium	C horizon <2 mm aqua regia	C horizon <2 mm total (XRF)	C horizon <2 mm total (INAA)		Lake water unfiltered (mg/l)
Median	-	-	-		-
Min	-	-	-		-
Max	-	-	-		-

CARBON C

Concentrations in Soils and Sediments (mg/kg)

Medium	Soil	Agricultural soil - Ap horizon	Agricultural soil - Top (0-25 cm)	Agricultural Soil - Bottom (50-75 cm)	Topsoil (0-15 cm)	Urban soil (0-2 cm)
	World	*Canada*	*Finland*	*Finland*	*England & Wales*	*Trondheim*
	<2 mm	*<2 mm*	*<2 mm*	*<2 mm*	*<2 mm*	*<2 mm*
	total	*aqua regia*	*aqua regia*	*aqua regia*	*aqua regia*	*aqua regia*
Median	-	-	-	-	-	-
Min	-	-	-	-	-	-
Max	-	-	-	-	-	-

*Estimated mean

Medium	Forest soil - Humus	Forest soil - B horizon	Forest soil - C horizon	Till (C horizon)	Till (C horizon)	Laterite (25 ±15 cm)
	Norway	*Norway*	*Norway*	*Finland*	*Finland*	*Australia*
	<2 mm	*<2 mm*	*<2 mm*	*<0.063 mm*	*<0.063 mm*	*0.45-2 mm*
	7N HNO$_3$	*7N HNO$_3$*	*7N HNO$_3$*	*aqua regia*	*total*	*total*
Median	-	-	-	-	-	-
Min	-	-	-	-	-	-
Max	-	-	-	-	-	-

Medium	Stream sediment	Stream sediment	Stream sediment	Overbank sediment	Overbank sediment	Organic stream sediment
		Southern Scotland	*Harz, Germany*	*Norway*	*Norway*	*Finland*
	Austria					
	<0.18 mm	*<0.10 mm*	*<0.063 mm*	*<0.063 mm*	*<0.063 mm*	*<2 mm*
		total	*total*	*total (XRF)*	*7N HNO$_3$*	*C,.HN-analyser*
Median	-	-	-	-	-	73,000
Min	-	-	-	-	-	14,000#
Max	-	-	-	-	-	305,000^

#2nd percentile ^98th percentile

Concentrations in Waters (mg/l)

Medium	Ocean water	Ocean water	Ocean water	Stream water	Stream water	Stream water
	World (1)	*World (2)*	*North Pacific*	*World*	*Nova Scotia, Canada*	*Finland*
					<0.45 µm	*<0.45 µm*
Median	28*	28*	27*	-	-	-
Min	-	-	-	-	-	-
Max	-	-	-	-	-	-

*Estimated mean   #2nd percentile   ^98th percentile

Medium	Stream water	Stream water	Stream water	Lake water	Ground water
	Romania	*Eastern India*	*Harz, Germany*	*Norway*	*Southern Norway*
	unfiltered	*<0.2 µm*	*unfiltered*	*unfiltered*	*unfiltered*
	ICP-MS	*ICP-MS*		*ICP-MS*	
Median	-	-	-	-	-
Min	-	-	-	-	-
Max	-	-	-	-	-

C — CARBON

Concentrations in Precipitation (mg/l)

Medium	Rain water	Rain water	Snow melt-water	Snow melt-water	Snow filter residue	Snow filter residue
	Kola, remote <0.45 μm	*Kola, polluted <0.45 μm*	*Kola, remote <0.45 μm*	*Kola, polluted <0.45 μm*	*Kola, remote >0.45 μm*	*Kola, polluted >0.45 μm*
Median	-	-	-	-	-	-
Min	-	-	-	-	-	-
Max	-	-	-	-	-	-

Medium	Rain water	Rain water				
	Coastal, Norway unfiltered ICP-MS	*Inland, Norway unfiltered ICP-MS*				
Median	-	-				
Min	-	-				
Max	-	-				

Concentrations in Air (ng/m³) and Yearly Deposition (kg/km²/yr)

Medium	Air	Air	Bulk deposition	Throughfall deposition	Bulk deposition	Bulk deposition
	World, remote	*World, polluted*	*West Germany*	*West Germany*	*Kola, remote*	*Kola, polluted*
Median	-	-	-	-	-	-
Min	-	-	-	-	-	-
Max	-	-	-	-	-	-

*Estimated value

Concentrations in Plants (mg/kg)

Medium	Moss	Moss	Crustose lichen	Lichen	Dandelion	Spruce bark
	Norway conc. HNO₃ (1990)	*Germany conc. HNO₃ + H₂O₂ (1995)*	*Germany conc. HNO₃*	*Northwest Territories oven-dried*	*Europe oven-dried*	*Canada ashed*
Median	-	-	-	-	-	-
Min	-	-	-	-	-	-
Max	-	-	-	-	-	-

*Mean #Geometric mean

Concentrations in Human Fluids (mg/l)

Medium	Human blood	Human serum	Human urine			
	Lombardy, Italy	*Lombardy, Italy*	*Lombardy, Italy*			
Mean	-	-	-			
Min	-	-	-			
Max	-	-	-			

CARBON   C

Environmental Geochemistry

Biological impacts	C-compounds can be essential to highly toxic
Uses	Countless uses from combustion (hydrocarbons), to chemical industry (organic chemical industry), to jewellery (diamonds)
Environmental pathways	Combustion, chemical industry. Natural emissions to the atmosphere are considerable
Environmental mobility	Oxidising conditions: - 　　　　　Acid conditions: - Reducing conditions: - 　　　　　Neutral to alkaline conditions: - Comments: Highly dependent on compound
Geochemical barriers	-
Natural association	C-H-O-(N)-(S) (organic matter)
Action levels	Many organic C compounds have action levels in drinking water
Remarks	Coal, petroleum, natural gas are all hydrocarbons. About 10 million known C compounds exist to date. CO_2 and CH_4 are the most important C compounds. HCO_3^- plays an important geochemical role in natural water. The dry mass of plants consists of about 45% C. Elemental C occurs as soot even in the atmosphere. Natural emissions of hydrocarbons are estimated at 4200 million tons per year; anthropogenic emissions are estimated at 200 million tons per year. CO is very toxic to mammals. CN^- is very toxic to most organisms. Organic matter tends to bind metals. Isotopic fractionation in nature is important
Suggested analytical method(s)	Elemental analyser (IR spectrometry)

World Yearly Production (t/y)	3,332,115,000 coal, 3,225,946,000 oil, 1,909,045,000 gas (as oil equivalent), 1,146,000,000 brown coal, 710,000 graphite, 20 diamond (all in 1995)
Price of Purest Form, in Small Quantities ($/kg)	81,875 (99.9%)
Market Price ($/kg)	0.85 (graphite, ca. 95% C), 15-25 $/barrel crude oil

Ca CALCIUM

Physico-Chemical Properties

Atomic number	20	Melting/boiling point (K)	1115 / 1757
Atomic mass	40.078(4)	Isotopes & isomers	6 stable + 13 unstable
Atomic radius (pm)	223	Acid/base of oxide	strong base
Main oxidation state(s)	+2	State (at 300 K, 1 atm.)	solid
Ionic radius (pm)	114-148	Metallic character	metal
Electronegativity (Pauling)	1	Element group(s)	alkaline earth, light metal
Density (g/cm^3)	1.55	Affinity	lithophile

Naturally occurring isotopes	Natural abundance (%)	Atomic mass	Half-life
^{40}Ca	96.941	39.962,59	stable
^{42}Ca	0.648	41.958,62	stable
^{43}Ca	0.136	42.958,77	stable
^{44}Ca	2.086	43.955,48	stable
^{46}Ca	0.004	45.953,69	stable
^{48}Ca	0.187	47.952,53	stable

Unstable isotopes	Longest half-life
^{35}Ca to ^{53}Ca	102,000 y

Concentrations in Crust/Rocks (mg/kg)

Bulk continental crust	38,500 / 41,500 / 52,900
Upper continental crust	29,450 / 30,000
Ultramafic rock	25,000
Ocean ridge basalt	84,000
Gabbro, basalt	74,000
Granite, granodiorite	9000
Sandstone	13,000
Greywacke	18,582
Shale, schist	22,000
Limestone	380,000
Coal	5000

Typical Minerals
calcite (CaCO$_3$), gypsum (CaSO$_4$.2H$_2$O), fluorite (CaF$_2$)

Possible Host Minerals
carbonates, feldspars, amphiboles, pyroxenes

Mass (kg) in

Continental crust	9.09x10^{+20}	Oceans	5.45x10^{+17}	Plants	1.84x10^{+13}

Concentrations in Media Sampled for the Kola Project (mg/kg)

Medium	Moss conc. HNO$_3$	Humus <2 mm conc. HNO$_3$	Humus <2 mm amm. acet.	Topsoil (0-5 cm) <2 mm total (INAA)	B horizon <2 mm aqua regia	B horizon <2 mm total (XRF)
Median	2620	2960	1680	<10,000	1320	19,368
Min	1680	460	223	<10,000	70	143
Max	9320	25,400	7090	40,000	7410	51,887

Medium	C horizon <2 mm aqua regia	C horizon <2 mm total (XRF)	C horizon <2 mm total (INAA)	Lake water unfiltered (mg/l)
Median	1900	21,700	20,000	2
Min	110	300	<10,000	0.2
Max	41,700	67,600	60,000	8.5

CALCIUM — Ca

Concentrations in Soils and Sediments (mg/kg)

Medium	Soil	Agricultural soil - Ap horizon	Agricultural soil - Top (0-25 cm)	Agricultural Soil - Bottom (50-75 cm)	Topsoil (0-15 cm)	Urban soil (0-2 cm)
	World	Canada	Finland	Finland	England & Wales	Trondheim
	<2 mm total	<2 mm	<2 mm aqua regia	<2 mm aqua regia	<2 mm aqua regia	<2 mm aqua regia
Median	14,000*	-	3280	2350	3278	5425
Min	-	-	<50	673	50	690
Max	-	-	20,000	18,600	339,630	106,000

*Estimated mean

Medium	Forest soil - Humus	Forest soil - B horizon	Forest soil - C horizon	Till (C horizon)	Till (C horizon)	Laterite (25 ±15 cm)
	Norway	Norway	Norway	Finland	Finland	Australia
	<2 mm 7N HNO$_3$	<2 mm 7N HNO$_3$	<2 mm 7N HNO$_3$	<0.063 mm aqua regia	<0.063 mm total	0.45-2 mm total
Median	2200	1600	4700	1700	18,000	214
Min	120	170	340	-	-	<70
Max	18,600	24,900	67,300	-	-	213,767

Medium	Stream sediment	Stream sediment	Stream sediment	Overbank sediment	Overbank sediment	Organic stream sediment
	Austria	Southern Scotland	Harz, Germany	Norway	Norway	Finland
	<0.18 mm "total" (ICP-AES)	<0.10 mm total (DCES)	<0.063 mm total (XRF)	<0.063 mm total (XRF)	<0.063 mm 7N HNO$_3$ ICP-AES	<2 mm Conc. HNO$_3$
Median	12,900	5718	4860	21,000	4100	5600
Min	-	<143	482	1200	390	2600#
Max	423,400	151,515	99,797	98,100	63,300	12,600^

#2nd percentile ^98th percentile

Concentrations in Waters (mg/l)

Medium	Ocean water	Ocean water	Ocean water	Stream water	Stream water	Stream water
	World (1)	World (2)	North Pacific	World	Nova Scotia, Canada	Finland
					<0.45 µm AAS	<0.45 µm ICP-AES
Median	412*	412*	412*	18*	3.6	4.06
Min	-	-	-	-	<0.1	1.46#
Max	-	-	-	-	355	27.2^

*Estimated mean   #2nd percentile   ^98th percentile

Medium	Stream water	Stream water	Stream water	Lake water	Ground water
	Romania	Eastern India	Harz, Germany	Norway	Southern Norway
	unfiltered ICP-MS	<0.2 µm ICP-MS	unfiltered ICP-AES	unfiltered ICP-MS	unfiltered ICP-AES
Median	56	-	5.4	-	25.7
Min	2	-	0.2	-	<0.2
Max	198	-	67	-	87.8

Ca　　　　　　　　　　　　　　　　　　　　　　　　　　CALCIUM

Concentrations in Precipitation (mg/l)

Medium	Rain water	Rain water	Snow melt-water	Snow melt-water	Snow filter residue	Snow filter residue
	Kola, remote <0.45 µm ICP-AES	Kola, polluted <0.45 µm ICP-AES	Kola, remote <0.45 µm ICP-AES	Kola, polluted <0.45 µm ICP-AES	Kola, remote >0.45 µm ICP-AES	Kola, polluted >0.45 µm ICP-AES
Median	0.05	0.1	<0.05	0.19	<0.1	0.12
Min	0.02	0.05	<0.05	0.15	<0.1	<0.1
Max	0.14	0.2	0.56	0.3	<0.1	0.84

Medium	Rain water	Rain water
	Coastal, Norway unfiltered ICP-MS	Inland, Norway unfiltered ICP-MS
Median	-	-
Min	-	-
Max	-	-

Concentrations in Air (ng/m^3) and Yearly Deposition (kg/km^2/yr)

Medium	Air	Air	Bulk deposition	Throughfall deposition	Bulk deposition	Bulk deposition
	World, remote	World, polluted	West Germany	West Germany	Kola, remote	Kola, polluted
Median	40 / 110	2100	-	-	60.3*	280*
Min	0.5	100	460	420	-	-
Max	820 (1600)	7000	2100	2700	-	-

*Estimated value

Concentrations in Plants (mg/kg)

Medium	Moss	Moss	Crustose lichen	Lichen	Dandelion	Spruce bark
	Norway conc. HNO$_3$ (1990)	Germany conc. HNO$_3$ + H$_2$O$_2$ (1995)	Germany conc. HNO$_3$	Northwest Territories oven-dried total (INAA)	Europe oven-dried total (INAA)	Canada ashed total (INAA)
Median	2800	3590	-	7626.9*	11,400#	16,300
Min	1300	810	-	-	-	1800
Max	12,000	20,600	-	-	-	35,300

*Mean　#Geometric mean

Concentrations in Human Fluids (mg/l)

Medium	Human blood	Human serum	Human urine			
	Lombardy, Italy	Lombardy, Italy	Lombardy, Italy			
Mean	-	-	-			
Min	-	-	-			
Max	-	-	-			

CALCIUM   Ca

Environmental Geochemistry

Biological impacts	Essential for most organisms. Non-toxic. Major nutrient.
Uses	Lime, cement, fertilisers, metallurgy (reducer, deoxidiser, desulphuriser)
Environmental pathways	Geogenic and anthropogenic dust, rock weathering, lime and cement factories, fertilisers, acidification. Generally, geogenic sources more important than anthropogenic ones
Environmental mobility	Oxidising conditions: high Acid conditions: high Reducing conditions: high Neutral to alkaline conditions: high Comments: -
Geochemical barriers	Incorporation into active organic matter, adsorption, pH decrease
Natural association	Ca-Sr (minerals), Cl-Ca-Na-Mg-Br-S (sea spray)
Action levels	Drinking water, recommended: 15-25 mg/l (Norway Shd)
Remarks	Fe availability can be inhibited in Ca-rich soils. Ca-arsenate used in agriculture and forestry
Suggested analytical method(s)	XRF, ICP-AES

World Yearly Production (t/y)	120,000,000 lime, 1,421,000,000 cement (all in 1995)
Price of Purest Form, in Small Quantities ($/kg)	13,950 (99.99%)
Market Price ($/kg)	0.08 (GCC-chalk, uncoated, ex-producer), 0.6 (PCC-chalk, uncoated, ex-producer)

Cd   CADMIUM

Physico-Chemical Properties

Atomic number	48	Melting/boiling point (K)	594.22 / 1040
Atomic mass	112.411(8)	Isotopes & isomers	8 stable + 34 unstable
Atomic radius (pm)	171	Acid/base of oxide	base
Main oxidation state(s)	+2	State (at 300 K, 1 atm.)	solid
Ionic radius (pm)	92-145	Metallic character	metal
Electronegativity (Pauling)	1.69	Element group(s)	heavy metal
Density (g/cm^3)	8.65	Affinity	chalcophile

Naturally occurring isotopes	Natural abundance (%)	Atomic mass	Half-life
^{106}Cd	1.25	105.906,46	stable
^{108}Cd	0.89	107.904,18	stable
^{110}Cd	12.49	109.903,01	stable
^{111}Cd	12.81	110.904,18	stable
^{112}Cd	24.13	111.902,67	stable
^{113}Cd	12.23	112.904,4	stable
^{114}Cd	28.73	113.903,36	stable
^{116}Cd	7.49	115.904,76	stable

Unstable isotopes	Longest half-life
^{97}Cd to ^{130}Cd	14.1 y

Concentrations in Crust/Rocks (mg/kg)

Bulk continental crust	0.1 / 0.15 / 0.098
Upper continental crust	0.102 / 0.098
Ultramafic rock	0.05
Ocean ridge basalt	0.1
Gabbro, basalt	0.2
Granite, granodiorite	0.1
Sandstone	<0.04
Greywacke	-
Shale, schist	0.25
Limestone	0.1
Coal	1

Typical Minerals
greenockite (CdS), octavite (CdCO$_3$), monteponite (CdO)

Possible Host Minerals
zinc ores, sphalerite (up to 0.5%), smithsonite (up to 5%), biotite, amphiboles

Mass (kg) in

Continental crust	2.36x10^{+15}	Oceans	1.32x10^{+11}	Plants	9.20x10^{+7}

Concentrations in Media Sampled for the Kola Project (mg/kg)

Medium	Moss conc. HNO$_3$	Humus <2 mm conc. HNO$_3$	Humus <2 mm amm. acet.	Topsoil (0-5 cm) <2 mm	B horizon <2 mm aqua regia	B horizon <2 mm total (XRF)
Median	0.089	0.303	0.22	-	0.024	-
Min	0.023	0.073	0.04	-	0.005	-
Max	1.23	1.39	1.02	-	0.236	-

Medium	C horizon <2 mm aqua regia	C horizon <2 mm total (XRF)	C horizon <2 mm total (INAA)	Lake water unfiltered (mg/l)
Median	0.024	-	-	<0.000,02
Min	0.007	-	-	<0.000,02
Max	0.221	-	-	0.000,81

CADMIUM   Cd

Concentrations in Soils and Sediments (mg/kg)

Medium	Soil	Agricultural soil - Ap horizon	Agricultural soil - Top (0-25 cm)	Agricultural Soil - Bottom (50-75 cm)	Topsoil (0-15 cm)	Urban soil (0-2 cm)
	World	Canada	Finland	Finland	England & Wales	Trondheim
	<2 mm	<2 mm	<2 mm	<2 mm	<2 mm	<2 mm
	total	total (AAS)	aqua regia	aqua regia	aqua regia	aqua regia
Median	0.3*	0.3	0.117	0.036	0.7	0.16
Min	-	<0.2	<0.01	<0.01	<0.2	0.01
Max	-	3.8	0.448	0.318	40.9	11.3

*Estimated mean

Medium	Forest soil - Humus	Forest soil - B horizon	Forest soil - C horizon	Till (C horizon)	Till (C horizon)	Laterite (25 ±15 cm)
	Norway	Norway	Norway	Finland	Finland	Australia
	<2 mm	<2 mm	<2 mm	<0.063 mm	<0.063 mm	0.45-2 mm
	7N HNO_3	7N HNO_3	7N HNO_3	aqua regia	total	total
Median	0.8	<1	<1	-	-	0.2
Min	<0.5	<1	<1	-	-	<0.1
Max	12.4	2.4	2.5	-	-	0.8

Medium	Stream sediment	Stream sediment	Stream sediment	Overbank sediment	Overbank sediment	Organic stream sediment
	Austria	Southern Scotland	Harz, Germany	Norway	Norway	Finland
	<0.18 mm	<0.10 mm	<0.063 mm	<0.063 mm	<0.063 mm	<2 mm
		total	total (ICP-MS)	total (XRF)	7N HNO_3	Conc. HNO_3
Median	-	-	1.8	-	-	0.08
Min	-	-	0.09	-	-	0.044#
Max	-	-	61	-	-	0.803^

#2nd percentile   ^98th percentile

Concentrations in Waters (mg/l)

Medium	Ocean water	Ocean water	Ocean water	Stream water	Stream water	Stream water
	World (1)	World (2)	North Pacific	World	Nova Scotia, Canada	Finland
					<0.45 μm	<0.45 μm
					ICP-MS	ICP-MS
Median	0.000,11*	0.000,1*	0.000,07*	0.000,02*	<0.000,05	<0.000,02
Min	-	-	-	-	<0.000,05	<0.000,02#
Max	-	-	-	-	0.000,59	0.000,077^

*Estimated mean   #2nd percentile   ^98th percentile

Medium	Stream water	Stream water	Stream water	Lake water	Ground water
	Romania	Eastern India	Harz, Germany	Norway	Southern Norway
	unfiltered	<0.2 μm	unfiltered	unfiltered	unfiltered
	ICP-MS	ICP-MS	ICP-MS	ICP-MS	ICP-MS
Median	0.000,03	0.000,04	0.000,29	<0.000,02	0.000,032
Min	0.000,002	0.000,02	<0.000,01	<0.000,02	<0.000,002
Max	0.009,6	0.000,46	0.006,7	0.000,255	0.005,54

Cd   CADMIUM

Concentrations in Precipitation (mg/l)

Medium	Rain water	Rain water	Snow melt-water	Snow melt-water	Snow filter residue	Snow filter residue
	Kola, remote <0.45 μm ICP-MS	*Kola, polluted <0.45 μm ICP-MS*	*Kola, remote <0.45 μm ICP-MS*	*Kola, polluted <0.45 μm ICP-MS*	*Kola, remote >0.45 μm ICP-AES*	*Kola, polluted >0.45 μm ICP-AES*
Median	0.000,06	0.000,89	<0.000,03	0.000,315	<0.000,05	0.000,095
Min	<0.000,02	0.000,32	<0.000,03	0.000,24	<0.000,05	0.000,05
Max	0.000,33	0.005,11	0.000,07	0.000,76	<0.000,05	0.000,34

Medium	Rain water	Rain water				
	Coastal, Norway unfiltered ICP-MS	*Inland, Norway unfiltered ICP-MS*				
Median	0.000,05	0.000,05				
Min	<0.000,01	<0.000,01				
Max	0.000,45	0.000,72				

Concentrations in Air (ng/m^3) and Yearly Deposition (kg/km^2/yr)

Medium	Air	Air	Bulk deposition	Throughfall deposition	Bulk deposition	Bulk deposition
	World, remote	*World, polluted*	*West Germany*	*West Germany*	*Kola, remote*	*Kola, polluted*
Median	<0.4 / 0.2	20	-	-	0.052,4*	0.387*
Min	0.002,5	0.5	0.05	0.08	-	-
Max	2.2	620	1.6	2	-	-

*Estimated value

Concentrations in Plants (mg/kg)

Medium	Moss	Moss	Crustose lichen	Lichen	Dandelion	Spruce bark
	Norway conc. HNO$_3$ (1990)	*Germany conc. HNO$_3$ + H$_2$O$_2$ (1995)*	*Germany conc. HNO$_3$*	*Northwest Territories oven-dried*	*Europe oven-dried conc. HNO$_3$ (AAS)*	*Canada ashed aqua regia*
Median	0.13	0.295	1.3	-	0.26#	3.35
Min	<0.02	0.05	0.8	-	-	<0.2
Max	3.4	1.8	13	-	-	63.4

*Mean   #Geometric mean

Concentrations in Human Fluids (mg/l)

Medium	Human blood	Human serum	Human urine			
	Lombardy, Italy	*Lombardy, Italy*	*Lombardy, Italy*			
Mean	0.000,6	0.000,2	0.000,86			
Min	0.000,1	0.000,09	0.000,15			
Max	0.005,5	0.000,66	0.002			

CADMIUM Cd

Environmental Geochemistry

Biological impacts	Seems to be essentials to some animals (e.g., rats) at very low concentrations. Toxic. Supposedly carcinogenic.
Uses	Electroplating, Ni-Cd batteries, pigments, stabilisers for plastics, solder, very low melting point alloys, television tubes
Environmental pathways	Coal combustion, smelters (esp. Zn and Cu), iron and steel mills, electroplating, fertilisers, traffic (tyre wear, exhaust), sewage sludge and waste water, waste incineration
Environmental mobility	Oxidising conditions: medium Acid conditions: medium Reducing conditions: very low Neutral to alkaline conditions: medium Comments: Some Cd components are able to leach through soil to ground water. Cd compounds bound to stream sediments are easily bioaccumulated or redissolved during sediment disturbance, e.g., flooding
Geochemical barriers	Forms chelates with humic acid, adsorbs onto clays
Natural association	Zn-Cd (all occurrences), Zn-Cd-Pb-Ba-F (Mississippi Valley type deposits)
Action levels	Drinking water, MAC: 0.005 mg/l (Norway Shd), 0.005 mg/l (US EPA), 0.005 mg/l (Canada CWQG); recommended: 0.003 mg/l (WHO); MAC: 0.001 mg/l (Russia MoH). Ground water, background: 0.000,4 mg/l; remediate: 0.006 mg/l (Netherlands VROM). Soil, background: 0.8 mg/kg; remediate: 12 mg/kg (Netherlands VROM). Agricultural soil, maximum tolerable concentration for disposal of sewage sludge of <10 mg/kg Cd: 1.5 mg/kg (Germany)
Remarks	Plants accumulate Cd via roots. Five mg Cd/kg in soil can result in reduced yield. Celery, carrot, salad, spinach, some mushrooms, blue poppy seeds, wheat concentrate Cd. Corn and potatoes have a low Cd uptake. Very soluble at low pH. Can substitute Ca^{2+}, Mn^{2+} (similar ionic radii). Generally released to atmosphere in very small particles (<2 micrometers). Practically all Cd is obtained as a by-product of processing Zn-Cu-Pb ores. Itaï Itaï disease, related to prolonged intake of Cd, is restricted to subsistence rice formers in the Far East who have Ca, Fe and Zn deficient diets. Cd can be released to drinking water through the corrosion of some galvanised plumbing and water main pipe materials. The US EPA suggests that drinking water exceeding action levels can lead to nausea, vomiting, diarrhea, muscle cramps, salivation, sensory disturbances, liver injury, convulsions, shock and renal failure in the short term, and to kidney, liver, bone and blood damage in the long term
Suggested analytical method(s)	GF-AAS, ICP-MS

World Yearly Production (t/y)	18,621 (in 1995)
Price of Purest Form, in Small Quantities ($/kg)	13,455 (99.999,9%)
Market Price ($/kg)	2.1

Ce CERIUM

Physico-Chemical Properties

Atomic number	58
Atomic mass	140.116(1)
Atomic radius (pm)	270
Main oxidation state(s)	+3 (+4)
Ionic radius (pm)	115-148 (+3), 101-128 (+4)
Electronegativity (Pauling)	1.12
Density (g/cm^3)	6.77

Melting/boiling point (K)	1071 / 3716
Isotopes & isomers	4 stable + 31 unstable
Acid/base of oxide	base
State (at 300 K, 1 atm.)	solid
Metallic character	metal
Element group(s)	heavy metal, REE
Affinity	lithophile

Naturally occurring isotopes	Natural abundance (%)	Atomic mass	Half-life
^{136}Ce	0.19	135.907,14	stable
^{138}Ce	0.25	137.905,99	stable
^{140}Ce	88.48	139.905,435	stable
^{142}Ce	11.08	141.901,241	stable

Unstable isotopes	Longest half-life
^{123}Ce to ^{152}Ce	284.6 d

Concentrations in Crust/Rocks (mg/kg)

Bulk continental crust	60 / 66.5 / 33
Upper continental crust	65.7 / 64
Ultramafic rock	3
Ocean ridge basalt	14
Gabbro, basalt	16
Granite, granodiorite	90
Sandstone	30
Greywacke	58
Shale, schist	80
Limestone	12
Coal	20

Typical Minerals
monazite ((Ce,La,Nd,Th)(PO$_4$,SiO$_4$)),
bastnaesite ((Ce,La)CO$_3$(F,OH)),
cerite ((Ce,La)$_9$(Mg,Fe)Si$_7$(O,OH,F)$_{28}$),
xenotime ((Y,Ce)PO$_4$)

Possible Host Minerals
feldspars, apatite, allanite, sphene, fluorite, zircon

Mass (kg) in

Continental crust 1.42x10^{+18}	Oceans 1.32x10^{+9}	Plants 9.20x10^{+8}

Concentrations in Media Sampled for the Kola Project (mg/kg)

Medium	Moss conc. HNO$_3$	Humus <2 mm amm. acet.	Humus <2 mm 	Topsoil (0-5 cm) <2 mm total (INAA)	B horizon <2 mm aqua regia	B horizon <2 mm total (XRF)
Median	-	-	-	17	-	-
Min	-	-	-	<3	-	-
Max	-	-	-	320	-	-

Medium	C horizon <2 mm aqua regia	C horizon <2 mm total (XRF)	C horizon <2 mm total (INAA)	Lake water unfiltered (mg/l)
Median	-	-	45	-
Min	-	-	12	-
Max	-	-	500	-

CERIUM   Ce

Concentrations in Soils and Sediments (mg/kg)

Medium	Soil	Agricultural soil - Ap horizon	Agricultural soil - Top (0-25 cm)	Agricultural Soil - Bottom (50-75 cm)	Topsoil (0-15 cm)	Urban soil (0-2 cm)
	World	Canada	Finland	Finland	England & Wales	Trondheim
	<2 mm	<2 mm	<2 mm	<2 mm	<2 mm	<2 mm
	total	total (INAA)	aqua regia	aqua regia	aqua regia	aqua regia
Median	65*	46	-	-	-	-
Min	-	2	-	-	-	-
Max	-	88	-	-	-	-

*Estimated mean

Medium	Forest soil - Humus	Forest soil - B horizon	Forest soil - C horizon	Till (C horizon)	Till (C horizon)	Laterite (25 ±15 cm)
	Norway	Norway	Norway	Finland	Finland	Australia
	<2 mm	<2 mm	<2 mm	<0.063 mm	<0.063 mm	0.45-2 mm
	7N HNO$_3$	7N HNO$_3$	7N HNO$_3$	aqua regia	total	total
Median	4.5	28.5	87.9	-	-	16
Min	<1.5	<3	<3	-	-	0.62
Max	204	498	1100	-	-	120

Medium	Stream sediment	Stream sediment	Stream sediment	Overbank sediment	Overbank sediment	Organic stream sediment
	Austria	Southern Scotland	Harz, Germany	Norway	Norway	Finland
	<0.18 mm	<0.10 mm	<0.063 mm	<0.063 mm	<0.063 mm	<2 mm
	"total" (ICP-AES)	total	total (XRF)	total (XRF)	7N HNO$_3$ ICP-AES	Conc. HNO$_3$
Median	96	-	86	-	50	-
Min	-	-	<15	-	6.5	-
Max	3590	-	517	-	514	-

#2nd percentile   ^98th percentile

Concentrations in Waters (mg/l)

Medium	Ocean water	Ocean water	Ocean water	Stream water	Stream water	Stream water
	World (1)	World (2)	North Pacific	World	Nova Scotia, Canada	Finland
					<0.45 µm	<0.45 µm
					ICP-MS	
Median	0.000,001,2*	0.000,001*	0.000,000,7*	0.000,06*	0.000,16	-
Min	-	-	-	-	<0.000,01	-
Max	-	-	-	-	0.001,34	-

*Estimated mean   #2nd percentile   ^98th percentile

Medium	Stream water	Stream water	Stream water	Lake water	Ground water
	Romania	Eastern India	Harz, Germany	Norway	Southern Norway
	unfiltered	<0.2 µm	unfiltered	unfiltered	unfiltered
	ICP-MS	ICP-MS	ICP-MS	ICP-MS	ICP-MS
Median	0.000,87	0.009,1	0.000,55	0.000,212	0.000,146,5
Min	0.000,02	0.001,1	0.000,01	0.000,007	<0.000,01
Max	0.241	0.110,7	0.009	0.004,52	0.232,076

Ce   CERIUM

Concentrations in Precipitation (mg/l)

Medium	Rain water Kola, remote <0.45 μm	Rain water Kola, polluted <0.45 μm	Snow melt- water Kola, remote <0.45 μm	Snow melt- water Kola, polluted <0.45 μm	Snow filter residue Kola, remote >0.45 μm	Snow filter residue Kola, polluted >0.45 μm
Median	-	-	-	-	-	-
Min	-	-	-	-	-	-
Max	-	-	-	-	-	-

Medium	Rain water Coastal, Norway unfiltered ICP-MS	Rain water Inland, Norway unfiltered ICP-MS				
Median	-	-				
Min	-	-				
Max	-	-				

Concentrations in Air (ng/m^3) and Yearly Deposition (kg/km^2/yr)

Medium	Air World, remote	Air World, polluted	Bulk deposition West Germany	Throughfall deposition West Germany	Bulk deposition Kola, remote	Bulk deposition Kola, polluted
Median	0.08 / 0.018	2.5	-	-	-	-
Min	0.002	0.02	-	-	-	-
Max	0.24 (0.6)	20	-	-	-	-

*Estimated value

Concentrations in Plants (mg/kg)

Medium	Moss Norway conc. HNO$_3$ (1990)	Moss Germany conc. HNO$_3$ + H$_2$O$_2$ (1995)	Crustose lichen Germany conc. HNO$_3$	Lichen Northwest Territories oven-dried	Dandelion Europe oven-dried total (INAA)	Spruce bark Canada ashed total (INAA)
Median	-	0.747	16	-	0.41#	28
Min	-	0.13	3.9	-	-	<3
Max	-	18	170	-	-	120

*Mean   #Geometric mean

Concentrations in Human Fluids (mg/l)

Medium	Human blood Lombardy, Italy	Human serum Lombardy, Italy	Human urine Lombardy, Italy			
Mean	0.003,1	-	0.003,1			
Min	0.000,7	-	0.000,15			
Max	0.01	-	0.02			

CERIUM   Ce

Environmental Geochemistry

Biological impacts	Considered non-essential. Generally low toxicity, but data to assess the health relevance of REE are scarce
Uses	Glass and ceramic industry, lighters, abrasives, steel alloys, automobile catalyst, hydrocarbon catalyst, television tubes, nuclear industry
Environmental pathways	Poorly understood. Probably mostly windblown dust, weathering of REE-bearing minerals. Generally, geogenic sources more important than anthropogenic ones
Environmental mobility	Oxidising conditions: very low Acid conditions: very low Reducing conditions: very low Neutral to alkaline conditions: very low Comments: -
Geochemical barriers	pH, mechanical
Natural association	REE-Li-Rb-Cs-Be-Nb-Ta-Zr-B-Th-U-F (pegmatites), REE-Th-P-Zr-Fe-Cu (monazite veins), REE-Th-Ba-Sr-P-F-N-C (carbonatites), REE-U-P-F (phosphorites), REE-Au-Ti-Sn-Sr-Th (placers)
Action levels	-
Remarks	Toxicity of REE decreases as atomic number increases. Inhaled REE probably cause pneumoconiosis. REE taken up orally (e.g., via drinking water) accumulate in the skeleton, teeth, lungs, liver and kidneys. Ce is the most abundant of all REE. All REE can replace Ca in biological processes. Some trees tend to accumulate Ce (e.g., hickory (*Carya*)). Unlike most other REE, Ce can occur in the +4 oxidation state. ^{144}Ce has been found in animal bones and clams, and in the lungs and lymph nodes of deceased persons having inhaled aerosols from nuclear explosions. In 1991, 7000-8600 t Ce-oxide were used in the glass industry, and 300-450 t in the ceramics industry
Suggested analytical method(s)	XRF, ICP-AES, ICP-MS (water)

World Yearly Production (t/y)	54,000 REE-minerals (in 1991)
Price of Purest Form, in Small Quantities ($/kg)	13,543 (99.99%)
Market Price ($/kg)	-

Cl   CHLORINE

Physico-Chemical Properties

Atomic number	17	Melting/boiling point (K)	171.65 / 239.11
Atomic mass	35.452,7(9)	Isotopes & isomers	2 stable + 17 unstable
Atomic radius (pm)	97	Acid/base of oxide	strong acid
Main oxidation state(s)	-1 (0,+1,+3,+4,+5,+7)	State (at 300 K, 1 atm.)	gas
Ionic radius (pm)	167 (-1), 26 (+5), 22-41 (+7)	Metallic character	non-metal
Electronegativity (Pauling)	3.16	Element group(s)	halogen, light non-metal
Density (g/cm^3)	0.003,214	Affinity	atmophile, lithophile

Naturally occurring isotopes	Natural abundance (%)	Atomic mass	Half-life
^{35}Cl	75.77	34.968,85	stable
^{37}Cl	24.23	36.965,903	stable

Unstable isotopes	Longest half-life
^{31}Cl to ^{47}Cl	301,000 y

Concentrations in Crust/Rocks (mg/kg)

Bulk continental crust	472 / 145 / -
Upper continental crust	640 / 150
Ultramafic rock	80
Ocean ridge basalt	-
Gabbro, basalt	130
Granite, granodiorite	200
Sandstone	10
Greywacke	-
Shale, schist	200
Limestone	150
Coal	1000

Typical Minerals
halite (NaCl), carnallite (KMgCl$_3$.6H$_2$O), sylvite (KCl), sodalite (Na$_4$Al$_3$Si$_3$O$_{12}$Cl), eudialyte (Na$_6$ZrSi$_6$O$_{18}$Cl)

Possible Host Minerals
biotite, amphiboles, apatite

Mass (kg) in

Continental crust	1.11x10^{+19}	Oceans	2.57x10^{+19}	Plants	3.68x10^{+12}

Concentrations in Media Sampled for the Kola Project (mg/kg)

Medium	Moss conc. HNO$_3$	Humus <2 mm	Humus <2 mm amm. acet.	Topsoil (0-5 cm) <2 mm	B horizon <2 mm aqua regia	B horizon <2 mm total (XRF)
Median	-	-	-	-	-	-
Min	-	-	-	-	-	-
Max	-	-	-	-	-	-

Medium	C horizon <2 mm aqua regia	C horizon <2 mm total (XRF)	C horizon <2 mm total (INAA)	Lake water unfiltered (mg/l)
Median	-	-	-	1.44
Min	-	-	-	0.4
Max	-	-	-	11.3

CHLORINE Cl

Concentrations in Soils and Sediments (mg/kg)

Medium	Soil	Agricultural soil - Ap horizon	Agricultural soil - Top (0-25 cm)	Agricultural Soil - Bottom (50-75 cm)	Topsoil (0-15 cm)	Urban soil (0-2 cm)
	World	Canada	Finland	Finland	England & Wales	Trondheim
	<2 mm total	<2 mm	<2 mm aqua regia	<2 mm aqua regia	<2 mm aqua regia	<2 mm aqua regia
Median	300*	-	-	-	-	-
Min	-	-	-	-	-	-
Max	-	-	-	-	-	-

*Estimated mean

Medium	Forest soil - Humus	Forest soil - B horizon	Forest soil - C horizon	Till (C horizon)	Till (C horizon)	Laterite (25 ±15 cm)
	Norway	Norway	Norway	Finland	Finland	Australia
	<2 mm 7N HNO$_3$	<2 mm 7N HNO$_3$	<2 mm 7N HNO$_3$	<0.063 mm aqua regia	<0.063 mm total	0.45-2 mm total
Median	-	-	-	-	-	-
Min	-	-	-	-	-	-
Max	-	-	-	-	-	-

Medium	Stream sediment	Stream sediment	Stream sediment	Overbank sediment	Overbank sediment	Organic stream sediment
	Austria	Southern Scotland	Harz, Germany	Norway	Norway	Finland
	<0.18 mm	<0.10 mm total	<0.063 mm total	<0.063 mm total (XRF)	<0.063 mm 7N HNO$_3$	<2 mm Conc. HNO$_3$
Median	-	-	-	155	-	-
Min	-	-	-	5	-	-
Max	-	-	-	2279	-	-

#2nd percentile ^98th percentile

Concentrations in Waters (mg/l)

Medium	Ocean water	Ocean water	Ocean water	Stream water	Stream water	Stream water
	World (1)	World (2)	North Pacific	World	Nova Scotia, Canada	Finland
					<0.45 µm IC	<0.45 µm IC
Median	19,400*	19,500*	19,350*	8*	5.32	1.4
Min	-	-	-	-	<0.05	0.5#
Max	-	-	-	-	526	25.6^

*Estimated mean   #2nd percentile   ^98th percentile

Medium	Stream water	Stream water	Stream water	Lake water	Ground water
	Romania	Eastern India	Harz, Germany	Norway	Southern Norway
	unfiltered ICP-MS	<0.2 µm ICP-MS	unfiltered IC	unfiltered ICP-MS	unfiltered
Median	-	-	3.6	-	-
Min	-	-	0.8	-	-
Max	-	-	84	-	-

Cl CHLORINE

Concentrations in Precipitation (mg/l)

Medium	Rain water	Rain water	Snow melt-water	Snow melt-water	Snow filter residue	Snow filter residue
	Kola, remote <0.45 µm IC	Kola, polluted <0.45 µm IC	Kola, remote <0.45 µm IC	Kola, polluted <0.45 µm IC	Kola, remote >0.45 µm	Kola, polluted >0.45 µm
Median	0.1	0.6	0.3	1.4	-	-
Min	<0.1	0.3	0.2	0.5	-	-
Max	0.3	2.1	0.3	1.7	-	-

Medium	Rain water	Rain water				
	Coastal, Norway unfiltered ICP-MS	Inland, Norway unfiltered ICP-MS				
Median	-	-				
Min	-	-				
Max	-	-				

Concentrations in Air (ng/m^3) and Yearly Deposition (kg/km^2/yr)

Medium	Air	Air	Bulk deposition	Throughfall deposition	Bulk deposition	Bulk deposition
	World, remote	World, polluted	West Germany	West Germany	Kola, remote	Kola, polluted
Median	230 / 60	2400	-	-	-	-
Min	2.6	9	650	1000	-	-
Max	4400 (9500)	6000	2500	5200	-	-

*Estimated value

Concentrations in Plants (mg/kg)

Medium	Moss	Moss	Crustose lichen	Lichen	Dandelion	Spruce bark
	Norway conc. HNO$_3$ (1977)	Germany conc. HNO$_3$ + H$_2$O$_2$ (1995)	Germany conc. HNO$_3$	Northwest Territories oven-dried total (INAA)	Europe oven-dried total (INAA)	Canada ashed
Median	160	-	-	339*	0.18#	-
Min	30	-	-	-	-	-
Max	810	-	-	-	-	-

*Mean #Geometric mean

Concentrations in Human Fluids (mg/l)

Medium	Human blood	Human serum	Human urine
	Lombardy, Italy	Lombardy, Italy	Lombardy, Italy
Mean	-	-	-
Min	-	-	-
Max	-	-	-

CHLORINE — Cl

Environmental Geochemistry

Biological impacts	Essential for humans and other organisms. Oxidised forms are toxic (Cl gas, hypochlorite, chlorates)
Uses	Chemical, paper, textile and petroleum industry, bleaching, water purification, antiseptic, dies, insecticide
Environmental pathways	Sea spray, volcano emissions (as Cl_2), potassium industry, paper mills, road salting
Environmental mobility	Oxidising conditions: very high Acid conditions: very high Reducing conditions: very high Neutral to alkaline conditions: very high Comments: -
Geochemical barriers	Evaporation
Natural association	Na-Cl-K-Mg-S (sea water), Li-Na-K-B-P-W-F-Br-Cl-I-SO_4-CO_3 (brines, evaporites)
Action levels	Drinking water, recommended value: 25 mg/l (Norway Shd); MAC: 250 mg/l (Canada CWQG). Separate action levels exist for many organic Cl compounds
Remarks	Many organic Cl compounds are carcinogenic. Many organic Cl compounds tend to bioaccumulate (e.g., DDT, PCBs, HFCs, dioxin). High Cl content in soils in arid and semi-arid climates and close to ocean. Irrigation with Cl-rich water can lead to salinisation of soils. Cl^- is an essential nutrient for many plants. Concentrations of 250 mg/l in drinking water may give rise to consumer complaints due to taste and corrosion
Suggested analytical method(s)	XRF, INAA, IC (water)

World Yearly Production (t/y)	40,000,000
Price of Purest Form, in Small Quantities ($/kg)	-
Market Price ($/kg)	-

Co COBALT

Physico-Chemical Properties

Atomic number	27	Melting/boiling point (K)	1768 / 3200
Atomic mass	58.933,200(9)	Isotopes & isomers	1 stable + 26 unstable
Atomic radius (pm)	167	Acid/base of oxide	amphoteric
Main oxidation state(s)	+2 (+3,+4)	State (at 300 K, 1 atm.)	solid
Ionic radius (pm)	72-104 (+2), 68.5-75 (+3), 54-67 (+4)	Metallic character	pred. metal
		Element group(s)	heavy metal
Electronegativity (Pauling)	1.88	Affinity	chalcophile, siderophile
Density (g/cm^3)	8.9		

Naturally occurring isotopes	Natural abundance (%)	Atomic mass	Half-life	Unstable isotopes	Longest half-life
^{59}Co	100	58.933,2	stable	^{50}Co to ^{71}Co	5.271 y

Concentrations in Crust/Rocks (mg/kg)

Bulk continental crust	24 / 25 / 29
Upper continental crust	11.6 / 10
Ultramafic rock	110
Ocean ridge basalt	50
Gabbro, basalt	45
Granite, granodiorite	4
Sandstone	0.3
Greywacke	15
Shale, schist	20
Limestone	0.1
Coal	10

Typical Minerals
smaltite ((Co,Ni)As$_{2-2.5}$), cobaltite ((Co,Fe)AsS), linnaeite ((Co,Ni)$_3$S$_4$), erythrite (Co$_3$(AsO$_4$)$_2$.8H$_2$O)

Possible Host Minerals
olivine, pyroxenes, amphiboles, micas, garnets, pyrite, sphalerite

Mass (kg) in

Continental crust	5.66x10^{+17}	Oceans	3.97x10^{+9}	Plants	3.68x10^{+8}

Concentrations in Media Sampled for the Kola Project (mg/kg)

Medium	Moss	Humus	Humus	Topsoil (0-5 cm)	B horizon	B horizon
		<2 mm	<2 mm	<2 mm	<2 mm	<2 mm
	conc. HNO$_3$	conc. HNO$_3$	amm. acet.	total (INAA)	aqua regia	total (XRF)
Median	0.395	1.57	0.29	5	5.7	-
Min	0.11	0.21	<0.02	<1	<0.2	-
Max	13.2	96	12.6	43	20.9	-

Medium	C horizon	C horizon	C horizon			Lake water
	<2 mm	<2 mm	<2 mm			unfiltered
	aqua regia	total (XRF)	total (INAA)			(mg/l)
Median	7	-	13			0.000,04
Min	1.2	-	<1			<0.000,02
Max	44.3	-	57			0.016,9

COBALT — Co

Concentrations in Soils and Sediments (mg/kg)

Medium	Soil	Agricultural soil - Ap horizon	Agricultural soil - Top (0-25 cm)	Agricultural Soil - Bottom (50-75 cm)	Topsoil (0-15 cm)	Urban soil (0-2 cm)
	World	Canada	Finland	Finland	England & Wales	Trondheim
	<2 mm total	<2 mm total (INAA)	<2 mm aqua regia	<2 mm aqua regia	<2 mm aqua regia	<2 mm aqua regia
Median	10*	9	4.45	4.98	9.8	13.5
Min	-	<4	<1	<1	0.2	1.6
Max	-	37	19.5	33.8	322	45

*Estimated mean

Medium	Forest soil - Humus	Forest soil - B horizon	Forest soil - C horizon	Till (C horizon)	Till (C horizon)	Laterite (25 ±15 cm)
	Norway	Norway	Norway	Finland	Finland	Australia
	<2 mm $7N\ HNO_3$	<2 mm $7N\ HNO_3$	<2 mm $7N\ HNO_3$	<0.063 mm aqua regia	<0.063 mm total	0.45-2 mm total
Median	1.9	9.7	12.6	6.7	14	3
Min	<0.5	2	1.1	-	-	<2
Max	29.8	80.3	469	-	-	52

Medium	Stream sediment	Stream sediment	Stream sediment	Overbank sediment	Overbank sediment	Organic stream sediment
	Austria	Southern Scotland	Harz, Germany	Norway	Norway	Finland
	<0.18 mm "total" (ICP-AES)	<0.10 mm total (DCES)	<0.063 mm total (XRF)	<0.063 mm total (XRF)	<0.063 mm $7N\ HNO_3$ ICP-AES	<2 mm Conc. HNO_3
Median	15	24	17	32	7	10.5
Min	-	<2	<7	6	1	2.4#
Max	226	659	220	92	65	48.5^

#2nd percentile ^98th percentile

Concentrations in Waters (mg/l)

Medium	Ocean water	Ocean water	Ocean water	Stream water	Stream water	Stream water
	World (1)	World (2)	North Pacific	World	Nova Scotia, Canada	Finland
					<0.45 µm ICP-MS	<0.45 µm ICP-MS
Median	0.000,02*	0.000,003*	0.000,001,2*	0.000,2*	<0.000,05	0.000,17
Min	-	-	-	-	<0.000,05	0.000,02#
Max	-	-	-	-	0.000,57	0.003,87^

*Estimated mean   #2nd percentile   ^98th percentile

Medium	Stream water	Stream water	Stream water	Lake water	Ground water
	Romania	Eastern India	Harz, Germany	Norway	Southern Norway
	unfiltered ICP-MS	<0.2 µm ICP-MS	unfiltered ICP-MS	unfiltered ICP-MS	unfiltered ICP-MS
Median	0.000,71	0.001,9	0.000,37	0.000,053	0.000,061,5
Min	0.000,03	0.000,4	0.000,03	<0.000,02	<0.000,005
Max	0.078	0.016,6	0.005,6	0.003,15	0.008,12

Co COBALT

Concentrations in Precipitation (mg/l)

Medium	Rain water	Rain water	Snow melt-water	Snow melt-water	Snow filter residue	Snow filter residue
	Kola, remote	Kola, polluted	Kola, remote	Kola, polluted	Kola, remote	Kola, polluted
	<0.45 μm	<0.45 μm	<0.45 μm	<0.45 μm	>0.45 μm	>0.45 μm
	ICP-MS	ICP-MS	ICP-MS	ICP-MS	ICP-AES	ICP-AES
Median	<0.000,02	0.011,8	<0.000,03	0.032,95	<0.000,05	0.046,45
Min	<0.000,02	0.002,17	<0.000,03	0.009,74	<0.000,05	0.030,4
Max	0.000,03	0.068,9	0.000,1	0.053	<0.000,05	0.104

Medium	Rain water	Rain water				
	Coastal, Norway unfiltered ICP-MS	Inland, Norway unfiltered ICP-MS				
Median	<0.000,01	0.000,04				
Min	<0.000,01	<0.000,01				
Max	0.000,44	0.000,46				

Concentrations in Air (ng/m^3) and Yearly Deposition (kg/km^2/yr)

Medium	Air	Air	Bulk deposition	Throughfall deposition	Bulk deposition	Bulk deposition
	World, remote	World, polluted	West Germany	West Germany	Kola, remote	Kola, polluted
Median	0.05 / 0.04	3	-	-	0.038,1*	60.3*
Min	0.000,4	0.13	-	0.13	-	-
Max	0.06 (1.0)	37	<0.1	0.27	-	-

*Estimated value

Concentrations in Plants (mg/kg)

Medium	Moss	Moss	Crustose lichen	Lichen	Dandelion	Spruce bark
	Norway conc. HNO$_3$ (1990)	Germany conc. HNO$_3$ + H$_2$O$_2$ (1995)	Germany conc. HNO$_3$	Northwest Territories oven-dried total (INAA)	Europe oven-dried	Canada ashed total (INAA)
Median	0.25	0.28	4.1	0.51*	-	8
Min	0.06	0.02	1.4	-	-	2
Max	14	3.4	14	-	-	27

*Mean   #Geometric mean

Concentrations in Human Fluids (mg/l)

Medium	Human blood	Human serum	Human urine			
	Lombardy, Italy	Lombardy, Italy	Lombardy, Italy			
Mean	0.000,39	0.000,21	0.000,57			
Min	0.000,1	0.000,08	0.000,12			
Max	0.004,2	0.000,52	0.002			

COBALT Co

Environmental Geochemistry

Biological impacts	Essential. Co is a central atom in vitamin B 12. Toxic to humans at doses of 25 mg/day or more. Toxic to rats at doses of >2 mg/l in drinking water. Co-dust is carcinogenic. Co deficiencies are more widespread than Co toxicity problems
Uses	Super-alloys, aerospace industry, stainless steel, electroplating, magnets, catalysts, paints, plastic hardener
Environmental pathways	Ni, Ag, Pb, Cu, Fe mining and processing, coal combustion, geogenic dust, weathering, fertilisers
Environmental mobility	Oxidising conditions: medium / Acid conditions: high / Reducing conditions: very low / Neutral to alkaline conditions: very low / Comments: -
Geochemical barriers	Presence of sulphide, adsorption, pH
Natural association	Ni-Co-Pt-Fe-Cu-Ag-Au-Se-Te-S (Ni-Cu massive sulphide deposits), Ni-Co-Ag-Fe-Cu-Pb-Zn-As-Sb-Bi-U (Cu-Co sulphide ores), Co-Au-Ag (some Au and Ag ores), Ni-Co-Fe-Mn-Cr (laterites), Mn-Ni-Cu-Zn-Co (deep-sea Mn nodules), U-Co-V-As-Mo (some U deposits)
Action levels	Drinking water: listed in Norwegian regulations, but no value given; MAC: 0.1 mg/l (Russia MoH). Ground water, background: 0.02 mg/l; remediate: 0.1 mg/l (Netherlands VROM). Soil, background: 20 mg/kg; remediate: 240 mg/kg (Netherlands VROM). Agricultural soil, maximum tolerable concentration: 50 mg/kg (Germany)
Remarks	Co depletion in agricultural soils is presently an increasing problem. Soil is considered Co-deficient at <5 mg/kg. Co fertilisers can increase yield. Excessive Co can result in Fe and Cu deficiencies. $CoCl_2$ used as a beer head stabiliser led to cases of Co poisoning in the 1960s. Some plants are thought to be indicators of Co deposits. ^{60}Co used in medicine
Suggested analytical method(s)	ICP-AES, ICP-MS

World Yearly Production (t/y)	18,000 (in 1995)
Price of Purest Form, in Small Quantities ($/kg)	8190 (99.9%)
Market Price ($/kg)	52

Cr CHROMIUM

Physico-Chemical Properties

Atomic number	24
Atomic mass	51.996,1(6)
Atomic radius (pm)	185
Main oxidation state(s)	+2,+3,+6 (+4,+5)
Ionic radius (pm)	87-94 (+2), 75.5 (+3), 55-69 (+4), 48.5-71 (+5), 40-58 (+6)
Electronegativity (Pauling)	1.66
Density (g/cm^3)	7.19

Melting/boiling point (K)	2180 / 2944
Isotopes & isomers	4 stable + 17 unstable
Acid/base of oxide	strong acid
State (at 300 K, 1 atm.)	solid
Metallic character	pred. non-metal
Element group(s)	heavy non-metal
Affinity	lithophile

Naturally occurring isotopes	Natural abundance (%)	Atomic mass	Half-life
^{50}Cr	4.345	49.946,05	stable
^{52}Cr	83.789	51.940,5	stable
^{53}Cr	9.501	52.940,7	stable
^{54}Cr	2.365	53.938,885	stable

Unstable isotopes	Longest half-life
^{43}Cr to ^{63}Cr	27.7 d

Concentrations in Crust/Rocks (mg/kg)

Bulk continental crust	126 / 102 / 185
Upper continental crust	35 / 35
Ultramafic rock	2300
Ocean ridge basalt	300
Gabbro, basalt	250
Granite, granodiorite	10
Sandstone	35
Greywacke	88
Shale, schist	100
Limestone	5
Coal	20

Typical Minerals
chromite (FeCr$_2$O$_4$), crocoite (PbCrO$_4$)

Possible Host Minerals
pyroxenes, amphiboles, micas, garnets, spinels

Mass (kg) in

Continental crust	2.97x10^{+18}	Oceans	3.97x10^{+11}	Plants	2.67x10^{+9}

Concentrations in Media Sampled for the Kola Project (mg/kg)

Medium	Moss	Humus	Humus	Topsoil (0-5 cm)	B horizon	B horizon
		<2 mm	<2 mm	<2 mm	<2 mm	<2 mm
	conc. HNO$_3$	conc. HNO$_3$	amm. acet.	total (INAA)	aqua regia	total (XRF)
Median	0.6	2.91	0.07	53	35.2	-
Min	<0.2	0.39	<0.02	<5	3.8	-
Max	14.4	109	2.96	670	413	-

Medium	C horizon	C horizon	C horizon	Lake water
	<2 mm	<2 mm	<2 mm	unfiltered
	aqua regia	total (XRF)	total (INAA)	(mg/l)
Median	28.2	-	99	0.000,29
Min	2.2	-	11	<0.000,2
Max	471	-	910	0.005,76

CHROMIUM   Cr

Concentrations in Soils and Sediments (mg/kg)

Medium	Soil	Agricultural soil - Ap horizon	Agricultural soil - Top (0-25 cm)	Agricultural Soil - Bottom (50-75 cm)	Topsoil (0-15 cm)	Urban soil (0-2 cm)
	World	Canada	Finland	Finland	England & Wales	Trondheim
	<2 mm	<2 mm	<2 mm	<2 mm	<2 mm	<2 mm
	total	total (INAA)	aqua regia	aqua regia	aqua regia	aqua regia
Median	80*	57	19.1	24.5	39.3	69.3
Min	-	10	<5	<5	0.2	7.9
Max	-	510	86	104	838	199

*Estimated mean

Medium	Forest soil - Humus	Forest soil - B horizon	Forest soil - C horizon	Till (C horizon)	Till (C horizon)	Laterite (25 ±15 cm)
	Norway	Norway	Norway	Finland	Finland	Australia
	<2 mm	<2 mm	<2 mm	<0.063 mm	<0.063 mm	0.45-2 mm
	7N HNO$_3$	7N HNO$_3$	7N HNO$_3$	aqua regia	total	total
Median	3	29.8	26.2	27.9	60	80.5
Min	<1	2.1	<2	-	-	<3
Max	188.6	840	388	-	-	4500

Medium	Stream sediment	Stream sediment	Stream sediment	Overbank sediment	Overbank sediment	Organic stream sediment
	Austria	Southern Scotland	Harz, Germany	Norway	Norway	Finland
	<0.18 mm	<0.10 mm	<0.063 mm	<0.063 mm	<0.063 mm	<2 mm
	"total" (ICP-AES)	total (DCES)	total (XRF)	total (XRF)	7N HNO$_3$ ICP-AES	Conc. HNO$_3$
Median	64	161	69	76	24	31.3
Min	-	<5	20	2	2.6	10.9#
Max	3176	114	448	4484	245	84.2^

#2nd percentile   ^98th percentile

Concentrations in Waters (mg/l)

Medium	Ocean water	Ocean water	Ocean water	Stream water	Stream water	Stream water
	World (1)	World (2)	North Pacific	World	Nova Scotia, Canada	Finland
					<0.45 μm	<0.45 μm
					ICP-MS	ICP-MS
Median	0.000,3*	0.000,3*	0.000,212*	0.000,7*	0.000,1	0.000,5
Min	-	-	-	-	<0.000,1	<0.000,2#
Max	-	-	-	-	0.001,1	0.001,61^

*Estimated mean   #2nd percentile   ^98th percentile

Medium	Stream water	Stream water	Stream water	Lake water	Ground water
	Romania	Eastern India	Harz, Germany	Norway	Southern Norway
	unfiltered	<0.2 μm	unfiltered	unfiltered	unfiltered
	ICP-MS	ICP-MS	ICP-MS	ICP-MS	ICP-MS
Median	0.001,13	0.009,7	0.000,41	<0.000,1	0.000,54
Min	0.000,05	0.004,9	<0.000,01	<0.000,1	<0.000,1
Max	0.179	0.058,8	0.007,1	0.004,85	0.005,86

Cr　　　　　　　　　　　　　　　　　　　　　　　　　CHROMIUM

Concentrations in Precipitation (mg/l)

Medium	Rain water	Rain water	Snow melt-water	Snow melt-water	Snow filter residue	Snow filter residue
	Kola, remote <0.45 μm ICP-MS	Kola, polluted <0.45 μm ICP-MS	Kola, remote <0.45 μm ICP-MS	Kola, polluted <0.45 μm ICP-MS	Kola, remote >0.45 μm ICP-AES	Kola, polluted >0.45 μm ICP-AES
Median	<0.000,2	0.000,47	<0.000,2	0.001,145	0.000,22	0.005,5
Min	<0.000,2	0.000,25	<0.000,2	0.000,32	0.000,09	0.003,44
Max	<0.000,2	0.000,95	0.000,31	0.001,31	0.000,33	0.020,7

Medium	Rain water	Rain water
	Coastal, Norway unfiltered ICP-MS	Inland, Norway unfiltered ICP-MS
Median	0.000,23	0.000,32
Min	<0.000,01	<0.000,01
Max	0.000,59	0.002,7

Concentrations in Air (ng/m^3) and Yearly Deposition (kg/km^2/yr)

Medium	Air	Air	Bulk deposition	Throughfall deposition	Bulk deposition	Bulk deposition
	World, remote	World, polluted	West Germany	West Germany	Kola, remote	Kola, polluted
Median	0.6 / 0.5	40	0.6	-	0.216*	5.24*
Min	0.005	1	-	0.61	-	-
Max	0.7 (7)	300	-	2.3	-	-

*Estimated value

Concentrations in Plants (mg/kg)

Medium	Moss	Moss	Crustose lichen	Lichen	Dandelion	Spruce bark
	Norway conc. HNO$_3$ (1990)	Germany conc. HNO$_3$ + H$_2$O$_2$ (1995)	Germany conc. HNO$_3$	Northwest Territories oven-dried total (INAA)	Europe oven-dried total (INAA)	Canada ashed total (INAA)
Median	0.9	1.393	33	1.6*	0.35#	31
Min	<0.4	0.05	16	-	-	4
Max	30	29	167	-	-	250

*Mean　#Geometric mean

Concentrations in Human Fluids (mg/l)

Medium	Human blood	Human serum	Human urine
	Lombardy, Italy	Lombardy, Italy	Lombardy, Italy
Mean	0.000,23	0.000,17	0.000,61
Min	0.000,09	0.000,04	0.000,04
Max	0.000,75	0.000,6	0.005,1

CHROMIUM — Cr

Environmental Geochemistry

Biological impacts	Essential for some organisms. Cr(3+) considered relatively harmless, Cr(6+) highly toxic; some Cr(6+) compounds are known to be carcinogenic
Uses	Stainless steel, many alloys, chrome plating, pigments, catalyst, dye, tanning, wood impregnation, refractory bricks, magnetic tapes
Environmental pathways	Geogenic dust, weathering, chemical industry, steel works, electrometallurgy, Cu smelting, combustion of natural gas, oil and coal, sewage sludge, waste incineration, some P-fertilisers
Environmental mobility	Oxidising conditions: very low Acid conditions: very low Reducing conditions: very low Neutral to alkaline conditions: very low Comments: Cr compounds are not likely to migrate to ground water through soils. Aquatic life can potentially accumulate Cr
Geochemical barriers	Very low mobility, mechanical
Natural association	Cr-Cu-Ni-Co-PGE
Action levels	Drinking water, MAC: 0.05 mg/l (Norway Shd), 0.1 mg/l (US EPA), 0.05 mg/l (Canada CWQG); recommended: 0.05 mg/l (WHO); MAC: 0.5 mg Cr^{3+}/l, 0.1 mg Cr^{6+}/l (Russia MoH). Ground water, background: 0.001 mg/l; remediate: 0.03 mg/l (Netherlands VROM). Soil, background: 100 mg/kg; remediate: 380 mg/kg (Netherlands VROM). Agricultural soil, maximum tolerable concentration: 100 mg/kg (Germany)
Remarks	Cr^{6+} is more mobile in soil than Cr^{3+}, but usually reduced to Cr^{3+} within a few weeks. Generally low uptake by most vegetables, except spinach. Some trees, lichen and moss can accumulate Cr. Rhubarb and geranium can tolerate very high Cr levels in soil. Cr deficiency has resulted in disturbed growth in test animals. The US EPA suggests that drinking water exceeding action levels can lead to skin irritation or ulceration in the short term, and to damage to liver, kidney circulatory and nerve tissues, and skin irritation in the long term
Suggested analytical method(s)	XRF, ICP-AES, ICP-MS (water)

World Yearly Production (t/y)	3,770,000 (in 1995)
Price of Purest Form, in Small Quantities ($/kg)	12,363 (99.998%)
Market Price ($/kg)	17.6

Cs   CESIUM

Physico-Chemical Properties

Atomic number	55	Melting/boiling point (K)	301.59 / 944
Atomic mass	132.905,45(2)	Isotopes & isomers	1 stable + 51 unstable
Atomic radius (pm)	334	Acid/base of oxide	strong base
Main oxidation state(s)	+1	State (at 300 K, 1 atm.)	liquid
Ionic radius (pm)	181-202	Metallic character	metal
Electronegativity (Pauling)	0.79	Element group(s)	alkali metal, light metal
Density (g/cm^3)	1.873	Affinity	lithophile

Naturally occurring isotopes	Natural abundance (%)	Atomic mass	Half-life	Unstable isotopes	Longest half-life
^{133}Cs	100	132.905,45	stable	^{112}Cs to ^{148}Cs	2,300,000 y

Concentrations in Crust/Rocks (mg/kg)

Bulk continental crust	3.4 / 3 / 1
Upper continental crust	5.8 / 3.7
Ultramafic rock	0.05
Ocean ridge basalt	0.2
Gabbro, basalt	0.8
Granite, granodiorite	4
Sandstone	1
Greywacke	2.2
Shale, schist	5
Limestone	0.5
Coal	1

Typical Minerals
pollucite ((Cs,Na)$_2$Al$_2$Si$_4$O$_{12}$.H$_2$O)

Possible Host Minerals
micas, K-feldspar

Mass (kg) in

Continental crust	8.02x10^{+16}	Oceans	5.29x10^{+11}	Plants	3.68x10^{+8}

Concentrations in Media Sampled for the Kola Project (mg/kg)

Medium	Moss	Humus <2 mm	Humus <2 mm	Topsoil (0-5 cm) <2 mm	B horizon <2 mm	B horizon <2 mm
	conc. HNO$_3$	amm. acet.	total (INAA)	aqua regia	total (XRF)	
Median	-	-	-	<1	-	-
Min	-	-	-	<1	-	-
Max	-	-	-	5	-	-

Medium	C horizon <2 mm	C horizon <2 mm	C horizon <2 mm			Lake water unfiltered
	aqua regia	total (XRF)	total (INAA)			(mg/l)
Median	-	-	<1			-
Min	-	-	<1			-
Max	-	-	8			-

CESIUM   Cs

Concentrations in Soils and Sediments (mg/kg)

Medium	Soil	Agricultural soil - Ap horizon	Agricultural soil - Top (0-25 cm)	Agricultural Soil - Bottom (50-75 cm)	Topsoil (0-15 cm)	Urban soil (0-2 cm)
	World	Canada	Finland	Finland	England & Wales	Trondheim
	<2 mm	<2 mm	<2 mm	<2 mm	<2 mm	<2 mm
	total	total (INAA)	aqua regia	aqua regia	aqua regia	aqua regia
Median	3*	2.1	-	-	-	-
Min	-	0.2	-	-	-	-
Max	-	7.1	-	-	-	-

*Estimated mean

Medium	Forest soil - Humus	Forest soil - B horizon	Forest soil - C horizon	Till (C horizon)	Till (C horizon)	Laterite (25 ±15 cm)
	Norway	Norway	Norway	Finland	Finland	Australia
	<2 mm	<2 mm	<2 mm	<0.063 mm	<0.063 mm	0.45-2 mm
	7N HNO$_3$	7N HNO$_3$	7N HNO$_3$	aqua regia	total	total
Median	-	-	-	-	2.1	-
Min	-	-	-	-	-	-
Max	-	-	-	-	-	-

Medium	Stream sediment	Stream sediment	Stream sediment	Overbank sediment	Overbank sediment	Organic stream sediment
	Austria	Southern Scotland	Harz, Germany	Norway	Norway	Finland
	<0.18 mm	<0.10 mm	<0.063 mm	<0.063 mm	<0.063 mm	<2 mm
		total	total (ICP-MS)	total (XRF)	7N HNO$_3$	Conc. HNO$_3$
Median	-	-	8.2	-	-	0.9
Min	-	-	3.3	-	-	0.12#
Max	-	-	52	-	-	4.89^

#2nd percentile   ^98th percentile

Concentrations in Waters (mg/l)

Medium	Ocean water	Ocean water	Ocean water	Stream water	Stream water	Stream water
	World (1)	World (2)	North Pacific	World	Nova Scotia, Canada	Finland
					<0.45 μm	<0.45 μm
					ICP-MS	
Median	0.000,3*	0.000,4*	0.000,306*	0.000,03*	<0.000,01	-
Min	-	-	-	-	<0.000,01	-
Max	-	-	-	-	0.000,4	-

*Estimated mean   #2nd percentile   ^98th percentile

Medium	Stream water	Stream water	Stream water	Lake water	Ground water
	Romania	Eastern India	Harz, Germany	Norway	Southern Norway
	unfiltered	<0.2 μm	unfiltered	unfiltered	unfiltered
	ICP-MS	ICP-MS	ICP-MS	ICP-MS	ICP-MS
Median	0.000,064	0.000,48	0.000,05	-	0.000,097
Min	0.000,001	0.000,04	<0.000,005	-	<0.000,01
Max	0.001,5	0.002,87	0.002,9	-	0.001,612

Cs   CESIUM

Concentrations in Precipitation (mg/l)

Medium	Rain water Kola, remote <0.45 µm	Rain water Kola, polluted <0.45 µm	Snow melt-water Kola, remote <0.45 µm	Snow melt-water Kola, polluted <0.45 µm	Snow filter residue Kola, remote >0.45 µm	Snow filter residue Kola, polluted >0.45 µm
Median	-	-	-	-	-	-
Min	-	-	-	-	-	-
Max	-	-	-	-	-	-

Medium	Rain water Coastal, Norway unfiltered ICP-MS	Rain water Inland, Norway unfiltered ICP-MS				
Median	-	-				
Min	-	-				
Max	-	-				

Concentrations in Air (ng/m^3) and Yearly Deposition (kg/km^2/yr)

Medium	Air World, remote	Air World, polluted	Bulk deposition West Germany	Throughfall deposition West Germany	Bulk deposition Kola, remote	Bulk deposition Kola, polluted
Median	0.045	0.4	-	-	-	-
Min	0.000,09	0.06	-	-	-	-
Max	0.6	1.4	-	-	-	-

*Estimated value

Concentrations in Plants (mg/kg)

Medium	Moss Norway conc. HNO$_3$ (1990)	Moss Germany conc. HNO$_3$ + H$_2$O$_2$ (1995)	Crustose lichen Germany conc. HNO$_3$	Lichen Northwest Territories oven-dried	Dandelion Europe oven-dried total (INAA)	Spruce bark Canada ashed total (INAA)
Median	0.18	0.22	1	-	0.1#	2.95
Min	0.03	0.04	0.3	-	-	<0.5
Max	2.5	9.8	7.4	-	-	14

*Mean   #Geometric mean

Concentrations in Human Fluids (mg/l)

Medium	Human blood Lombardy, Italy	Human serum Lombardy, Italy	Human urine Lombardy, Italy			
Mean	0.003	0.001,5	0.008,1			
Min	0.000,5	0.000,11	0.001,1			
Max	0.008,5	0.006,8	0.022			

CESIUM — Cs

Environmental Geochemistry

Biological impacts	Considered non-essential. ^{137}Cs is radiotoxic
Uses	Catalyst, rocket propellant ($CsBH_4$), photovoltaic cells, atomic clocks
Environmental pathways	Geogenic dust, rock weathering, nuclear bomb testing, nuclear accidents
Environmental mobility	Oxidising conditions: low Acid conditions: low Reducing conditions: very low Neutral to alkaline conditions: low Comments: -
Geochemical barriers	-
Natural association	K-Cs (silicates)
Action levels	-
Remarks	Least electronegative and most alkaline element. ^{134}Cs and ^{137}Cs are very abundant in nuclear reactors. In contrast to ^{137}Cs ^{134}Cs is not present in detectable amounts in fallout from nuclear explosions
Suggested analytical method(s)	ICP-MS, INAA, HPGe detector (radioisotopes)

World Yearly Production (t/y)	30
Price of Purest Form, in Small Quantities ($/kg)	62,875 (99.98%)
Market Price ($/kg)	-

Cu   COPPER

Physico-Chemical Properties

Atomic number	29	Melting/boiling point (K)	1357.77 / 2835
Atomic mass	63.546(3)	Isotopes & isomers	2 stable + 25 unstable
Atomic radius (pm)	157	Acid/base of oxide	base
Main oxidation state(s)	+2 (+1)	State (at 300 K, 1 atm.)	solid
Ionic radius (pm)	60-91 (+1), 71-87 (+2)	Metallic character	metal
Electronegativity (Pauling)	1.9	Element group(s)	heavy metal
Density (g/cm^3)	8.96	Affinity	chalcophile

Naturally occurring isotopes	Natural abundance (%)	Atomic mass	Half-life
^{63}Cu	69.17	62.929,6	stable
^{65}Cu	30.83	64.927,79	stable

Unstable isotopes	Longest half-life
^{55}Cu to ^{79}Cu	2.58 d

Concentrations in Crust/Rocks (mg/kg)

Bulk continental crust	25 / 60 / 75
Upper continental crust	14.3 / 25
Ultramafic rock	40
Ocean ridge basalt	80
Gabbro, basalt	90
Granite, granodiorite	12
Sandstone	2
Greywacke	24
Shale, schist	45
Limestone	6
Coal	20

Typical Minerals
chalcopyrite (CuFeS$_2$), bornite (Cu$_5$FeS$_4$), chalcosite (Cu$_2$S), malachite (Cu$_2$CO$_3$(OH)$_2$), covellite (CuS), digenite (Cu$_9$S$_5$), tetrahedrite (Cu$_{12}$Sb$_4$S$_{13}$), native Cu

Possible Host Minerals
micas (biotite), pyroxenes, amphiboles, magnetite

Mass (kg) in

Continental crust	5.90x10^{+17}	Oceans	1.32x10^{+11}	Plants	1.84x10^{+10}

Concentrations in Media Sampled for the Kola Project (mg/kg)

Medium	Moss	Humus <2 mm	Humus <2 mm	Topsoil (0-5 cm) <2 mm	B horizon <2 mm	B horizon <2 mm
	conc. HNO$_3$	conc. HNO$_3$	amm. acet.		aqua regia	total (XRF)
Median	7.16	9.69	0.2	-	10.5	-
Min	2.63	2.69	<0.02	-	0.9	-
Max	355	4080	1160	-	126	-

Medium	C horizon <2 mm aqua regia	C horizon <2 mm total (XRF)	C horizon <2 mm total (INAA)	Lake water unfiltered (mg/l)
Median	16.1	-	-	0.000,935
Min	2	-	-	0.000,17
Max	149	-	-	0.152

COPPER — Cu

Concentrations in Soils and Sediments (mg/kg)

Medium	Soil	Agricultural soil - Ap horizon	Agricultural soil - Top (0-25 cm)	Agricultural Soil - Bottom (50-75 cm)	Topsoil (0-15 cm)	Urban soil (0-2 cm)
	World	Canada	Finland	Finland	England & Wales	Trondheim
	<2 mm	<2 mm	<2 mm	<2 mm	<2 mm	<2 mm
	total	total (AAS)	aqua regia	aqua regia	aqua regia	aqua regia
Median	25*	19	14.7	12.8	18.1	34.7
Min	-	5	<1	1.38	1.2	1.7
Max	-	221	54.8	78.2	1508	706

*Estimated mean

Medium	Forest soil - Humus	Forest soil - B horizon	Forest soil - C horizon	Till (C horizon)	Till (C horizon)	Laterite (25 ±15 cm)
	Norway	Norway	Norway	Finland	Finland	Australia
	<2 mm	<2 mm	<2 mm	<0.063 mm	<0.063 mm	0.45-2 mm
	7N HNO$_3$	7N HNO$_3$	7N HNO$_3$	aqua regia	total	total
Median	7	13	26.3	21.1	20	14
Min	1.6	1.1	2.6	-	-	<2
Max	1300	298	324	-	-	145

Medium	Stream sediment	Stream sediment	Stream sediment	Overbank sediment	Overbank sediment	Organic stream sediment
	Austria	Southern Scotland	Harz, Germany	Norway	Norway	Finland
	<0.18 mm	<0.10 mm	<0.063 mm	<0.063 mm	<0.063 mm	<2 mm
	"total" (ICP-AES)	total (DCES)	total (XRF)	total (XRF)	7N HNO$_3$ AAS	Conc. HNO$_3$
Median	19	23	23	24	21	12.4
Min	-	<3	<10	<1	0.1	4.1#
Max	6400	104	2440	383	288	45.2^

#2nd percentile   ^98th percentile

Concentrations in Waters (mg/l)

Medium	Ocean water	Ocean water	Ocean water	Stream water	Stream water	Stream water
	World (1)	World (2)	North Pacific	World	Nova Scotia, Canada	Finland
					<0.45 µm	<0.45 µm
					ICP-MS	ICP-MS
Median	0.000,25*	0.000,1*	0.000,15*	0.003*	0.000,4	0.000,64
Min	-	-	-	-	<0.000,1	0.000,13#
Max	-	-	-	-	0.002,3	0.003,71^

*Estimated mean   #2nd percentile   ^98th percentile

Medium	Stream water	Stream water	Stream water	Lake water	Ground water
	Romania	Eastern India	Harz, Germany	Norway	Southern Norway
	unfiltered	<0.2 µm	unfiltered	unfiltered	unfiltered
	ICP-MS	ICP-MS	ICP-MS	ICP-MS	ICP-MS
Median	0.002,2	0.005,7	0.001,2	0.000,413	0.011,75
Min	0.000,1	0.002,5	<0.000,01	<0.000,2	0.000,4
Max	0.111	0.057,6	0.138	0.037,7	1.332

Cu COPPER

Concentrations in Precipitation (mg/l)

Medium	Rain water	Rain water	Snow melt-water	Snow melt-water	Snow filter residue	Snow filter residue
	Kola, remote <0.45 μm ICP-MS	Kola, polluted <0.45 μm ICP-MS	Kola, remote <0.45 μm ICP-MS	Kola, polluted <0.45 μm ICP-MS	Kola, remote >0.45 μm ICP-AES	Kola, polluted >0.45 μm ICP-AES
Median	0.000,51	0.231	0.000,23	0.555	0.000,11	0.135,5
Min	0.000,24	0.086	0.000,12	0.186	<0.000,05	0.066
Max	0.001,75	0.848	0.006,76	2.19	0.000,53	0.426

Medium	Rain water	Rain water				
	Coastal, Norway unfiltered ICP-MS	Inland, Norway unfiltered ICP-MS				
Median	0.000,16	0.001,1				
Min	<0.000,01	0.000,05				
Max	0.004,6	0.012				

Concentrations in Air (ng/m^3) and Yearly Deposition (kg/km^2/yr)

Medium	Air	Air	Bulk deposition	Throughfall deposition	Bulk deposition	Bulk deposition
	World, remote	World, polluted	West Germany	West Germany	Kola, remote	Kola, polluted
Median	11 / 2.6	310	-	-	0.626*	494*
Min	0.014	5	0.9	2.6	-	-
Max	20 (110)	4900	10	9.6	-	-

*Estimated value

Concentrations in Plants (mg/kg)

Medium	Moss	Moss	Crustose lichen	Lichen	Dandelion	Spruce bark
	Norway conc. HNO$_3$ (1990)	Germany conc. HNO$_3$ + H$_2$O$_2$ (1995)	Germany conc. HNO$_3$	Northwest Territories oven-dried total (XRF)	Europe oven-dried conc. HNO$_3$ (AAS)	Canada ashed aqua regia
Median	5.2	9.57	24	8.5*	14.8#	164
Min	2.1	3	10	-	-	27
Max	240	57	220	-	-	701

*Mean #Geometric mean

Concentrations in Human Fluids (mg/l)

Medium	Human blood	Human serum	Human urine			
	Lombardy, Italy	Lombardy, Italy	Lombardy, Italy			
Mean	1.225	0.985	0.023			
Min	0.535	0.6	0.004,2			
Max	1.94	1.76	0.075			

COPPER   Cu

Environmental Geochemistry

Biological impacts	Essential for all organisms. Toxic at high doses (deadly poisoning of small children from drinking water has been reported)
Uses	Electrical industry, water piping, pigments, alloys, coins, algaecide, bactericide, molluscicide, fungicide, insecticide
Environmental pathways	Cu mining and smelting, other non-ferrous smelters, plastic industry, steel works, agriculture, geogenic dust, rock weathering, sewage sludge (pig farming)
Environmental mobility	Oxidising conditions: medium Acid conditions: high Reducing conditions: very low Neutral to alkaline conditions: very low Comments: All waters are corrosive towards Cu to some degree, especially very acidic waters
Geochemical barriers	Presence of sulphide, pH increase, adsorption
Natural association	Ni-Cu-Pt-Cr (ultrabasic Pt deposits), Cu-Pb-Zn-Cd-Ag-Fe-As-Sb (massive sulphide deposits), Cu-Mo-Re-Fe (porphyry Cu deposits), Ag-Zn-Pb-Mo-Co (Cu shale deposits)
Action levels	Drinking water, recommended: 0.3 mg/l (Norway Shd); MAC: 1.3 mg/l (US EPA), 1 mg/l (Canada CWQG); recommended: 2 mg/l (WHO); MAC: 1 mg/l (Russia MoH). Ground water, background: 0.015 mg/l; remediate: 0.075 mg/l (Netherlands VROM). Soil, background: 36 mg/kg; remediate: 190 mg/kg (Netherlands VROM). Agricultural soil, maximum tolerable concentration: 100 mg/kg (Germany).
Remarks	Excess Cu can lead to Zn deficiencies, and vice-versa. Excess Mo can lead to Cu deficiencies. Cu deficiencies for plants occur at <5 mg Cu/kg in soils. Plants can tolerate very high Cu content in soil when organic C level is high. Cu mobility is strongly dependent on C_{org}, and Cu is especially stable at pH 5 to 6. High uptake by carrot, low uptake by corn and celery. Humans, pigs are very tolerant to high Cu levels. High Cu leads to increased growth rate. $CuSO_4$ frequently used as food supplement in pig farming, leading to high Cu levels on farmland fertilised with sludge from pig farms. In contrast, cows and sheep are very vulnerable to Cu poisoning. Cu has been mined for more than 5000 y. Concentrations of 1 mg/l in drinking water may give rise to consumer complaints due to staining of laundry and sanitary ware. The US EPA suggests that drinking water exceeding action levels can lead to stomach and intestinal distress, liver and kidney damage, and anaemia in the short term
Suggested analytical method(s)	ICP-AES, ICP-MS, (XRF)

World Yearly Production (t/y)	9,988,000 (in 1995)
Price of Purest Form, in Small Quantities ($/kg)	866 (99.999,8%)
Market Price ($/kg)	2.5

Dy DYSPROSIUM

Physico-Chemical Properties

Atomic number	66	Melting/boiling point (K)	1685 / 2840
Atomic mass	162.50(3)	Isotopes & isomers	7 stable + 26 unstable
Atomic radius (pm)	249	Acid/base of oxide	weak base
Main oxidation state(s)	+2,+3,+4	State (at 300 K, 1 atm.)	solid
Ionic radius (pm)	121-133 (+2), 105-122 (+3)	Metallic character	metal
		Element group(s)	heavy metal, REE
Electronegativity (Pauling)	1.22	Affinity	lithophile
Density (g/cm^3)	8.551		

Naturally occurring isotopes	Natural abundance (%)	Atomic mass	Half-life
^{156}Dy	0.06	155.924,28	stable
^{158}Dy	0.1	157.924,41	stable
^{160}Dy	2.34	159.925,19	stable
^{161}Dy	18.9	160.926,93	stable
^{162}Dy	25.5	161.926,795	stable
^{163}Dy	24.9	162.928,73	stable
^{164}Dy	28.2	163.929,17	stable

Unstable isotopes	Longest half-life
^{141}Dy to ^{169}Dy	3,000,000 y

Concentrations in Crust/Rocks (mg/kg)

Bulk continental crust	3.8 / 5.2 / 3.7
Upper continental crust	2.9 / 3.5
Ultramafic rock	-
Ocean ridge basalt	-
Gabbro, basalt	-
Granite, granodiorite	-
Sandstone	-
Greywacke	3.4
Shale, schist	-
Limestone	-
Coal	3

Typical Minerals
monazite ((Ce,La,Nd,Dy,Th)(PO$_4$,SiO$_4$)), bastnaesite ((Ce,La,Dy)CO$_3$(F,OH))

Possible Host Minerals
feldspars, apatite, allanite, sphene, fluorite, zircon

Mass (kg) in

Continental crust	8.97x10^{+16}	Oceans	1.19x10^{+9}	Plants	5.52x10^{+7}

Concentrations in Media Sampled for the Kola Project (mg/kg)

Medium	Moss conc. HNO$_3$	Humus <2 mm amm. acet.	Humus <2 mm	Topsoil (0-5 cm) <2 mm	B horizon <2 mm aqua regia	B horizon <2 mm total (XRF)
Median	-	-	-	-	-	-
Min	-	-	-	-	-	-
Max	-	-	-	-	-	-

Medium	C horizon <2 mm aqua regia	C horizon <2 mm total (XRF)	C horizon <2 mm total (INAA)	Lake water unfiltered (mg/l)
Median	-	-	-	-
Min	-	-	-	-
Max	-	-	-	-

DYSPROSIUM   Dy

Concentrations in Soils and Sediments (mg/kg)

Medium	Soil	Agricultural soil - Ap horizon	Agricultural soil - Top (0-25 cm)	Agricultural Soil - Bottom (50-75 cm)	Topsoil (0-15 cm)	Urban soil (0-2 cm)
	World	Canada	Finland	Finland	England & Wales	Trondheim
	<2 mm	<2 mm	<2 mm	<2 mm	<2 mm	<2 mm
	total		aqua regia	aqua regia	aqua regia	aqua regia
Median	-	-	-	-	-	-
Min	-	-	-	-	-	-
Max	-	-	-	-	-	-

*Estimated mean

Medium	Forest soil - Humus	Forest soil - B horizon	Forest soil - C horizon	Till (C horizon)	Till (C horizon)	Laterite (25 ±15 cm)
	Norway	Norway	Norway	Finland	Finland	Australia
	<2 mm	<2 mm	<2 mm	<0.063 mm	<0.063 mm	0.45-2 mm
	7N HNO₃	7N HNO₃	7N HNO₃	aqua regia	total	total
Median	-	-	-	-	-	-
Min	-	-	-	-	-	-
Max	-	-	-	-	-	-

Medium	Stream sediment	Stream sediment	Stream sediment	Overbank sediment	Overbank sediment	Organic stream sediment
		Southern Scotland	Harz, Germany	Norway	Norway	Finland
	Austria					
	<0.18 mm	<0.10 mm	<0.063 mm	<0.063 mm	<0.063 mm	<2 mm
	total	total	total	total (XRF)	7N HNO₃	Conc. HNO₃
Median	-	-	-	-	-	-
Min	-	-	-	-	-	-
Max	-	-	-	-	-	-

#2nd percentile ^98th percentile

Concentrations in Waters (mg/l)

Medium	Ocean water	Ocean water	Ocean water	Stream water	Stream water	Stream water
	World (1)	World (2)	North Pacific	World	Nova Scotia, Canada	Finland
					<0.45 μm	<0.45 μm
					ICP-MS	
Median	0.000,000,91*	0.000,000,9*	0.000,001,1*	-	0.000,028	-
Min	-	-	-	-	<0.000,005	-
Max	-	-	-	-	0.000,164	-

*Estimated mean   #2nd percentile   ^98th percentile

Medium	Stream water	Stream water	Stream water	Lake water	Ground water
	Romania	Eastern India	Harz, Germany	Norway	Southern Norway
	unfiltered	<0.2 μm	unfiltered	unfiltered	unfiltered
	ICP-MS	ICP-MS	ICP-MS	ICP-MS	ICP-MS
Median	-	0.000,73	0.000,11	0.000,018	0.000,021
Min	-	0.000,08	<0.000,002	<0.000,01	<0.000,001
Max	-	0.010,63	0.001,4	0.000,43	0.001,66

Dy   DYSPROSIUM

Concentrations in Precipitation (mg/l)

Medium	Rain water Kola, remote <0.45 µm	Rain water Kola, polluted <0.45 µm	Snow melt-water Kola, remote <0.45 µm	Snow melt-water Kola, polluted <0.45 µm	Snow filter residue Kola, remote >0.45 µm	Snow filter residue Kola, polluted >0.45 µm
Median	-	-	-	-	-	-
Min	-	-	-	-	-	-
Max	-	-	-	-	-	-

Medium	Rain water Coastal, Norway unfiltered ICP-MS	Rain water Inland, Norway unfiltered ICP-MS				
Median	-	-				
Min	-	-				
Max	-	-				

Concentrations in Air (ng/m³) and Yearly Deposition (kg/km²/yr)

Medium	Air World, remote	Air World, polluted	Bulk deposition West Germany	Throughfall deposition West Germany	Bulk deposition Kola, remote	Bulk deposition Kola, polluted
Median	-	-	-	-	-	-
Min	-	-	-	-	-	-
Max	-	-	-	-	-	-

*Estimated value

Concentrations in Plants (mg/kg)

Medium	Moss Norway conc. HNO$_3$ (1990)	Moss Germany conc. HNO$_3$ + H$_2$O$_2$ (1995)	Crustose lichen Germany conc. HNO$_3$	Lichen Northwest Territories oven-dried	Dandelion Europe oven-dried	Spruce bark Canada ashed
Median	-	-	-	-	-	-
Min	-	-	-	-	-	-
Max	-	-	-	-	-	-

*Mean   #Geometric mean

Concentrations in Human Fluids (mg/l)

Medium	Human blood Lombardy, Italy	Human serum Lombardy, Italy	Human urine Lombardy, Italy			
Mean	-	-	-			
Min	-	-	-			
Max	-	-	-			

DYSPROSIUM Dy

Environmental Geochemistry

Biological impacts	Considered non-essential. Data to assess the toxicity of REE are scarce
Uses	Magnets, magneto-optical materials, nuclear industry
Environmental pathways	Poorly understood. Probably mostly windblown dust, weathering of REE-bearing minerals. Generally, geogenic sources more important than anthropogenic ones
Environmental mobility	Oxidising conditions: very low Acid conditions: very low Reducing conditions: very low Neutral to alkaline conditions: very low Comments: -
Geochemical barriers	Mechanical
Natural association	REE-Li-Rb-Cs-Be-Nb-Ta-Zr-B-Th-U-F (pegmatites), REE-Th-P-Zr-Fe-Cu (monazite veins), REE-Th-Ba-Sr-P-F-N-C (carbonatites), REE-U-P-F (phosphorites), REE-Au-Ti-Sn-Sr-Th (placers)
Action levels	-
Remarks	Toxicity of REE decreases as atomic number increases. Inhaled REE probably cause pneumoconiosis. REE taken up orally (e.g., via drinking water) accumulate in the skeleton, teeth, lungs, liver and kidneys. Some trees tend to accumulate Dy (e.g., hickory (*Carya*))
Suggested analytical method(s)	ICP-MS

World Yearly Production (t/y)	54,000 REE-minerals (in 1991)
Price of Purest Form, in Small Quantities ($/kg)	47,838 (99.99%)
Market Price ($/kg)	-

Er — ERBIUM

Physico-Chemical Properties

Atomic number	68	Melting/boiling point (K)	1802 / 3141
Atomic mass	167.26(3)	Isotopes & isomers	6 stable + 24 unstable
Atomic radius (pm)	245	Acid/base of oxide	weak base
Main oxidation state(s)	+3	State (at 300 K, 1 atm.)	solid
Ionic radius (pm)	103-120	Metallic character	metal
Electronegativity (Pauling)	1.24	Element group(s)	heavy metal, REE
Density (g/cm^3)	9.066	Affinity	lithophile

Naturally occurring isotopes	Natural abundance (%)	Atomic mass	Half-life
^{162}Er	0.14	161.928,78	stable
^{164}Er	1.61	163.929,197	stable
^{166}Er	33.6	165.930,29	stable
^{167}Er	22.95	166.932,05	stable
^{168}Er	26.8	167.932,37	stable
^{170}Er	14.9	169.935,46	stable

Unstable isotopes	Longest half-life
^{147}Er to ^{174}Er	9.4 d

Concentrations in Crust/Rocks (mg/kg)

Bulk continental crust	2.1 / 3.5 / 2.2
Upper continental crust	- / 2.3
Ultramafic rock	-
Ocean ridge basalt	-
Gabbro, basalt	-
Granite, granodiorite	-
Sandstone	-
Greywacke	2.2
Shale, schist	-
Limestone	-
Coal	0.1

Typical Minerals
monazite ((Ce,La,Nd,Er,Th)(PO$_4$,SiO$_4$)), bastnaesite ((Ce,La,Er)CO$_3$(F,OH))

Possible Host Minerals
feldspars, apatite, allanite, sphene, fluorite, zircon

Mass (kg) in

Continental crust	4.96x10^{+16}	Oceans	1.06x10^{+9}	Plants	3.68x10^{+7}

Concentrations in Media Sampled for the Kola Project (mg/kg)

Medium	Moss conc. HNO$_3$	Humus <2 mm	Humus <2 mm amm. acet.	Topsoil (0-5 cm) <2 mm	B horizon <2 mm aqua regia	B horizon <2 mm total (XRF)
Median	-	-	-	-	-	-
Min	-	-	-	-	-	-
Max	-	-	-	-	-	-

Medium	C horizon <2 mm aqua regia	C horizon <2 mm total (XRF)	C horizon <2 mm total (INAA)	Lake water unfiltered (mg/l)
Median	-	-	-	-
Min	-	-	-	-
Max	-	-	-	-

ERBIUM   Er

Concentrations in Soils and Sediments (mg/kg)

Medium	Soil World <2 mm total	Agricultural soil - Ap horizon Canada <2 mm	Agricultural soil - Top (0-25 cm) Finland <2 mm aqua regia	Agricultural Soil - Bottom (50-75 cm) Finland <2 mm aqua regia	Topsoil (0-15 cm) England & Wales <2 mm aqua regia	Urban soil (0-2 cm) Trondheim <2 mm aqua regia
Median	-	-	-	-	-	-
Min	-	-	-	-	-	-
Max	-	-	-	-	-	-

*Estimated mean

Medium	Forest soil - Humus Norway <2 mm 7N HNO$_3$	Forest soil - B horizon Norway <2 mm 7N HNO$_3$	Forest soil - C horizon Norway <2 mm 7N HNO$_3$	Till (C horizon) Finland <0.063 mm aqua regia	Till (C horizon) Finland <0.063 mm total	Laterite (25 ±15 cm) Australia 0.45-2 mm total
Median	-	-	-	-	-	-
Min	-	-	-	-	-	-
Max	-	-	-	-	-	-

Medium	Stream sediment Austria <0.18 mm	Stream sediment Southern Scotland <0.10 mm total	Stream sediment Harz, Germany <0.063 mm total	Overbank sediment Norway <0.063 mm total (XRF)	Overbank sediment Norway <0.063 mm 7N HNO$_3$	Organic stream sediment Finland <2 mm Conc. HNO$_3$
Median	-	-	-	-	-	-
Min	-	-	-	-	-	-
Max	-	-	-	-	-	-

#2nd percentile ^98th percentile

Concentrations in Waters (mg/l)

Medium	Ocean water World (1)	Ocean water World (2)	Ocean water North Pacific	Stream water World	Stream water Nova Scotia, Canada <0.45 µm ICP-MS	Stream water Finland <0.45 µm
Median	0.000,000,87*	0.000,000,8*	0.000,001,2*	-	0.000,016	-
Min	-	-	-	-	<0.000,005	-
Max	-	-	-	-	0.000,122	-

*Estimated mean   #2nd percentile   ^98th percentile

Medium	Stream water Romania unfiltered ICP-MS	Stream water Eastern India <0.2 µm ICP-MS	Stream water Harz, Germany unfiltered ICP-MS	Lake water Norway unfiltered ICP-MS	Ground water Southern Norway unfiltered ICP-MS
Median	-	0.000,2	0.000,06	0.000,01	0.000,015,5
Min	-	0.000,02	<0.000,002	<0.000,004	<0.000,001
Max	-	0.002,92	0.000,71	0.000,286	0.001,33

Er — ERBIUM

Concentrations in Precipitation (mg/l)

Medium	Rain water *Kola, remote <0.45 μm*	Rain water *Kola, polluted <0.45 μm*	Snow melt-water *Kola, remote <0.45 μm*	Snow melt-water *Kola, polluted <0.45 μm*	Snow filter residue *Kola, remote >0.45 μm*	Snow filter residue *Kola, polluted >0.45 μm*
Median	-	-	-	-	-	-
Min	-	-	-	-	-	-
Max	-	-	-	-	-	-

Medium	Rain water *Coastal, Norway unfiltered ICP-MS*	Rain water *Inland, Norway unfiltered ICP-MS*				
Median	-	-				
Min	-	-				
Max	-	-				

Concentrations in Air (ng/m³) and Yearly Deposition (kg/km²/yr)

Medium	Air *World, remote*	Air *World, polluted*	Bulk deposition *West Germany*	Throughfall deposition *West Germany*	Bulk deposition *Kola, remote*	Bulk deposition *Kola, polluted*
Median	-	-	-	-	-	-
Min	-	-	-	-	-	-
Max	-	-	-	-	-	-

*Estimated value

Concentrations in Plants (mg/kg)

Medium	Moss *Norway conc. HNO₃ (1990)*	Moss *Germany conc. HNO₃ + H₂O₂ (1995)*	Crustose lichen *Germany conc. HNO₃*	Lichen *Northwest Territories oven-dried*	Dandelion *Europe oven-dried*	Spruce bark *Canada ashed*
Median	-	-	-	-	-	-
Min	-	-	-	-	-	-
Max	-	-	-	-	-	-

*Mean   #Geometric mean

Concentrations in Human Fluids (mg/l)

Medium	Human blood *Lombardy, Italy*	Human serum *Lombardy, Italy*	Human urine *Lombardy, Italy*			
Mean	-	-	-			
Min	-	-	-			
Max	-	-	-			

ERBIUM — Er

Environmental Geochemistry

Biological impacts	Considered non-essential. Data to assess the toxicity of REE are scarce
Uses	Glass industry, alloys (e.g., with V), nuclear industry
Environmental pathways	Poorly understood. Probably mostly windblown dust, weathering of REE-bearing minerals. Generally, geogenic sources more important than anthropogenic ones
Environmental mobility	Oxidising conditions: very low · Acid conditions: very low Reducing conditions: very low · Neutral to alkaline conditions: very low Comments: -
Geochemical barriers	Mechanical
Natural association	REE-Li-Rb-Cs-Be-Nb-Ta-Zr-B-Th-U-F (pegmatites), REE-Th-P-Zr-Fe-Cu (monazite veins), REE-Th-Ba-Sr-P-F-N-C (carbonatites), REE-U-P-F (phosphorites), REE-Au-Ti-Sn-Sr-Th (placers)
Action levels	-
Remarks	Toxicity of REE decreases as atomic number increases. Inhaled REE probably cause pneumoconiosis. REE taken up orally (e.g., via drinking water) accumulate in the skeleton, teeth, lungs, liver and kidneys. Some trees tend to accumulate Er (e.g., hickory (*Carya*)). In 1991, 20 t Er-oxide were used in the glass industry
Suggested analytical method(s)	ICP-MS

World Yearly Production (t/y)	54,000 REE-minerals (in 1991)
Price of Purest Form, in Small Quantities ($/kg)	33,856 (99.99%)
Market Price ($/kg)	-

Eu — EUROPIUM

Physico-Chemical Properties

Atomic number	63	Melting/boiling point (K)	1095 / 1869
Atomic mass	151.964(1)	Isotopes & isomers	2 stable + 35 unstable
Atomic radius (pm)	256	Acid/base of oxide	base
Main oxidation state(s)	+3 (+2)	State (at 300 K, 1 atm.)	solid
Ionic radius (pm)	131-149 (+2), 109-126 (+3)	Metallic character	metal
Electronegativity (Pauling)	1.2	Element group(s)	heavy metal, REE
Density (g/cm^3)	5.244	Affinity	lithophile

Naturally occurring isotopes	Natural abundance (%)	Atomic mass	Half-life
^{151}Eu	47.8	150.919,85	stable
^{153}Eu	52.2	152.921,23	stable

Unstable isotopes	Longest half-life
^{134}Eu to ^{162}Eu	36 y

Concentrations in Crust/Rocks (mg/kg)

Bulk continental crust	1.3 / 2 / 1.1
Upper continental crust	0.95 / 0.88
Ultramafic rock	-
Ocean ridge basalt	-
Gabbro, basalt	-
Granite, granodiorite	-
Sandstone	-
Greywacke	1.2
Shale, schist	-
Limestone	-
Coal	0.5

Typical Minerals
monazite ((Ce,La,Nd,Eu,Th)(PO$_4$,SiO$_4$)), bastnaesite ((Ce,La,Eu)CO$_3$(F,OH))

Possible Host Minerals
feldspars, apatite, allanite, sphene, fluorite, zircon

Mass (kg) in

Continental crust	3.07x10^{+16}	Oceans	1.32x10^{+7}	Plants	1.47x10^{+7}

Concentrations in Media Sampled for the Kola Project (mg/kg)

Medium	Moss conc. HNO$_3$	Humus <2 mm amm. acet.	Humus <2 mm amm. acet.	Topsoil (0-5 cm) <2 mm total (INAA)	B horizon <2 mm aqua regia	B horizon <2 mm total (XRF)
Median	-	-	-	0.5	-	-
Min	-	-	-	<0.2	-	-
Max	-	-	-	7.3	-	-

Medium	C horizon <2 mm aqua regia	C horizon <2 mm total (XRF)	C horizon <2 mm total (INAA)	Lake water unfiltered (mg/l)
Median	-	-	1.05	-
Min	-	-	0.3	-
Max	-	-	14.3	-

EUROPIUM — Eu

Concentrations in Soils and Sediments (mg/kg)

Medium	Soil World <2 mm total	Agricultural soil - Ap horizon Canada <2 mm aqua regia	Agricultural soil - Top (0-25 cm) Finland <2 mm aqua regia	Agricultural Soil - Bottom (50-75 cm) Finland <2 mm aqua regia	Topsoil (0-15 cm) England & Wales <2 mm aqua regia	Urban soil (0-2 cm) Trondheim <2 mm aqua regia
Median	-	-	-	-	-	-
Min	-	-	-	-	-	-
Max	-	-	-	-	-	-

*Estimated mean

Medium	Forest soil - Humus Norway <2 mm 7N HNO₃	Forest soil - B horizon Norway <2 mm 7N HNO₃	Forest soil - C horizon Norway <2 mm 7N HNO₃	Till (C horizon) Finland <0.063 mm aqua regia	Till (C horizon) Finland <0.063 mm total	Laterite (25 ±15 cm) Australia 0.45-2 mm total
Median	-	-	-	-	-	-
Min	-	-	-	-	-	-
Max	-	-	-	-	-	-

Medium	Stream sediment Austria <0.18 mm total	Stream sediment Southern Scotland <0.10 mm total	Stream sediment Harz, Germany <0.063 mm total	Overbank sediment Norway <0.063 mm total (XRF)	Overbank sediment Norway <0.063 mm 7N HNO₃	Organic stream sediment Finland <2 mm Conc. HNO₃
Median	-	-	-	-	-	-
Min	-	-	-	-	-	-
Max	-	-	-	-	-	-

#2nd percentile ^98th percentile

Concentrations in Waters (mg/l)

Medium	Ocean water World (1)	Ocean water World (2)	Ocean water North Pacific	Stream water World	Stream water Nova Scotia, Canada <0.45 μm ICP-MS	Stream water Finland <0.45 μm
Median	0.000,000,13*	0.000,000,01*	0.000,000,17*	-	0.000,006	-
Min	-	-	-	-	<0.000,005	-
Max	-	-	-	-	0.000,047	-

*Estimated mean   #2nd percentile   ^98th percentile

Medium	Stream water Romania unfiltered ICP-MS	Stream water Eastern India <0.2 μm ICP-MS	Stream water Harz, Germany unfiltered ICP-MS	Lake water Norway unfiltered ICP-MS	Ground water Southern Norway unfiltered ICP-MS
Median	-	0.000,27	0.000,02	0.000,005	0.000,011
Min	-	0.000,08	<0.000,002	<0.000,004	<0.000,001
Max	-	0.003,17	0.000,08	0.000,062	0.001

Eu EUROPIUM

Concentrations in Precipitation (mg/l)

Medium	Rain water	Rain water	Snow melt-water	Snow melt-water	Snow filter residue	Snow filter residue
	Kola, remote <0.45 µm	Kola, polluted <0.45 µm	Kola, remote <0.45 µm	Kola, polluted <0.45 µm	Kola, remote >0.45 µm	Kola, polluted >0.45 µm
Median	-	-	-	-	-	-
Min	-	-	-	-	-	-
Max	-	-	-	-	-	-

Medium	Rain water	Rain water				
	Coastal, Norway unfiltered ICP-MS	Inland, Norway unfiltered ICP-MS				
Median	-	-				
Min	-	-				
Max	-	-				

Concentrations in Air (ng/m^3) and Yearly Deposition (kg/km^2/yr)

Medium	Air	Air	Bulk deposition	Throughfall deposition	Bulk deposition	Bulk deposition
	World, remote	World, polluted	West Germany	West Germany	Kola, remote	Kola, polluted
Median	-	-	-	-	-	-
Min	-	-	-	-	-	-
Max	-	-	-	-	-	-

*Estimated value

Concentrations in Plants (mg/kg)

Medium	Moss	Moss	Crustose lichen	Lichen	Dandelion	Spruce bark
	Norway conc. HNO$_3$ (1990)	Germany conc. HNO$_3$ + H$_2$O$_2$ (1995)	Germany conc. HNO$_3$	Northwest Territories oven-dried	Europe oven-dried total (INAA)	Canada ashed total (INAA)
Median	-	-	-	-	0.004,6#	0.495
Min	-	-	-	-	-	<0.01
Max	-	-	-	-	-	2.45

*Mean   #Geometric mean

Concentrations in Human Fluids (mg/l)

Medium	Human blood	Human serum	Human urine			
	Lombardy, Italy	Lombardy, Italy	Lombardy, Italy			
Mean	0.000,21	-	0.000,11			
Min	0.000,005	-	0.000,003			
Max	0.000,662	-	0.000,4			

EUROPIUM   Eu

Environmental Geochemistry

Biological impacts	Considered non-essential. Data to assess the toxicity of REE are scarce
Uses	Nuclear research (neutron absorber), laser, colour television screen (with Y)
Environmental pathways	Poorly understood. Probably mostly windblown dust, weathering of REE-bearing minerals. Generally, geogenic sources more important than anthropogenic ones
Environmental mobility	Oxidising conditions: very low — Acid conditions: very low Reducing conditions: very low — Neutral to alkaline conditions: very low Comments: -
Geochemical barriers	Mechanical
Natural association	REE-Li-Rb-Cs-Be-Nb-Ta-Zr-B-Th-U-F (pegmatites), REE-Th-P-Zr-Fe-Cu (monazite veins), REE-Th-Ba-Sr-P-F-N-C (carbonatites), REE-U-P-F (phosphorites), REE-Au-Ti-Sn-Sr-Th (placers)
Action levels	-
Remarks	Toxicity of REE decreases as atomic number increases. Inhaled REE probably cause pneumoconiosis. REE taken up orally (e.g., via drinking water) accumulate in the skeleton, teeth, lungs, liver and kidneys. Some trees tend to accumulate Eu (e.g., hickory (*Carya*)). Unlike most other REE, Eu can occur in the +2 oxidation state, which implies different geochemical substitution and behaviour. Eu substitutes more commonly into feldspars than other REE
Suggested analytical method(s)	ICP-MS

World Yearly Production (t/y)	54,000 REE-minerals (in 1991)
Price of Purest Form, in Small Quantities ($/kg)	221,975 (99.99%)
Market Price ($/kg)	-

F | FLUORINE

Physico-Chemical Properties

Atomic number	9	Melting/boiling point (K)	53.53 / 85.03
Atomic mass	18.998,403,2(5)	Isotopes & isomers	1 stable + 12 unstable
Atomic radius (pm)	57	Acid/base of oxide	-
Main oxidation state(s)	-1	State (at 300 K, 1 atm.)	gas
Ionic radius (pm)	114.5-119	Metallic character	non-metal
Electronegativity (Pauling)	3.98	Element group(s)	halogen, light non-metal
Density (g/cm^3)	1.696	Affinity	lithophile

Naturally occurring isotopes	Natural abundance (%)	Atomic mass	Half-life
^{19}F	100	18.998,4	stable

Unstable isotopes	Longest half-life
^{15}F to ^{27}F	1.83 h

Concentrations in Crust/Rocks (mg/kg)

Bulk continental crust	525 / 585 / -
Upper continental crust	611 / 650
Ultramafic rock	20
Ocean ridge basalt	-
Gabbro, basalt	300
Granite, granodiorite	800
Sandstone	200
Greywacke	-
Shale, schist	700
Limestone	300
Coal	50

Typical Minerals
fluorite (CaF$_2$), topaz (Al$_2$SiO$_4$(F,OH)$_2$), cryolite (Na$_2$AlF$_6$)

Possible Host Minerals
micas, amphiboles, tourmaline, apatite

Mass (kg) in

Continental crust 1.24x10^{+19}	Oceans 1.72x10^{+15}	Plants 3.68x10^{+9}

Concentrations in Media Sampled for the Kola Project (mg/kg)

Medium	Moss	Humus <2 mm	Humus <2 mm	Topsoil (0-5 cm) <2 mm	B horizon <2 mm	B horizon <2 mm
	conc. HNO$_3$	amm. acet.		aqua regia	total (XRF)	
Median	-	-	-	-	-	-
Min	-	-	-	-	-	-
Max	-	-	-	-	-	-

Medium	C horizon <2 mm aqua regia	C horizon <2 mm total (XRF)	C horizon <2 mm total (INAA)		Lake water unfiltered (mg/l)
Median	-	-	-		0.05
Min	-	-	-		<0.03
Max	-	-	-		0.23

FLUORINE   F

Concentrations in Soils and Sediments (mg/kg)

Medium	Soil	Agricultural soil - Ap horizon	Agricultural soil - Top (0-25 cm)	Agricultural Soil - Bottom (50-75 cm)	Topsoil (0-15 cm)	Urban soil (0-2 cm)
	World	Canada	Finland	Finland	England & Wales	Trondheim
	<2 mm	<2 mm	<2 mm	<2 mm	<2 mm	<2 mm
	total	aqua regia	aqua regia	aqua regia	aqua regia	aqua regia
Median	400*	-	-	-	-	-
Min	-	-	-	-	-	-
Max	-	-	-	-	-	-

*Estimated mean

Medium	Forest soil - Humus	Forest soil - B horizon	Forest soil - C horizon	Till (C horizon)	Till (C horizon)	Laterite (25 ±15 cm)
	Norway	Norway	Norway	Finland	Finland	Australia
	<2 mm	<2 mm	<2 mm	<0.063 mm	<0.063 mm	0.45-2 mm
	7N HNO$_3$	7N HNO$_3$	7N HNO$_3$	aqua regia	total	total
Median	-	-	-	-	-	<100
Min	-	-	-	-	-	<100
Max	-	-	-	-	-	1200

Medium	Stream sediment	Stream sediment	Stream sediment	Overbank sediment	Overbank sediment	Organic stream sediment
	Austria	Southern Scotland	Harz, Germany	Norway	Norway	Finland
	<0.18 mm	<0.10 mm	<0.063 mm	<0.063 mm	<0.063 mm	<2 mm
		total	total	total (XRF)	7N HNO$_3$	Conc. HNO$_3$
Median	-	-	-	-	-	-
Min	-	-	-	-	-	-
Max	-	-	-	-	-	-

#2nd percentile   ^98th percentile

Concentrations in Waters (mg/l)

Medium	Ocean water	Ocean water	Ocean water	Stream water	Stream water	Stream water
	World (1)	World (2)	North Pacific	World	Nova Scotia, Canada	Finland
					<0.45 µm	<0.45 µm
					Dionex	IC
Median	1.3*	1.3*	1.3*	0.001*	<0.05	0.008
Min	-	-	-	-	<0.05	0.3#
Max	-	-	-	-	0.837	0.99^

*Estimated mean   #2nd percentile   ^98th percentile

Medium	Stream water	Stream water	Stream water	Lake water	Ground water
	Romania	Eastern India	Harz, Germany	Norway	Southern Norway
	unfiltered	<0.2 µm	unfiltered	unfiltered	unfiltered
	ICP-MS	ICP-MS		ICP-MS	Electrode
Median	-	-	-	-	0.33
Min	-	-	-	-	0.03
Max	-	-	-	-	9.18

FLUORINE

Concentrations in Precipitation (mg/l)

Medium	Rain water	Rain water	Snow melt-water	Snow melt-water	Snow filter residue	Snow filter residue
	Kola, remote <0.45 μm IC	Kola, polluted <0.45 μm IC	Kola, remote <0.45 μm IC	Kola, polluted <0.45 μm IC	Kola, remote >0.45 μm	Kola, polluted >0.45 μm
Median	<0.05	<0.05	<0.05	<0.05	-	-
Min	<0.05	<0.05	<0.05	<0.05	-	-
Max	<0.05	0.08	<0.05	<0.05	-	-

Medium	Rain water	Rain water				
	Coastal, Norway unfiltered ICP-MS	Inland, Norway unfiltered ICP-MS				
Median	-	-				
Min	-	-				
Max	-	-				

Concentrations in Air (ng/m^3) and Yearly Deposition (kg/km^2/yr)

Medium	Air	Air	Bulk deposition	Throughfall deposition	Bulk deposition	Bulk deposition
	World, remote	World, polluted	West Germany	West Germany	Kola, remote	Kola, polluted
Median	-	1.5	-	-	-	-
Min	-	1.5	50	260	-	-
Max	-	400	90	450	-	-

*Estimated value

Concentrations in Plants (mg/kg)

Medium	Moss	Moss	Crustose lichen	Lichen	Dandelion	Spruce bark
	Norway conc. HNO$_3$ (1990)	Germany conc. HNO$_3$ + H$_2$O$_2$ (1995)	Germany conc. HNO$_3$	Northwest Territories oven-dried	Europe oven-dried	Canada ashed
Median	-	-	-	-	-	-
Min	-	-	-	-	-	-
Max	-	-	-	-	-	-

*Mean #Geometric mean

Concentrations in Human Fluids (mg/l)

Medium	Human blood	Human serum	Human urine			
	Lombardy, Italy	Lombardy, Italy	Lombardy, Italy			
Mean	-	-	-			
Min	-	-	-			
Max	-	-	-			

FLUORINE F

Environmental Geochemistry

Biological impacts	Essential for some organisms. F and many F compounds are highly toxic
Uses	U production, plastic, glass etching, flux (fluorite), has been studied as rocket propellant
Environmental pathways	Al-industry, cement and brick industry, added to drinking water in some countries, geogenic sources, sea spray
Environmental mobility	Oxidising conditions: high Acid conditions: high Reducing conditions: high Neutral to alkaline conditions: high Comments: -
Geochemical barriers	Adsorption on Al oxide, precipitation in presence of Ca (as CaF_2)
Natural association	F-Ca-Fe-S-Si-Ba-Sr-Pb-Zn-Cu (veins and stockworks), F-Al-Ca-Sn-Mo-W (greisens), Nb-Ta-P-F-Ti-REE (carbonatites), F-U-V-Se-As-REE (phosphorites), Pb-Zn-Ba-F (MVT deposits)
Action levels	Drinking water, MAC: 1.5 mg/l, or 0.7 mg/l (at T >25°C) (Norway Shd), 4 mg/l (US EPA), 1.5 mg/l (Canada CWQG); recommended: 1.5 mg/l (WHO); MAC: 1.5 mg/l (Russia MoH). Agricultural soil, maximum tolerable concentration: 200 mg/kg (Germany)
Remarks	Most electronegative and reactive of all elements. Many organic compounds. Some P fertilisers can contain high F contents. F is important for strong teeth in humans (around 0.5 mg F/l in drinking water), but too much F leads to fluorosis (starting at 1.5 mg F/l, serious at >6 mg F/l in drinking water)
Suggested analytical method(s)	Ion selective electrode, IC

World Yearly Production (t/y)	1,960,000 (in 1995, from 4,023,000 CaF_2)
Price of Purest Form, in Small Quantities ($/kg)	-
Market Price ($/kg)	0.33 (acidspar, dry, 97% CaF_2)

Fe — IRON

Physico-Chemical Properties

Atomic number	26	Melting/boiling point (K)	1811 / 3134
Atomic mass	55.845(2)	Isotopes & isomers	4 stable + 22 unstable
Atomic radius (pm)	172	Acid/base of oxide	amphoteric
Main oxidation state(s)	+3 (+2,+4,+6)	State (at 300 K, 1 atm.)	solid
Ionic radius (pm)	77-108 (+2), 63-92 (+3), 72.5 (+4), 39 (+6)	Metallic character	pred. metal
Electronegativity (Pauling)	1.83	Element group(s)	heavy metal
Density (g/cm^3)	7.874	Affinity	chalcophile, siderophile

Naturally occurring isotopes	Natural abundance (%)	Atomic mass	Half-life
^{54}Fe	5.845	53.939,62	stable
^{56}Fe	91.754	55.934,94	stable
^{57}Fe	2.119	56.935,398	stable
^{58}Fe	0.282	57.933,28	stable

Unstable isotopes	Longest half-life
^{46}Fe to ^{69}Fe	1,500,000 y

Concentrations in Crust/Rocks (mg/kg)

Bulk continental crust	43,200 / 56,300 / 70,700
Upper continental crust	30,890 / 35,000
Ultramafic rock	94,000
Ocean ridge basalt	71,000
Gabbro, basalt	86,000
Granite, granodiorite	20,000
Sandstone	10,000
Greywacke	41,265
Shale, schist	55,000
Limestone	5000
Coal	10,000

Typical Minerals

magnetite (Fe$_3$O$_4$), hematite (Fe$_2$O$_3$), goethite/limonite (FeO.OH), siderite (FeCO$_3$), pyrite (FeS$_2$) and many major rock forming

Possible Host Minerals

minerals, e.g., olivine, pyroxenes, amphiboles, micas, garnets

Mass (kg) in

Continental crust	1.02x10^{+21}	Oceans	2.64x10^{+12}	Plants	2.67x10^{+11}

Concentrations in Media Sampled for the Kola Project (mg/kg)

Medium	Moss	Humus <2 mm	Humus <2 mm	Topsoil (0-5 cm) <2 mm	B horizon <2 mm	B horizon <2 mm
	conc. HNO$_3$	conc. HNO$_3$	amm. acet.	total (INAA)	aqua regia	total (XRF)
Median	212	1970	9.8	14,800	23,500	38,817
Min	46.5	430	1.3	700	4980	5455
Max	5140	44,800	2630	66,400	121,000	107,778

Medium	C horizon <2 mm	C horizon <2 mm	C horizon <2 mm			Lake water unfiltered
	aqua regia	total (XRF)	total (INAA)			(mg/l)
Median	14,700	34,300	35,750			0.07
Min	3310	5900	6800			<0.02
Max	79,200	123,500	119,000			1.22

IRON   Fe

Concentrations in Soils and Sediments (mg/kg)

Medium	Soil	Agricultural soil -	Agricultural soil - Top	Agricultural Soil - Bottom	Topsoil (0-15 cm)	Urban soil (0-2 cm)
	World	Ap horizon Canada	(0-25 cm) Finland	(50-75 cm) Finland	England & Wales	Trondheim
	<2 mm total	<2 mm total (INAA)	<2 mm aqua regia	<2 mm aqua regia	<2 mm aqua regia	<2 mm aqua regia
Median	35,000*	22,000	12,100	12,000	26,786	31,050
Min	-	1000	1140	560	395	3360
Max	-	59,000	51,000	63,800	264,405	84,900

*Estimated mean

Medium	Forest soil - Humus	Forest soil - B horizon	Forest soil - C horizon	Till (C horizon)	Till (C horizon)	Laterite (25 ±15 cm)
	Norway	Norway	Norway	Finland	Finland	Australia
	<2 mm 7N HNO$_3$	<2 mm 7N HNO$_3$	<2 mm 7N HNO$_3$	<0.063 mm aqua regia	<0.063 mm total	0.45-2 mm total
Median	2100	38,400	24,050	17,000	31,000	22,241
Min	320	800	600	-	-	4686
Max	49,200	98,500	94,800	-	-	333,544

Medium	Stream sediment	Stream sediment	Stream sediment	Overbank sediment	Overbank sediment	Organic stream sediment
	Austria	Southern Scotland	Harz, Germany	Norway	Norway	Finland
	<0.18 mm "total" (ICP-AES)	<0.10 mm total (DCES)	<0.063 mm total (XRF)	<0.063 mm total (XRF)	<0.063 mm 7N HNO$_3$ ICP-AES	<2 mm Conc. HNO$_3$
Median	37,200	54,553	25,598	50,300	10,000	26,000
Min	-	2098	6784	23,400	890	6500#
Max	190,500	428,032	137,152	195,600	71,400	111,000^

#2nd percentile   ^98th percentile

Concentrations in Waters (mg/l)

Medium	Ocean water	Ocean water	Ocean water	Stream water	Stream water	Stream water
	World (1)	World (2)	North Pacific	World	Nova Scotia, Canada	Finland
					<0.45 µm ICP-MS	<0.45 µm ICP-MS
Median	0.002*	0.002*	0.000,03*	0.04*	0.076	0.68
Min	-	-	-	-	<0.005	0.3#
Max	-	-	-	-	1.113	3.6^

*Estimated mean   #2nd percentile   ^98th percentile

Medium	Stream water	Stream water	Stream water	Lake water	Ground water
	Romania	Eastern India	Harz, Germany	Norway	Southern Norway
	unfiltered ICP-MS	<0.2 µm ICP-MS	unfiltered ICP-MS	unfiltered ICP-MS	unfiltered ICP-MS
Median	0.743	-	0.285	0.060,7	0.025
Min	0.017	-	<0.01	<0.015	<0.005
Max	32	-	11.8	7.68	5.323

Fe — IRON

Concentrations in Precipitation (mg/l)

Medium	Rain water	Rain water	Snow melt-water	Snow melt-water	Snow filter residue	Snow filter residue
	Kola, remote	*Kola, polluted*	*Kola, remote*	*Kola, polluted*	*Kola, remote*	*Kola, polluted*
	<0.45 µm	*<0.45 µm*	*<0.45 µm*	*<0.45 µm*	*>0.45 µm*	*>0.45 µm*
	ICP-MS	*ICP-MS*	*ICP-MS*	*ICP-MS*	*ICP-AES*	*ICP-AES*
Median	<0.01	0.04	<0.015	0.08	<0.01	1.11
Min	<0.01	<0.01	<0.015	0.02	<0.01	0.5
Max	0.01	0.09	<0.015	0.1	0.03	4.15

Medium	Rain water	Rain water				
	Coastal, Norway unfiltered ICP-MS	*Inland, Norway unfiltered ICP-MS*				
Median	0.011	0.021				
Min	<0.002	<0.002				
Max	0.8	0.57				

Concentrations in Air (ng/m^3) and Yearly Deposition (kg/km^2/yr)

Medium	Air	Air	Bulk deposition	Throughfall deposition	Bulk deposition	Bulk deposition
	World, remote	*World, polluted*	*West Germany*	*West Germany*	*Kola, remote*	*Kola, polluted*
Median	60 / 79	2500	-	-	10.5*	1144*
Min	0.25	130	9	13	-	-
Max	90 (660)	14,000	90	300	-	-

*Estimated value

Concentrations in Plants (mg/kg)

Medium	Moss	Moss	Crustose lichen	Lichen	Dandelion	Spruce bark
	Norway conc. HNO$_3$ (1990)	*Germany conc. HNO$_3$ + H$_2$O$_2$ (1995)*	*Germany conc. HNO$_3$*	*Northwest Territories oven-dried total (INAA)*	*Europe oven-dried total (INAA)*	*Canada ashed total (INAA)*
Median	470	444	-	478.8*	272#	20,300
Min	130	118	-	-	-	3200
Max	18,000	6780	-	-	-	22,200

*Mean   #Geometric mean

Concentrations in Human Fluids (mg/l)

Medium	Human blood	Human serum	Human urine			
	Lombardy, Italy	*Lombardy, Italy*	*Lombardy, Italy*			
Mean	-	-	-			
Min	-	-	-			
Max	-	-	-			

IRON   Fe

Environmental Geochemistry

Biological impacts	Essential for all organisms. Toxic to humans at drinking water level >200 mg/l. Iron deficiencies widespread
Uses	Steel, building and construction, transport industry, pigments, sewage treatment
Environmental pathways	Rock weathering, geogenic dust, Fe and steel industry
Environmental mobility	Oxidising conditions: very low Acid conditions: low Reducing conditions: low Neutral to alkaline conditions: low Comments: -
Geochemical barriers	Oxidation, pH increase, precipitation as Fe oxides, hydroxides, oxy-hydroxides, co-precipitating many other metals
Natural association	Fe-Mg-Mn-V-Ti-Sc-S (many silicates, sulphides), Fe-Mn (seafloor polymetallic nodules)
Action levels	Drinking water, MAC: 0.2 mg/l (Norway Shd), 0.3 mg/l (Canada CWQG), 0.3 mg/l (Russia MoH)
Remarks	Plant growth sensitive to soil changes affecting Fe availability. $FeSO_4 \cdot 7H_2O$ used as fertiliser and herbicide. Availability in soils depending on pH, phosphate content, content of other metals (e.g., Co). Fe deficiency widespread on Ca-rich soils even at high total Fe concentrations. Some plants (e.g., bean, pea, potato, tomato) are able to reduce Fe in soils, if needed. Concentrations of 0.3 mg/l in drinking water may give rise to consumer complaints due to staining of laundry and sanitary ware
Suggested analytical method(s)	XRF, ICP-AES, ICP-MS (water)

World Yearly Production (t/y)	568,128,000 (in 1995)
Price of Purest Form, in Small Quantities ($/kg)	5633 (99.99%)
Market Price ($/kg)	0.04 (Fe, in iron ore), 0.12 (Fe-oxide pigment), 0.14 (scrap metal)

Fr FRANCIUM

Physico-Chemical Properties

Atomic number	87	Melting/boiling point (K)	300 / 950
Atomic mass	[223]	Isotopes & isomers	0 stable + 35 unstable
Atomic radius (pm)	-	Acid/base of oxide	strong base
Main oxidation state(s)	+1	State (at 300 K, 1 atm.)	liquid
Ionic radius (pm)	194	Metallic character	metal
Electronegativity (Pauling)	0.7	Element group(s)	alkaline earth, heavy metal
Density (g/cm^3)	-	Affinity	-

Naturally occurring isotopes	Natural abundance (%)	Atomic mass	Half-life
-	-	-	-

Unstable isotopes	Longest half-life
^{201}Fr to ^{232}Fr	22 min

Concentrations in Crust/Rocks (mg/kg)

Bulk continental crust	- / - / -
Upper continental crust	- / -
Ultramafic rock	-
Ocean ridge basalt	-
Gabbro, basalt	-
Granite, granodiorite	-
Sandstone	-
Greywacke	-
Shale, schist	-
Limestone	-
Coal	-

Typical Minerals
-

Possible Host Minerals
uraninite/pitchblende, brannerite, carnotite

Mass (kg) in

Continental crust	-	Oceans	-	Plants	-

Concentrations in Media Sampled for the Kola Project (mg/kg)

Medium	Moss conc. HNO$_3$	Humus <2 mm	Humus <2 mm amm. acet.	Topsoil (0-5 cm) <2 mm	B horizon <2 mm aqua regia	B horizon <2 mm total (XRF)
Median	-	-	-	-	-	-
Min	-	-	-	-	-	-
Max	-	-	-	-	-	-

Medium	C horizon <2 mm aqua regia	C horizon <2 mm total (XRF)	C horizon <2 mm total (INAA)		Lake water unfiltered (mg/l)
Median	-	-	-		-
Min	-	-	-		-
Max	-	-	-		-

FRANCIUM — Fr

Environmental Geochemistry

Biological impacts	Extremely rare; probably no biological implications
Uses	No viable quantity ever produced
Environmental pathways	-
Environmental mobility	Oxidising conditions: - Acid conditions: - Reducing conditions: - Neutral to alkaline conditions: - Comments: -
Geochemical barriers	-
Natural association	U-Ac-Fr (U deposits)
Action levels	-
Remarks	There is only about 50 g of Fr in the earth crust. Chemical properties should resemble Cs
Suggested analytical method(s)	-

World Yearly Production (t/y) -
Price of Purest Form, in Small Quantities ($/kg) -
Market Price ($/kg) -

Ga   GALLIUM

Physico-Chemical Properties

Atomic number	31	Melting/boiling point (K)	302.91 / 2477
Atomic mass	69.723(1)	Isotopes & isomers	2 stable + 23 unstable
Atomic radius (pm)	181	Acid/base of oxide	amphoteric
Main oxidation state(s)	+3 (+1,+2)	State (at 300 K, 1 atm.)	liquid
Ionic radius (pm)	61-76 (+3)	Metallic character	pred. metal
Electronegativity (Pauling)	1.81	Element group(s)	heavy metal
Density (g/cm^3)	5.904	Affinity	chalcophile

Naturally occurring isotopes	Natural abundance (%)	Atomic mass	Half-life
^{69}Ga	60.108	68.925,581	stable
^{71}Ga	39.892	70.924,71	stable

Unstable isotopes	Longest half-life
^{61}Ga to ^{84}Ga	3.26 d

Concentrations in Crust/Rocks (mg/kg)

Bulk continental crust	15 / 19 / 18
Upper continental crust	14 / 17
Ultramafic rock	0.5
Ocean ridge basalt	17
Gabbro, basalt	17
Granite, granodiorite	18
Sandstone	8
Greywacke	1.6
Shale, schist	20
Limestone	1
Coal	5

Typical Minerals
söhngeite (Ga(OH)$_3$), gallite (CuGaS$_2$)

Possible Host Minerals
feldspars, amphiboles, micas, magnetite

Mass (kg) in

Continental crust	3.54x10^{+17}	Oceans	2.64x10^{+9}	Plants	1.84x10^{+8}

Concentrations in Media Sampled for the Kola Project (mg/kg)

Medium	Moss	Humus	Humus	Topsoil (0-5 cm)	B horizon	B horizon
	<2 mm	<2 mm	<2 mm	<2 mm	<2 mm	<2 mm
	conc. HNO$_3$		amm. acet.		aqua regia	total (XRF)
Median	-	-	-	-	-	-
Min	-	-	-	-	-	-
Max	-	-	-	-	-	-

Medium	C horizon	C horizon	C horizon	Lake water
	<2 mm	<2 mm	<2 mm	unfiltered
	aqua regia	total (XRF)	total (INAA)	(mg/l)
Median	-	-	-	-
Min	-	-	-	-
Max	-	-	-	-

GALLIUM   Ga

Concentrations in Soils and Sediments (mg/kg)

Medium	Soil	Agricultural soil - Ap horizon	Agricultural soil - Top (0-25 cm)	Agricultural Soil - Bottom (50-75 cm)	Topsoil (0-15 cm)	Urban soil (0-2 cm)
	World	Canada	Finland	Finland	England & Wales	Trondheim
	<2 mm	<2 mm	<2 mm	<2 mm	<2 mm	<2 mm
	total		aqua regia	aqua regia	aqua regia	aqua regia
Median	-	-	-	-	-	-
Min	-	-	-	-	-	-
Max	-	-	-	-	-	-

*Estimated mean

Medium	Forest soil - Humus	Forest soil - B horizon	Forest soil - C horizon	Till (C horizon)	Till (C horizon)	Laterite (25 ±15 cm)
	Norway	Norway	Norway	Finland	Finland	Australia
	<2 mm	<2 mm	<2 mm	<0.063 mm	<0.063 mm	0.45-2 mm
	7N HNO$_3$	7N HNO$_3$	7N HNO$_3$	aqua regia	total	total
Median	-	-	-	-	-	11
Min	-	-	-	-	-	1.35
Max	-	-	-	-	-	44

Medium	Stream sediment	Stream sediment	Stream sediment	Overbank sediment	Overbank sediment	Organic stream sediment
		Southern Scotland	Harz, Germany	Norway	Norway	Finland
	Austria	<0.10 mm	<0.063 mm	<0.063 mm	<0.063 mm	<2 mm
	<0.18 mm	total (DCES)	total (XRF)	total (XRF)	7N HNO$_3$	Conc. HNO$_3$
	"total" (ICP-AES)					
Median	21	18.2	14	-	-	-
Min	-	<4	<5	-	-	-
Max	91	49	32	-	-	-

#2nd percentile ^98th percentile

Concentrations in Waters (mg/l)

Medium	Ocean water	Ocean water	Ocean water	Stream water	Stream water	Stream water
	World (1)	World (2)	North Pacific	World	Nova Scotia, Canada	Finland
					<0.45 µm	<0.45 µm
Median	0.000,03*	0.000,002*	0.000,001,2*	0.000,1*	-	-
Min	-	-	-	-	-	-
Max	-	-	-	-	-	-

*Estimated mean   #2nd percentile   ^98th percentile

Medium	Stream water	Stream water	Stream water	Lake water	Ground water
	Romania	Eastern India	Harz, Germany	Norway	Southern Norway
	unfiltered	<0.2 µm	unfiltered	unfiltered	unfiltered
	ICP-MS	ICP-MS	ICP-MS	ICP-MS	ICP-MS
Median	0.000,14	0.001,39	0.000,04	0.000,118	0.000,027,5
Min	0.000,01	0.000,22	<0.000,01	<0.000,02	0.000,004
Max	0.02	0.027,88	0.002,3	0.005,39	0.003,06

Ga GALLIUM

Concentrations in Precipitation (mg/l)

Medium	Rain water Kola, remote <0.45 µm	Rain water Kola, polluted <0.45 µm	Snow melt- water Kola, remote <0.45 µm	Snow melt- water Kola, polluted <0.45 µm	Snow filter residue Kola, remote >0.45 µm	Snow filter residue Kola, polluted >0.45 µm
Median	-	-	-	-	-	-
Min	-	-	-	-	-	-
Max	-	-	-	-	-	-

Medium	Rain water Coastal, Norway unfiltered ICP-MS	Rain water Inland, Norway unfiltered ICP-MS				
Median	0.000,03	0.000,04				
Min	<0.000,01	<0.000,01				
Max	0.000,19	0.000,16				

Concentrations in Air (ng/m^3) and Yearly Deposition (kg/km^2/yr)

Medium	Air World, remote	Air World, polluted	Bulk deposition West Germany	Throughfall deposition West Germany	Bulk deposition Kola, remote	Bulk deposition Kola, polluted
Median	-	-	-	-	-	-
Min	-	0.23	-	-	-	-
Max	<0.000,1	1	-	-	-	-

*Estimated value

Concentrations in Plants (mg/kg)

Medium	Moss Norway conc. HNO$_3$ (1990)	Moss Germany conc. HNO$_3$ + H$_2$O$_2$ (1995)	Crustose lichen Germany conc. HNO$_3$	Lichen Northwest Territories oven-dried	Dandelion Europe oven-dried	Spruce bark Canada ashed
Median	0.2	0.157	-	-	-	-
Min	0.02	0.04	-	-	-	-
Max	2	3	-	-	-	-

*Mean   #Geometric mean

Concentrations in Human Fluids (mg/l)

Medium	Human blood Lombardy, Italy	Human serum Lombardy, Italy	Human urine Lombardy, Italy			
Mean	0.000,26	-	<0.000,5			
Min	0.000,1	-	-			
Max	0.000,38	-	-			

GALLIUM   Ga

Environmental Geochemistry

Biological impacts	Considered non-essential. Toxicity considered low. Ga nitrate is anti-carcinogenic
Uses	Electronics (semi-conductors, transistors), low melting alloys, chemotherapy
Environmental pathways	Geogenic sources, coal combustion, Al-Mn-Cr-Fe production
Environmental mobility	Oxidising conditions: - Acid conditions: - Reducing conditions: - Neutral to alkaline conditions: - Comments: -
Geochemical barriers	-
Natural association	Ag-Zn-Cd-Cu-Ba-Sr-V-Cr-Mn-Fe-Ga-In-Ta-Ge-Sn-As-Sb-Bi-Se-Hg-Te-Pb (sulphide deposits), Nb-Ta-Ti-Ga-Be-Al (bauxite developed on alkaline rocks)
Action levels	Agricultural soil, maximum tolerable concentration: 10 mg/kg (Germany)
Remarks	Widest temperature range in liquid state of any metals. Ga expands upon solidification. Chemical and biochemical behaviour similar to Fe. Ga is invariably associated with Zn in nature (e.g., in sphalerite). Ga recovered as a side-product of Zn and Al production. Flue dust of coal fired power plants can contain up to 1.5% Ga. Ga in drinking water (5 mg/l) reported to cause severe health problems in mice. Clinical use of Ga as anti-carcinogenic limited by nephrotoxicity. Can lead to DNA damage. Production of Ga is bound to increase due to its importance in the semi-conductor industry
Suggested analytical method(s)	ICP-MS, INAA

World Yearly Production (t/y)	20
Price of Purest Form, in Small Quantities ($/kg)	10,606 (99.999,99%)
Market Price ($/kg)	-

Gd GADOLINIUM

Physico-Chemical Properties

Atomic number	64	Melting/boiling point (K)	1586 / 3546
Atomic mass	157.25(3)	Isotopes & isomers	7 stable + 23 unstable
Atomic radius (pm)	254	Acid/base of oxide	base
Main oxidation state(s)	+3	State (at 300 K, 1 atm.)	solid
Ionic radius (pm)	108-125	Metallic character	metal
Electronegativity (Pauling)	1.2	Element group(s)	heavy metal, REE
Density (g/cm^3)	7.901	Affinity	lithophile

Naturally occurring isotopes	Natural abundance (%)	Atomic mass	Half-life
^{152}Gd	0.2	151.919,79	stable
^{154}Gd	2.18	153.920,86	stable
^{155}Gd	14.8	154.922,62	stable
^{156}Gd	20.47	155.922,12	stable
^{157}Gd	15.65	156.923,96	stable
^{158}Gd	24.84	157.924,101	stable
^{160}Gd	21.86	159.927,05	stable

Unstable isotopes	Longest half-life
^{137}Gd to ^{164}Gd	1,800,000 y

Concentrations in Crust/Rocks (mg/kg)

Bulk continental crust	4 / 6.2 / 3.3
Upper continental crust	2.8 / 3.8
Ultramafic rock	-
Ocean ridge basalt	-
Gabbro, basalt	-
Granite, granodiorite	-
Sandstone	-
Greywacke	4
Shale, schist	-
Limestone	-
Coal	0.2

Typical Minerals
monazite ((Ce,La,Nd,Gd,Th)(PO$_4$,SiO$_4$)),
bastnaesite ((Ce,La,Gd)CO$_3$(F,OH)),
gadolinite (Be$_2$Fe(Y,Gd)$_2$Y$_2$Si$_2$O$_{10}$)

Possible Host Minerals
feldspars, apatite, allanite, sphene, fluorite, zircon

Mass (kg) in

Continental crust	9.44x10^{+16}	Oceans	9.25x10^{+8}	Plants	7.36x10^{+7}

Concentrations in Media Sampled for the Kola Project (mg/kg)

Medium	Moss	Humus <2 mm amm. acet.	Humus <2 mm amm. acet.	Topsoil (0-5 cm) <2 mm	B horizon <2 mm aqua regia	B horizon <2 mm total (XRF)
	conc. HNO$_3$					
Median	-	-	-	-	-	-
Min	-	-	-	-	-	-
Max	-	-	-	-	-	-

Medium	C horizon <2 mm aqua regia	C horizon <2 mm total (XRF)	C horizon <2 mm total (INAA)		Lake water unfiltered (mg/l)
Median	-	-	-		-
Min	-	-	-		-
Max	-	-	-		-

GADOLINIUM — Gd

Concentrations in Soils and Sediments (mg/kg)

Medium	Soil	Agricultural soil - Ap horizon	Agricultural soil - Top (0-25 cm)	Agricultural Soil - Bottom (50-75 cm)	Topsoil (0-15 cm)	Urban soil (0-2 cm)
	World	Canada	Finland	Finland	England & Wales	Trondheim
	<2 mm total	<2 mm	<2 mm aqua regia	<2 mm aqua regia	<2 mm aqua regia	<2 mm aqua regia
Median	-	-	-	-	-	-
Min	-	-	-	-	-	-
Max	-	-	-	-	-	-

*Estimated mean

Medium	Forest soil - Humus	Forest soil - B horizon	Forest soil - C horizon	Till (C horizon)	Till (C horizon)	Laterite (25 ±15 cm)
	Norway	Norway	Norway	Finland	Finland	Australia
	<2 mm 7N HNO$_3$	<2 mm 7N HNO$_3$	<2 mm 7N HNO$_3$	<0.063 mm aqua regia	<0.063 mm total	0.45-2 mm total
Median	-	-	-	-	-	-
Min	-	-	-	-	-	-
Max	-	-	-	-	-	-

Medium	Stream sediment	Stream sediment	Stream sediment	Overbank sediment	Overbank sediment	Organic stream sediment
	Austria	Southern Scotland	Harz, Germany	Norway	Norway	Finland
	<0.18 mm	<0.10 mm total	<0.063 mm total	<0.063 mm total (XRF)	<0.063 mm 7N HNO$_3$	<2 mm Conc. HNO$_3$
Median	-	-	-	-	-	-
Min	-	-	-	-	-	-
Max	-	-	-	-	-	-

#2nd percentile ^98th percentile

Concentrations in Waters (mg/l)

Medium	Ocean water	Ocean water	Ocean water	Stream water	Stream water	Stream water
	World (1)	World (2)	North Pacific	World	Nova Scotia, Canada	Finland
					<0.45 μm ICP-MS	<0.45 μm
Median	0.000,000,7*	0.000,000,7*	0.000,000,9*	-	0.000,033	-
Min	-	-	-	-	<0.000,005	-
Max	-	-	-	-	0.000,195	-

*Estimated mean   #2nd percentile   ^98th percentile

Medium	Stream water	Stream water	Stream water	Lake water	Ground water
	Romania	Eastern India	Harz, Germany	Norway	Southern Norway
	unfiltered ICP-MS	<0.2 μm ICP-MS	unfiltered ICP-MS	unfiltered ICP-MS	unfiltered ICP-MS
Median	-	0.001,23	0.000,11	0.000,026	0.000,029
Min	-	0.000,16	<0.000,002	<0.000,015	<0.000,001
Max	-	0.017,07	0.001,1	0.000,5	0.003,9

Gd GADOLINIUM

Concentrations in Precipitation (mg/l)

Medium	Rain water	Rain water	Snow melt-water	Snow melt-water	Snow filter residue	Snow filter residue
	Kola, remote <0.45 μm	Kola, polluted <0.45 μm	Kola, remote <0.45 μm	Kola, polluted <0.45 μm	Kola, remote >0.45 μm	Kola, polluted >0.45 μm
Median	-	-	-	-	-	-
Min	-	-	-	-	-	-
Max	-	-	-	-	-	-

Medium	Rain water	Rain water				
	Coastal, Norway unfiltered ICP-MS	Inland, Norway unfiltered ICP-MS				
Median	-	-				
Min	-	-				
Max	-	-				

Concentrations in Air (ng/m^3) and Yearly Deposition (kg/km^2/yr)

Medium	Air	Air	Bulk deposition	Throughfall deposition	Bulk deposition	Bulk deposition
	World, remote	World, polluted	West Germany	West Germany	Kola, remote	Kola, polluted
Median	-	-	-	-	-	-
Min	-	-	-	-	-	-
Max	-	-	-	-	-	-

*Estimated value

Concentrations in Plants (mg/kg)

Medium	Moss	Moss	Crustose lichen	Lichen	Dandelion	Spruce bark
	Norway conc. HNO$_3$ (1990)	Germany conc. HNO$_3$ + H$_2$O$_2$ (1995)	Germany conc. HNO$_3$	Northwest Territories oven-dried	Europe oven-dried	Canada ashed
Median	-	-	-	-	-	-
Min	-	-	-	-	-	-
Max	-	-	-	-	-	-

*Mean   #Geometric mean

Concentrations in Human Fluids (mg/l)

Medium	Human blood	Human serum	Human urine
	Lombardy, Italy	Lombardy, Italy	Lombardy, Italy
Mean	-	-	<0.001
Min	-	-	-
Max	-	-	-

GADOLINIUM — Gd

Environmental Geochemistry

Biological impacts	Considered non-essential. Data to assess the toxicity of REE are scarce
Uses	Glass additives, magnets, Gd-Y garnets (microwave), electrical ceramics
Environmental pathways	Poorly understood. Probably mostly windblown dust, weathering of REE-bearing minerals. Generally, geogenic sources more important than anthropogenic ones
Environmental mobility	Oxidising conditions: very low — Acid conditions: very low Reducing conditions: very low — Neutral to alkaline conditions: very low Comments: -
Geochemical barriers	Mechanical
Natural association	REE-Li-Rb-Cs-Be-Nb-Ta-Zr-B-Th-U-F (pegmatites), REE-Th-P-Zr-Fe-Cu (monazite veins), REE-Th-Ba-Sr-P-F-N-C (carbonatites), REE-U-P-F (phosphorites), REE-Au-Ti-Sn-Sr-Th (placers)
Action levels	-
Remarks	Toxicity of REE decreases as atomic number increases. Inhaled REE probably cause pneumoconiosis. REE taken up orally (e.g., via drinking water) accumulate in the skeleton, teeth, lungs, liver and kidneys. Some trees tend to accumulate Gd (e.g., hickory (*Carya*)). In 1991, 25 t Gd-oxide were used in the glass industry, and 10 t in the ceramics industry
Suggested analytical method(s)	ICP-MS

World Yearly Production (t/y) 54,000 REE-minerals (in 1991)
Price of Purest Form, in Small Quantities ($/kg) 90,656 (99.99%)
Market Price ($/kg) -

Ge GERMANIUM

Physico-Chemical Properties

Atomic number	32	Melting/boiling point (K)	1211.4 / 3106
Atomic mass	72.61(2)	Isotopes & isomers	5 stable + 24 unstable
Atomic radius (pm)	152	Acid/base of oxide	amphoteric
Main oxidation state(s)	+4 (+2)	State (at 300 K, 1 atm.)	solid
Ionic radius (pm)	87 (+2), 53-67 (+4)	Metallic character	pred. metal
Electronegativity (Pauling)	2.01	Element group(s)	heavy metal
Density (g/cm^3)	5.323	Affinity	siderophile

Naturally occurring isotopes	Natural abundance (%)	Atomic mass	Half-life	Unstable isotopes	Longest half-life
^{70}Ge	21.23	69.924,25	stable	^{61}Ge to ^{85}Ge	270.8 d
^{72}Ge	27.66	71.922,08	stable		
^{73}Ge	7.73	72.923,46	stable		
^{74}Ge	35.94	73.921,18	stable		
^{76}Ge	7.44	75.921,4	stable		

Concentrations in Crust/Rocks (mg/kg)

Bulk continental crust	1.4 / 1.5 / 1.6
Upper continental crust	1.4 / 1.6
Ultramafic rock	0.9
Ocean ridge basalt	1.2
Gabbro, basalt	1.3
Granite, granodiorite	1.6
Sandstone	1.4
Greywacke	-
Shale, schist	1.9
Limestone	0.15
Coal	1

Typical Minerals
argyrodite (Ag$_8$GeS$_6$),
germanite (Cu$_3$(Ge,Ga,Fe)(S,As)$_4$)

Possible Host Minerals
olivine, amphiboles, pyroxenes, micas, feldspars, sphalerite, Pb-Zn-Cu sulphide ore

Mass (kg) in

Continental crust	3.30x10^{+16}	Oceans	6.61x10^{+10}	Plants	1.84x10^{+7}

Concentrations in Media Sampled for the Kola Project (mg/kg)

Medium	Moss	Humus	Humus	Topsoil (0-5 cm)	B horizon	B horizon
	<2 mm conc. HNO$_3$	<2 mm	<2 mm amm. acet.	<2 mm	<2 mm aqua regia	<2 mm total (XRF)
Median	-	-	-	-	-	-
Min	-	-	-	-	-	-
Max	-	-	-	-	-	-

Medium	C horizon	C horizon	C horizon	Lake water
	<2 mm aqua regia	<2 mm total (XRF)	<2 mm total (INAA)	unfiltered (mg/l)
Median	-	-	-	-
Min	-	-	-	-
Max	-	-	-	-

GERMANIUM / Ge

Concentrations in Soils and Sediments (mg/kg)

Medium	Soil	Agricultural soil - Ap horizon	Agricultural soil - Top (0-25 cm)	Agricultural Soil - Bottom (50-75 cm)	Topsoil (0-15 cm)	Urban soil (0-2 cm)
	World	Canada	Finland	Finland	England & Wales	Trondheim
	<2 mm total	<2 mm	<2 mm aqua regia	<2 mm aqua regia	<2 mm aqua regia	<2 mm aqua regia
Median	2.1*	-	-	-	-	-
Min	-	-	-	-	-	-
Max	-	-	-	-	-	-

*Estimated mean

Medium	Forest soil - Humus	Forest soil - B horizon	Forest soil - C horizon	Till (C horizon)	Till (C horizon)	Laterite (25 ±15 cm)
	Norway	Norway	Norway	Finland	Finland	Australia
	<2 mm 7N HNO₃	<2 mm 7N HNO₃	<2 mm 7N HNO₃	<0.063 mm aqua regia	<0.063 mm total	0.45-2 mm total
Median	-	-	-	-	-	-
Min	-	-	-	-	-	-
Max	-	-	-	-	-	-

Medium	Stream sediment	Stream sediment	Stream sediment	Overbank sediment	Overbank sediment	Organic stream sediment
	Austria	Southern Scotland	Harz, Germany	Norway	Norway	Finland
	<0.18 mm	<0.10 mm total	<0.063 mm total	<0.063 mm total (XRF)	<0.063 mm 7N HNO₃	<2 mm Conc. HNO₃
Median	-	-	-	-	-	-
Min	-	-	-	-	-	-
Max	-	-	-	-	-	-

#2nd percentile   ^98th percentile

Concentrations in Waters (mg/l)

Medium	Ocean water	Ocean water	Ocean water	Stream water	Stream water	Stream water
	World (1)	World (2)	North Pacific	World	Nova Scotia, Canada	Finland
					<0.45 µm	<0.45 µm
Median	0.000,05*	0.000,05*	0.000,005,5*	0.000,05*	-	-
Min	-	-	-	-	-	-
Max	-	-	-	-	-	-

*Estimated mean   #2nd percentile   ^98th percentile

Medium	Stream water	Stream water	Stream water	Lake water	Ground water
	Romania	Eastern India	Harz, Germany	Norway	Southern Norway
	unfiltered	<0.2 µm	unfiltered	unfiltered	unfiltered
	ICP-MS	ICP-MS	ICP-MS	ICP-MS	ICP-MS
Median	0.000,021	0.000,08	0.000,01	0.000,1	0.000,018
Min	0.000,005	0.000,03	<0.000,01	<0.000,05	<0.000,002
Max	0.000,33	0.007,85	0.000,75	0.007,18	0.001,14

Ge GERMANIUM

Concentrations in Precipitation (mg/l)

Medium	Rain water	Rain water	Snow melt-water	Snow melt-water	Snow filter residue	Snow filter residue
	Kola, remote <0.45 µm	Kola, polluted <0.45 µm	Kola, remote <0.45 µm	Kola, polluted <0.45 µm	Kola, remote >0.45 µm	Kola, polluted >0.45 µm
Median	-	-	-	-	-	-
Min	-	-	-	-	-	-
Max	-	-	-	-	-	-

Medium	Rain water	Rain water				
	Coastal, Norway unfiltered ICP-MS	Inland, Norway unfiltered ICP-MS				
Median	-	-				
Min	-	-				
Max	-	-				

Concentrations in Air (ng/m^3) and Yearly Deposition (kg/km^2/yr)

Medium	Air	Air	Bulk deposition	Throughfall deposition	Bulk deposition	Bulk deposition
	World, remote	World, polluted	West Germany	West Germany	Kola, remote	Kola, polluted
Median	-	-	-	-	-	-
Min	-	-	-	-	-	-
Max	-	-	-	-	-	-

*Estimated value

Concentrations in Plants (mg/kg)

Medium	Moss	Moss	Crustose lichen	Lichen	Dandelion	Spruce bark
	Norway conc. HNO$_3$ (1990)	Germany conc. HNO$_3$ + H$_2$O$_2$ (1995)	Germany conc. HNO$_3$	Northwest Territories oven-dried	Europe oven-dried	Canada ashed
Median	-	0.027	-	-	-	-
Min	-	0.002	-	-	-	-
Max	-	0.39	-	-	-	-

*Mean   #Geometric mean

Concentrations in Human Fluids (mg/l)

Medium	Human blood	Human serum	Human urine			
	Lombardy, Italy	Lombardy, Italy	Lombardy, Italy			
Mean	-	-	-			
Min	-	-	-			
Max	-	-	-			

GERMANIUM   Ge

Environmental Geochemistry

Biological impacts	Considered non-essential. Low toxicity. Anti-carcinogenic
Uses	Glass (lenses, prisms, IR spectroscopy), catalyst, alloys (Ag, Cu), semi-conductors, transistors, pharmaceutical industry
Environmental pathways	Coal combustion, Pb, Zn, Cu smelters, geogenic dust, weathering
Environmental mobility	Oxidising conditions: low Acid conditions: low Reducing conditions: very low Neutral to alkaline conditions: low Comments: -
Geochemical barriers	-
Natural association	Ag-Zn-Cd-Cu-Ba-Sr-V-Cr-Mn-Fe-Ga-In-Ta-Ge-Sn-As-Sb-Bi-Se-Hg-Te-Pb (sulphide deposits)
Action levels	-
Remarks	Ge is invariably associated with Zn in nature (e.g., in sphalerite). Frequently obtained as by-product from Zn smelting. Ge compounds used as anti-tumour agents (but use limited by neurotoxicity). Bactericide. Green tea can contain high Ge concentrations. Geochemical similarity to Si, resulting in occasional replacements of Si by Ge in organisms
Suggested analytical method(s)	ICP-MS

World Yearly Production (t/y)	53 (in 1995)
Price of Purest Form, in Small Quantities ($/kg)	13,729 (99.999,9%)
Market Price ($/kg)	1730

H HYDROGEN

Physico-Chemical Properties

Atomic number	1	Melting/boiling point (K)	13.81 / 20.28
Atomic mass	1.007,94(7)	Isotopes & isomers	2 stable + 1 unstable
Atomic radius (pm)	79	Acid/base of oxide	amphoteric
Main oxidation state(s)	0,+1 (-1)	State (at 300 K, 1 atm.)	gas
Ionic radius (pm)	-	Metallic character	pred. metal
Electronegativity (Pauling)	2.2	Element group(s)	alkali metal, light metal
Density (g/cm^3)	0.000,089,88	Affinity	atmophile, biophile, lithophile

Naturally occurring isotopes	Natural abundance (%)	Atomic mass	Half-life
^{1}H	99.985	1.007,825	stable
^{2}H	0.015	2.014,102	stable

Unstable isotopes	Longest half-life
^{3}H	12.32 y

Concentrations in Crust/Rocks (mg/kg)

Bulk continental crust	- / 1400 / -
Upper continental crust	- / -
Ultramafic rock	-
Ocean ridge basalt	-
Gabbro, basalt	-
Granite, granodiorite	-
Sandstone	-
Greywacke	-
Shale, schist	-
Limestone	-
Coal	57,000

Typical Minerals
ice (H$_2$O)

Possible Host Minerals
H-containing groups in many minerals

Mass (kg) in

Continental crust	3.30x10^{+19}	Oceans	1.43x10^{+20}	Plants	1.20x10^{+14}

Concentrations in Media Sampled for the Kola Project (mg/kg)

| Medium | Moss | Humus <2 mm | Humus <2 mm | Topsoil (0-5 cm) <2 mm | B horizon <2 mm | B horizon <2 mm |
	conc. HNO$_3$	CHN-anal.	amm. acet.		aqua regia	total (XRF)
Median	-	61,000	-	-	-	-
Min	-	22,000	-	-	-	-
Max	-	71,000	-	-	-	-

Medium	C horizon <2 mm aqua regia	C horizon <2 mm total (XRF)	C horizon <2 mm total (INAA)			Lake water unfiltered (mg/l)
Median	-	-	-			-
Min	-	-	-			-
Max	-	-	-			-

HYDROGEN H

Concentrations in Soils and Sediments (mg/kg)

Medium	Soil	Agricultural soil - Ap horizon	Agricultural soil - Top (0-25 cm)	Agricultural Soil - Bottom (50-75 cm)	Topsoil (0-15 cm)	Urban soil (0-2 cm)
	World	Canada	Finland	Finland	England & Wales	Trondheim
	<2 mm total	<2 mm	<2 mm aqua regia	<2 mm aqua regia	<2 mm aqua regia	<2 mm aqua regia
Median	-	-	-	-	-	-
Min	-	-	-	-	-	-
Max	-	-	-	-	-	-

*Estimated mean

Medium	Forest soil - Humus	Forest soil - B horizon	Forest soil - C horizon	Till (C horizon)	Till (C horizon)	Laterite (25 ±15 cm)
	Norway	Norway	Norway	Finland	Finland	Australia
	<2 mm 7N HNO₃	<2 mm 7N HNO₃	<2 mm 7N HNO₃	<0.063 mm aqua regia	<0.063 mm total	0.45-2 mm total
Median	-	-	-	-	-	-
Min	-	-	-	-	-	-
Max	-	-	-	-	-	-

Medium	Stream sediment	Stream sediment	Stream sediment	Overbank sediment	Overbank sediment	Organic stream sediment
	Austria	Southern Scotland	Harz, Germany	Norway	Norway	Finland
	<0.18 mm	<0.10 mm total	<0.063 mm total	<0.063 mm total (XRF)	<0.063 mm 7N HNO₃	<2 mm C, H, N-analyser
Median	-	-	-	-	-	15,000
Min	-	-	-	-	-	4000#
Max	-	-	-	-	-	49,000^

#2nd percentile   ^98th percentile

Concentrations in Waters (mg/l)

Medium	Ocean water	Ocean water	Ocean water	Stream water	Stream water	Stream water
	World (1)	World (2)	North Pacific	World	Nova Scotia, Canada	Finland
					<0.45 μm	<0.45 μm
Median	108,000*	-	-	-	-	-
Min	-	-	-	-	-	-
Max	-	-	-	-	-	-

*Estimated mean   #2nd percentile   ^98th percentile

Medium	Stream water	Stream water	Stream water	Lake water	Ground water
	Romania	Eastern India	Harz, Germany	Norway	Southern Norway
	unfiltered ICP-MS	<0.2 μm ICP-MS	unfiltered	unfiltered ICP-MS	unfiltered
Median	-	-	-	-	-
Min	-	-	-	-	-
Max	-	-	-	-	-

H   HYDROGEN

Concentrations in Precipitation (mg/l)

Medium	Rain water	Rain water	Snow melt-water	Snow melt-water	Snow filter residue	Snow filter residue
	Kola, remote <0.45 μm	Kola, polluted <0.45 μm	Kola, remote <0.45 μm	Kola, polluted <0.45 μm	Kola, remote >0.45 μm	Kola, polluted >0.45 μm
Median	-	-	-	-	-	-
Min	-	-	-	-	-	-
Max	-	-	-	-	-	-

Medium	Rain water	Rain water				
	Coastal, Norway unfiltered ICP-MS	Inland, Norway unfiltered ICP-MS				
Median	-	-				
Min	-	-				
Max	-	-				

Concentrations in Air (ng/m^3) and Yearly Deposition (kg/km^2/yr)

Medium	Air	Air	Bulk deposition	Throughfall deposition	Bulk deposition	Bulk deposition
	World, remote	World, polluted	West Germany	West Germany	Kola, remote	Kola, polluted
Median	-	-	-	-	-	-
Min	-	-	-	-	-	-
Max	-	-	-	-	-	-

*Estimated value

Concentrations in Plants (mg/kg)

Medium	Moss	Moss	Crustose lichen	Lichen	Dandelion	Spruce bark
	Norway conc. HNO$_3$ (1990)	Germany conc. HNO$_3$ + H$_2$O$_2$ (1995)	Germany conc. HNO$_3$	Northwest Territories oven-dried	Europe oven-dried	Canada ashed
Median	-	-	-	-	-	-
Min	-	-	-	-	-	-
Max	-	-	-	-	-	-

*Mean   #Geometric mean

Concentrations in Human Fluids (mg/l)

Medium	Human blood	Human serum	Human urine			
	Lombardy, Italy	Lombardy, Italy	Lombardy, Italy			
Mean	-	-	-			
Min	-	-	-			
Max	-	-	-			

HYDROGEN   H

Environmental Geochemistry

Biological impacts	Essential for all organisms (e.g., as H_2O). Toxicity varies with compound (e.g., D_2O)
Uses	Ammonia production, methanol production, hydrogenation of fats and oils, welding, rocket fuel, etc.
Environmental pathways	Countless
Environmental mobility	Oxidising conditions: - Acid conditions: - Reducing conditions: - Neutral to alkaline conditions: - Comments: Highly dependent on compound
Geochemical barriers	-
Natural association	C-H-O-(N)-(S) (organic matter), H-O (water)
Action levels	Drinking water, recommended pH (= -log [H^+]) range: 6.5 to 8.5 (Norway Shd); max pH: 9.5 (Norway Shd)
Remarks	H is the most abundant element in the universe. Coal, petroleum, natural gas are all hydrocarbons. ^2H (deuterium) is toxic. ^3H (tritium) released during the 1960s atmospheric nuclear bomb testing, leading to ^3H enrichments in surface waters. ^3H used in hydrological tracer tests. Air contains 0.5 mg H_2/kg. Isotopic fractionation in nature is important. Consumer complaints related to drinking water may arise at low pH due to corrosion, at high pH due to taste and soapy feel. Drinking water pH should generally be lower than 8 for effective disinfection with chlorine
Suggested analytical method(s)	Ion selective electrode (pH), elemental analyser (IR spectrometry)

World Yearly Production (t/y)	85,000,000 m^3 (USA only)
Price of Purest Form, in Small Quantities ($/kg)	-
Market Price ($/kg)	-

He — HELIUM

Physico-Chemical Properties

Atomic number	2	Melting/boiling point (K)	<1 (26 atm) / 4.22
Atomic mass	4.002,602(2)	Isotopes & isomers	2 stable + 6 unstable
Atomic radius (pm)	49	Acid/base of oxide	-
Main oxidation state(s)	0	State (at 300 K, 1 atm.)	gas
Ionic radius (pm)	-	Metallic character	-
Electronegativity (Pauling)	-	Element group(s)	noble gas
Density (g/cm^3)	0.000,178,5 (273.15 K, 1 atm)	Affinity	atmophile

Naturally occurring isotopes	Natural abundance (%)	Atomic mass	Half-life
^{3}He	0.000,137	3.016,03	stable
^{4}He	99.999,863	4.002,603	stable

Unstable isotopes	Longest half-life
^{5}He to ^{10}He	0.8 s

Concentrations in Crust/Rocks (mg/kg)

Bulk continental crust	- / 0.008 / -
Upper continental crust	- / -
Ultramafic rock	-
Ocean ridge basalt	-
Gabbro, basalt	-
Granite, granodiorite	-
Sandstone	-
Greywacke	-
Shale, schist	-
Limestone	-
Coal	-

Typical Minerals
-

Possible Host Minerals
zeolite

Mass (kg) in

Continental crust	1.89x10^{+14}	Oceans	8.99x10^{+9}	Plants	-

Concentrations in Media Sampled for the Kola Project (mg/kg)

Medium	Moss	Humus	Humus	Topsoil (0-5 cm)	B horizon	B horizon
	<2 mm conc. HNO$_3$	<2 mm amm. acet.	<2 mm	<2 mm	<2 mm aqua regia	<2 mm total (XRF)
Median	-	-	-	-	-	-
Min	-	-	-	-	-	-
Max	-	-	-	-	-	-

Medium	C horizon	C horizon	C horizon			Lake water
	<2 mm aqua regia	<2 mm total (XRF)	<2 mm total (INAA)			unfiltered (mg/l)
Median	-	-	-			-
Min	-	-	-			-
Max	-	-	-			-

HELIUM He

Concentrations in Soils and Sediments (mg/kg)

Medium	Soil	Agricultural soil - Ap horizon	Agricultural soil - Top (0-25 cm)	Agricultural Soil - Bottom (50-75 cm)	Topsoil (0-15 cm)	Urban soil (0-2 cm)
	World	Canada	Finland	Finland	England & Wales	Trondheim
	<2 mm	<2 mm	<2 mm	<2 mm	<2 mm	<2 mm
	total		aqua regia	aqua regia	aqua regia	aqua regia
Median	-	-	-	-	-	-
Min	-	-	-	-	-	-
Max	-	-	-	-	-	-

*Estimated mean

Medium	Forest soil - Humus	Forest soil - B horizon	Forest soil - C horizon	Till (C horizon)	Till (C horizon)	Laterite (25 ±15 cm)
	Norway	Norway	Norway	Finland	Finland	Australia
	<2 mm	<2 mm	<2 mm	<0.063 mm	<0.063 mm	0.45-2 mm
	7N HNO$_3$	7N HNO$_3$	7N HNO$_3$	aqua regia	total	total
Median	-	-	-	-	-	-
Min	-	-	-	-	-	-
Max	-	-	-	-	-	-

Medium	Stream sediment	Stream sediment	Stream sediment	Overbank sediment	Overbank sediment	Organic stream sediment
		Southern Scotland	Harz, Germany			
	Austria			Norway	Norway	Finland
	<0.18 mm	<0.10 mm	<0.063 mm	<0.063 mm	<0.063 mm	<2 mm
		total	total	total (XRF)	7N HNO$_3$	Conc. HNO$_3$
Median	-	-	-	-	-	-
Min	-	-	-	-	-	-
Max	-	-	-	-	-	-

#2nd percentile ^98th percentile

Concentrations in Waters (mg/l)

Medium	Ocean water	Ocean water	Ocean water	Stream water	Stream water	Stream water
	World (1)	World (2)	North Pacific	World	Nova Scotia, Canada	Finland
					<0.45 µm	<0.45 µm
Median	0.000,007*	0.000,006,8*	0.000,007,6*	-	-	-
Min	-	-	-	-	-	-
Max	-	-	-	-	-	-

*Estimated mean   #2nd percentile   ^98th percentile

Medium	Stream water	Stream water	Stream water	Lake water	Ground water
	Romania	Eastern India	Harz, Germany	Norway	Southern Norway
	unfiltered	<0.2 µm	unfiltered	unfiltered	unfiltered
	ICP-MS	ICP-MS		ICP-MS	
Median	-	-	-	-	-
Min	-	-	-	-	-
Max	-	-	-	-	-

He HELIUM

Concentrations in Precipitation (mg/l)

Medium	Rain water Kola, remote <0.45 μm	Rain water Kola, polluted <0.45 μm	Snow melt-water Kola, remote <0.45 μm	Snow melt-water Kola, polluted <0.45 μm	Snow filter residue Kola, remote >0.45 μm	Snow filter residue Kola, polluted >0.45 μm
Median	-	-	-	-	-	-
Min	-	-	-	-	-	-
Max	-	-	-	-	-	-

Medium	Rain water Coastal, Norway unfiltered ICP-MS	Rain water Inland, Norway unfiltered ICP-MS				
Median	-	-				
Min	-	-				
Max	-	-				

Concentrations in Air (ng/m^3) and Yearly Deposition (kg/km^2/yr)

Medium	Air World, remote	Air World, polluted	Bulk deposition West Germany	Throughfall deposition West Germany	Bulk deposition Kola, remote	Bulk deposition Kola, polluted
Median	-	-	-	-	-	-
Min	-	-	-	-	-	-
Max	-	-	-	-	-	-

*Estimated value

Concentrations in Plants (mg/kg)

Medium	Moss Norway conc. HNO$_3$ (1990)	Moss Germany conc. HNO$_3$ + H$_2$O$_2$ (1995)	Crustose lichen Germany conc. HNO$_3$	Lichen Northwest Territories oven-dried	Dandelion Europe oven-dried	Spruce bark Canada ashed
Median	-	-	-	-	-	-
Min	-	-	-	-	-	-
Max	-	-	-	-	-	-

*Mean   #Geometric mean

Concentrations in Human Fluids (mg/l)

Medium	Human blood Lombardy, Italy	Human serum Lombardy, Italy	Human urine Lombardy, Italy			
Mean	-	-	-			
Min	-	-	-			
Max	-	-	-			

HELIUM — He

Environmental Geochemistry

Biological impacts	Considered non-essential
Uses	Balloons, welding, coolant in nuclear reactors, cryogenic research, inert gas in research laboratories and technological applications
Environmental pathways	Mostly natural sources. Natural trace component of atmosphere
Environmental mobility	Oxidising conditions: very high　　　Acid conditions: very high Reducing conditions: very high　　　Neutral to alkaline conditions: very high Comments: Inert gas
Geochemical barriers	-
Natural association	-
Action levels	-
Remarks	He is the second most abundant element in the universe. It has the lowest melting point of all elements. He occurs together with U minerals (0.5 - 9% He have been measured in naturally degassing faults). He is a decay product of U and Th. Occurs in atmosphere at low concentrations (0.9 mg/m^3), together with the other noble gasses
Suggested analytical method(s)	-

World Yearly Production (t/y)	-
Price of Purest Form, in Small Quantities ($/kg)	141 $/l
Market Price ($/kg)	8 $/m^3

Hf HAFNIUM

Physico-Chemical Properties

Atomic number	72	Melting/boiling point (K)	2506 / 4876
Atomic mass	178.49(2)	Isotopes & isomers	6 stable + 34 unstable
Atomic radius (pm)	216	Acid/base of oxide	amphoteric
Main oxidation state(s)	+4	State (at 300 K, 1 atm.)	solid
Ionic radius (pm)	72-94	Metallic character	metal
Electronegativity (Pauling)	1.3	Element group(s)	heavy metal
Density (g/cm^3)	13.31	Affinity	lithophile

Naturally occurring isotopes	Natural abundance (%)	Atomic mass	Half-life
^{174}Hf	0.162	173.940,04	stable
^{176}Hf	5.206	175.941,4	stable
^{177}Hf	18.606	176.943,22	stable
^{178}Hf	27.297	177.943,698	stable
^{179}Hf	13.629	178.945,82	stable
^{180}Hf	35.1	179.946,55	stable

Unstable isotopes	Longest half-life
^{154}Hf to ^{185}Hf	2x10^{+15} y

Concentrations in Crust/Rocks (mg/kg)

Bulk continental crust	4.9 / 3 / 3
Upper continental crust	5.8 / 5.8
Ultramafic rock	0.5
Ocean ridge basalt	2.5
Gabbro, basalt	3
Granite, granodiorite	5
Sandstone	6
Greywacke	3.5
Shale, schist	4
Limestone	0.4
Coal	1

Typical Minerals
zircon ((Zr,Hf)SiO$_4$),
baddeleyite ((Zr,Hf,Ti,Fe,Th)O$_2$)

Possible Host Minerals
pyroxenes, garnets, biotite

Mass (kg) in

Continental crust	1.16x10^{+17}	Oceans	9.25x10^{+9}	Plants	9.20x10^{+7}

Concentrations in Media Sampled for the Kola Project (mg/kg)

Medium	Moss <2 mm conc. HNO$_3$	Humus <2 mm	Humus <2 mm amm. acet.	Topsoil (0-5 cm) <2 mm total (INAA)	B horizon <2 mm aqua regia	B horizon <2 mm total (XRF)
Median	-	-	-	5	-	-
Min	-	-	-	<1	-	-
Max	-	-	-	81	-	-

Medium	C horizon <2 mm aqua regia	C horizon <2 mm total (XRF)	C horizon <2 mm total (INAA)	Lake water unfiltered (mg/l)
Median	-	-	6	-
Min	-	-	2	-
Max	-	-	120	-

HAFNIUM Hf

Concentrations in Soils and Sediments (mg/kg)

Medium	Soil	Agricultural soil - Ap horizon	Agricultural soil - Top (0-25 cm)	Agricultural Soil - Bottom (50-75 cm)	Topsoil (0-15 cm)	Urban soil (0-2 cm)
	World	Canada	Finland	Finland	England & Wales	Trondheim
	<2 mm total	<2 mm total (INAA)	<2 mm aqua regia	<2 mm aqua regia	<2 mm aqua regia	<2 mm aqua regia
Median	5*	5	-	-	-	-
Min	-	1	-	-	-	-
Max	-	10	-	-	-	-

*Estimated mean

Medium	Forest soil - Humus	Forest soil - B horizon	Forest soil - C horizon	Till (C horizon)	Till (C horizon)	Laterite (25 ±15 cm)
	Norway	Norway	Norway	Finland	Finland	Australia
	<2 mm 7N HNO₃	<2 mm 7N HNO₃	<2 mm 7N HNO₃	<0.063 mm aqua regia	<0.063 mm total	0.45-2 mm total
Median	-	-	-	-	-	-
Min	-	-	-	-	-	-
Max	-	-	-	-	-	-

Medium	Stream sediment	Stream sediment	Stream sediment	Overbank sediment	Overbank sediment	Organic stream sediment
	Austria	Southern Scotland	Harz, Germany	Norway	Norway	Finland
	<0.18 mm	<0.10 mm total	<0.063 mm total (ICP-MS)	<0.063 mm total (XRF)	<0.063 mm 7N HNO₃	<2 mm Conc. HNO₃
Median	-	-	16	-	-	-
Min	-	-	2	-	-	-
Max	-	-	38	-	-	-

#2nd percentile ^98th percentile

Concentrations in Waters (mg/l)

Medium	Ocean water	Ocean water	Ocean water	Stream water	Stream water	Stream water
	World (1)	World (2)	North Pacific	World	Nova Scotia, Canada	Finland
					<0.45 μm	<0.45 μm
Median	0.000,007*	0.000,007*	0.000,003,4*	<0.000,008*	-	-
Min	-	-	-	-	-	-
Max	-	-	-	-	-	-

*Estimated mean #2nd percentile ^98th percentile

Medium	Stream water	Stream water	Stream water	Lake water	Ground water
	Romania	Eastern India	Harz, Germany	Norway	Southern Norway
	unfiltered	<0.2 μm	unfiltered	unfiltered	unfiltered
	ICP-MS	ICP-MS	ICP-MS	ICP-MS	ICP-MS
Median	0.000,009	0.000,02	0.000,01	<0.000,03	0.000,01
Min	0.000,003	0.000,01	<0.000,01	<0.000,03	<0.000,002
Max	0.000,22	0.000,4	0.000,51	<0.000,03	0.000,75

Hf   HAFNIUM

Concentrations in Precipitation (mg/l)

Medium	Rain water Kola, remote <0.45 μm	Rain water Kola, polluted <0.45 μm	Snow melt-water Kola, remote <0.45 μm	Snow melt-water Kola, polluted <0.45 μm	Snow filter residue Kola, remote >0.45 μm	Snow filter residue Kola, polluted >0.45 μm
Median	-	-	-	-	-	-
Min	-	-	-	-	-	-
Max	-	-	-	-	-	-

Medium	Rain water Coastal, Norway unfiltered ICP-MS	Rain water Inland, Norway unfiltered ICP-MS				
Median	-	-				
Min	-	-				
Max	-	-				

Concentrations in Air (ng/m³) and Yearly Deposition (kg/km²/yr)

Medium	Air World, remote	Air World, polluted	Bulk deposition West Germany	Throughfall deposition West Germany	Bulk deposition Kola, remote	Bulk deposition Kola, polluted
Median	-	-	-	-	-	-
Min	-	-	-	-	-	-
Max	-	-	-	-	-	-

*Estimated value

Concentrations in Plants (mg/kg)

Medium	Moss Norway conc. HNO₃ (1990)	Moss Germany conc. HNO₃ + H₂O₂ (1995)	Crustose lichen Germany conc. HNO₃	Lichen Northwest Territories oven-dried	Dandelion Europe oven-dried	Spruce bark Canada ashed total (INAA)
Median	-	-	-	-	-	3.6
Min	-	-	-	-	-	<0.5
Max	-	-	-	-	-	11

*Mean   #Geometric mean

Concentrations in Human Fluids (mg/l)

Medium	Human blood Lombardy, Italy	Human serum Lombardy, Italy	Human urine Lombardy, Italy			
Mean	0.000,21	-	0.000,49			
Min	0.000,012	-	0.000,01			
Max	0.000,6	-	0.001,4			

HAFNIUM   Hf

Environmental Geochemistry

Biological impacts	Considered non-essential
Uses	Alloys, nuclear industry (reactor control rods), bulbs for photographic flashes
Environmental pathways	Geogenic sources much more important than anthropogenic sources
Environmental mobility	Oxidising conditions: - Acid conditions: - Reducing conditions: - Neutral to alkaline conditions: - Comments: -
Geochemical barriers	-
Natural association	Nb-Ta-Sn-W-Li-Be-Ti-Rb-Cs-U-Th-B-Zr-Hf-P-F-REE (granites and syenitic pegmatites)
Action levels	-
Remarks	Hf is invariably associated with Zr in nature (e.g., in zircon). Yearly demand in the USA is >50 tons
Suggested analytical method(s)	ICP-MS

World Yearly Production (t/y)	-
Price of Purest Form, in Small Quantities ($/kg)	25,638 (99.9+%)
Market Price ($/kg)	-

Hg   MERCURY

Physico-Chemical Properties

Atomic number	80	Melting/boiling point (K)	234.32 / 629.88
Atomic mass	200.59(2)	Isotopes & isomers	7 stable + 34 unstable
Atomic radius (pm)	176	Acid/base of oxide	base
Main oxidation state(s)	+2 (0,+1,+3)	State (at 300 K, 1 atm.)	liquid
Ionic radius (pm)	83-128 (+2), 111 (+3), 133 (+6)	Metallic character	metal
Electronegativity (Pauling)	2	Element group(s)	heavy metal
Density (g/cm^3)	13.546	Affinity	chalcophile

Naturally occurring isotopes	Natural abundance (%)	Atomic mass	Half-life
^{196}Hg	0.15	195.965,81	stable
^{198}Hg	9.97	197.966,75	stable
^{199}Hg	16.87	198.968,26	stable
^{200}Hg	23.1	199.968,31	stable
^{201}Hg	13.18	200.970,29	stable
^{202}Hg	29.86	201.970,63	stable
^{204}Hg	6.87	203.973,48	stable

Unstable isotopes	Longest half-life
^{175}Hg to ^{208}Hg	520 y

Concentrations in Crust/Rocks (mg/kg)

Bulk continental crust	0.04 / 0.085 / -
Upper continental crust	0.056 / 0.02
Ultramafic rock	0.004
Ocean ridge basalt	0.01
Gabbro, basalt	0.01
Granite, granodiorite	0.03
Sandstone	0.01
Greywacke	-
Shale, schist	0.18
Limestone	0.02
Coal	0.1

Typical Minerals
cinnabar (HgS), native Hg

Possible Host Minerals
amphiboles, astrophyllite, aegirine, sphene, sphalerite, tetrahedrite, other sulphides

Mass (kg) in

Continental crust	9.44x10^{+14}	Oceans	3.97x10^{+10}	Plants	1.84x10^{+8}

Concentrations in Media Sampled for the Kola Project (mg/kg)

Medium	Moss conc. HNO$_3$	Humus <2 mm conc. HNO$_3$	Humus <2 mm amm. acet.	Topsoil (0-5 cm) <2 mm	B horizon <2 mm aqua regia	B horizon <2 mm total (XRF)
Median	0.05	0.227	-	-	<0.05	-
Min	0.02	0.094	-	-	<0.05	-
Max	0.155	0.974	-	-	0.15	-

Medium	C horizon <2 mm aqua regia	C horizon <2 mm total (XRF)	C horizon <2 mm total (INAA)	Lake water unfiltered (mg/l)
Median	<0.02	-	-	-
Min	<0.02	-	-	-
Max	0.16	-	-	-

MERCURY   Hg

Concentrations in Soils and Sediments (mg/kg)

Medium	Soil	Agricultural soil - Ap horizon	Agricultural soil - Top (0-25 cm)	Agricultural Soil - Bottom (50-75 cm)	Topsoil (0-15 cm)	Urban soil (0-2 cm)
	World	*Canada*	*Finland*	*Finland*	*England & Wales*	*Trondheim*
	<2 mm	<2 mm	<2 mm	<2 mm	<2 mm	<2 mm
	total	total (AAS)	aqua regia	aqua regia	aqua regia	aqua regia
Median	0.05*	0.04	-	-	-	0.13
Min	-	0.005	-	-	-	<0.02
Max	-	0.13	-	-	-	4.49

*Estimated mean

Medium	Forest soil - Humus	Forest soil - B horizon	Forest soil - C horizon	Till (C horizon)	Till (C horizon)	Laterite (25 ±15 cm)
	Norway	*Norway*	*Norway*	*Finland*	*Finland*	*Australia*
	<2 mm	<2 mm	<2 mm	<0.063 mm	<0.063 mm	0.45-2 mm
	7N HNO$_3$	7N HNO$_3$	7N HNO$_3$	aqua regia	total	total
Median	-	-	-	-	-	-
Min	-	-	-	-	-	-
Max	-	-	-	-	-	-

Medium	Stream sediment	Stream sediment	Stream sediment	Overbank sediment	Overbank sediment	Organic stream sediment
		Southern Scotland	*Harz, Germany*	*Norway*	*Norway*	*Finland*
	Austria					
	<0.18 mm	<0.10 mm	<0.063 mm	<0.063 mm	<0.063 mm	<2 mm
		total	total (ICP-MS)	total (XRF)	7N HNO$_3$	Conc. HNO$_3$
Median	-	-	0.09	-	-	0.04
Min	-	-	<0.01	-	-	0.01#
Max	-	-	3.3	-	-	0.13^

#2nd percentile   ^98th percentile

Concentrations in Waters (mg/l)

Medium	Ocean water	Ocean water	Ocean water	Stream water	Stream water	Stream water
	World (1)	*World (2)*	*North Pacific*	*World*	*Nova Scotia, Canada*	*Finland*
					<0.45 µm	<0.45 µm
					ICP-MS-Hydride	
Median	0.000,03*	0.000,03*	0.000,000,14*	0.000,05*	0.000,007	-
Min	-	-	-	-	<0.000,004	-
Max	-	-	-	-	0.000,219	-

*Estimated mean   #2nd percentile   ^98th percentile

Medium	Stream water	Stream water	Stream water	Lake water	Ground water
	Romania	*Eastern India*	*Harz, Germany*	*Norway*	*Southern Norway*
	unfiltered	<0.2 µm	unfiltered	unfiltered	unfiltered
	ICP-MS	ICP-MS	ICP-MS	ICP-MS	ICP-MS
Median	0.000,005	<0.000,1	0.000,01	-	0.000,034
Min	0.000,005	<0.000,1	<0.000,01	-	<0.000,005
Max	0.000,11	<0.000,1	0.000,18	-	0.001,16

Hg   MERCURY

Concentrations in Precipitation (mg/l)

Medium	Rain water Kola, remote <0.45 µm	Rain water Kola, polluted <0.45 µm	Snow melt- water Kola, remote <0.45 µm	Snow melt- water Kola, polluted <0.45 µm	Snow filter residue Kola, remote >0.45 µm	Snow filter residue Kola, polluted >0.45 µm
Median	-	-	-	-	-	-
Min	-	-	-	-	-	-
Max	-	-	-	-	-	-

Medium	Rain water Coastal, Norway unfiltered ICP-MS	Rain water Inland, Norway unfiltered ICP-MS				
Median	-	-				
Min	-	-				
Max	-	-				

Concentrations in Air (ng/m^3) and Yearly Deposition (kg/km^2/yr)

Medium	Air World, remote	Air World, polluted	Bulk deposition West Germany	Throughfall deposition West Germany	Bulk deposition Kola, remote	Bulk deposition Kola, polluted
Median	<0.04 / 2	-	-	-	-	-
Min	0.01	<0.09	-	-	-	-
Max	0.06 (3.4)	38	-	-	-	-

*Estimated value

Concentrations in Plants (mg/kg)

Medium	Moss Norway conc. HNO$_3$ (1990)	Moss Germany conc. HNO$_3$ + H$_2$O$_2$ (1995)	Crustose lichen Germany conc. HNO$_3$	Lichen Northwest Territories oven-dried	Dandelion Europe oven-dried total (INAA)	Spruce bark Canada ashed
Median	0.06	0.044	-	-	0.14#	-
Min	0.02	<0.001	-	-	-	-
Max	0.88	0.4	-	-	-	-

*Mean   #Geometric mean

Concentrations in Human Fluids (mg/l)

Medium	Human blood Lombardy, Italy	Human serum Lombardy, Italy	Human urine Lombardy, Italy			
Mean	0.005,3	0.002,1	0.003,5			
Min	0.000,5	0.000,39	0.000,3			
Max	0.017,3	0.004,8	0.016,5			

MERCURY   Hg

Environmental Geochemistry

Biological impacts	Considered non-essential. Very toxic. Teratogenic. No carcinogenic effect proven yet (probably too toxic)
Uses	Caustic soda and chlorine production, gold ore processing (amalgamation), batteries, dentistry, wood impregnation, thermometers, barometers, detonators, Hg-vapour lamps
Environmental pathways	Windblown dust, sea spray, volcanic dust, open-cast mining (detonators), crude oil and coal combustion, chemical industry, electric lamp production, paper mills, sewage sludge
Environmental mobility	Oxidising conditions: medium Acid conditions: high Reducing conditions: very low Neutral to alkaline conditions: very low Comments: Hg can evaporate when released to water or soil. Microbes can convert inorganic Hg to organic forms, which can accumulate in aquatic life
Geochemical barriers	Sulphide adsorption, organic matter
Natural association	Hg-Sb-As-(Sn)-(W), Hg-Au or Hg-Ag (some Au, quartz and Ag vein deposits, e.g., Cobalt-type deposits), Hg-Zn (some Zn vein deposits)
Action levels	Drinking water, MAC: 0.000,5 mg/l (Norway Shd), 0.002 mg/l (US EPA), 0.001 mg/l (Canada CWQG); recommended: 0.001 mg/l (WHO); MAC: 0.000,5 mg/l (Russia MoH). Ground water, background: 0.000,05 mg/l; remediate: 0.000,3 mg/l (Netherlands VROM). Soil, background: 0.3 mg/kg; remediate: 10 mg/kg (Netherlands VROM). Agricultural soil, maximum tolerable concentration for disposal of sewage sludge of <8 mg Hg/kg: 1 mg/kg (Germany)
Remarks	Hg use can be traced back >3000 y. Alkyl-Hg fungicide and seed treatment has led to serious poisonings. Fish accumulate methyl-Hg. Minamata disease caused by uptake of methyl-Hg via contaminated aquatic organisms. Se can act as antagonist to Hg. Estimates of natural and anthropogenic global Hg fluxes vary considerably. The US EPA suggests that drinking water exceeding action levels can lead to kidney damage in the short term
Suggested analytical method(s)	CV-AAS

World Yearly Production (t/y)	2900 (in 1995)
Price of Purest Form, in Small Quantities ($/kg)	927 (99.999,995%)
Market Price ($/kg)	5

Ho   HOLMIUM

Physico-Chemical Properties

Atomic number	67	Melting/boiling point (K)	1747 / 2973
Atomic mass	164.930,32(2)	Isotopes & isomers	1 stable + 47 unstable
Atomic radius (pm)	247	Acid/base of oxide	weak base
Main oxidation state(s)	+3	State (at 300 K, 1 atm.)	solid
Ionic radius (pm)	104-126	Metallic character	metal
Electronegativity (Pauling)	1.23	Element group(s)	heavy metal, REE
Density (g/cm^3)	8.795	Affinity	lithophile

Naturally occurring isotopes	Natural abundance (%)	Atomic mass	Half-life
^{165}Ho	100	164.930,3	stable

Unstable isotopes	Longest half-life
^{144}Ho to ^{172}Ho	4570 y

Concentrations in Crust/Rocks (mg/kg)

Bulk continental crust	0.8 / 1.3 / 0.78
Upper continental crust	0.62 / 0.8
Ultramafic rock	-
Ocean ridge basalt	-
Gabbro, basalt	-
Granite, granodiorite	-
Sandstone	-
Greywacke	0.78
Shale, schist	-
Limestone	-
Coal	0.1

Typical Minerals
monazite ((Ce,La,Nd,Ho,Th)(PO$_4$,SiO$_4$)),
bastnaesite ((Ce,La,Ho)CO$_3$(F,OH))

Possible Host Minerals
feldspars, apatite, allanite, sphene, fluorite, zircon

Mass (kg) in

Continental crust	1.89x10^{+16}	Oceans	2.64x10^{+8}	Plants	1.47x10^{+7}

Concentrations in Media Sampled for the Kola Project (mg/kg)

Medium	Moss	Humus	Humus	Topsoil (0-5 cm)	B horizon	B horizon
		<2 mm	<2 mm	<2 mm	<2 mm	<2 mm
	conc. HNO$_3$		amm. acet.		aqua regia	total (XRF)
Median	-	-	-	-	-	-
Min	-	-	-	-	-	-
Max	-	-	-	-	-	-

Medium	C horizon	C horizon	C horizon			Lake water
	<2 mm	<2 mm	<2 mm			unfiltered
	aqua regia	total (XRF)	total (INAA)			(mg/l)
Median	-	-	-			-
Min	-	-	-			-
Max	-	-	-			-

HOLMIUM   Ho

Concentrations in Soils and Sediments (mg/kg)

Medium	Soil	Agricultural soil - Ap horizon	Agricultural soil - Top (0-25 cm)	Agricultural Soil - Bottom (50-75 cm)	Topsoil (0-15 cm)	Urban soil (0-2 cm)
	World	Canada	Finland	Finland	England & Wales	Trondheim
	<2 mm total	<2 mm	<2 mm aqua regia	<2 mm aqua regia	<2 mm aqua regia	<2 mm aqua regia
Median	-	-	-	-	-	-
Min	-	-	-	-	-	-
Max	-	-	-	-	-	-

*Estimated mean

Medium	Forest soil - Humus	Forest soil - B horizon	Forest soil - C horizon	Till (C horizon)	Till (C horizon)	Laterite (25 ±15 cm)
	Norway	Norway	Norway	Finland	Finland	Australia
	<2 mm 7N HNO$_3$	<2 mm 7N HNO$_3$	<2 mm 7N HNO$_3$	<0.063 mm aqua regia	<0.063 mm total	0.45-2 mm total
Median	-	-	-	-	-	-
Min	-	-	-	-	-	-
Max	-	-	-	-	-	-

Medium	Stream sediment	Stream sediment	Stream sediment	Overbank sediment	Overbank sediment	Organic stream sediment
	Austria	Southern Scotland	Harz, Germany	Norway	Norway	Finland
	<0.18 mm	<0.10 mm total	<0.063 mm total	<0.063 mm total (XRF)	<0.063 mm 7N HNO$_3$	<2 mm Conc. HNO$_3$
Median	-	-	-	-	-	-
Min	-	-	-	-	-	-
Max	-	-	-	-	-	-

#2nd percentile   ^98th percentile

Concentrations in Waters (mg/l)

Medium	Ocean water	Ocean water	Ocean water	Stream water	Stream water	Stream water
	World (1)	World (2)	North Pacific	World	Nova Scotia, Canada	Finland
					<0.45 µm ICP-MS	<0.45 µm
Median	0.000,000,22*	0.000,000,2*	0.000,000,36*	-	0.000,005	-
Min	-	-	-	-	<0.000,005	-
Max	-	-	-	-	0.000,039	-

*Estimated mean   #2nd percentile   ^98th percentile

Medium	Stream water	Stream water	Stream water	Lake water	Ground water
	Romania	Eastern India	Harz, Germany	Norway	Southern Norway
	unfiltered ICP-MS	<0.2 µm ICP-MS	unfiltered ICP-MS	unfiltered ICP-MS	unfiltered ICP-MS
Median	-	0.000,12	0.000,02	0.000,003	0.000,004,5
Min	-	0.000,02	<0.000,002	<0.000,002	<0.000,001
Max	-	0.001,96	0.000,24	0.000,07	0.000,36

Ho — HOLMIUM

Concentrations in Precipitation (mg/l)

Medium	Rain water Kola, remote <0.45 µm	Rain water Kola, polluted <0.45 µm	Snow melt-water Kola, remote <0.45 µm	Snow melt-water Kola, polluted <0.45 µm	Snow filter residue Kola, remote >0.45 µm	Snow filter residue Kola, polluted >0.45 µm
Median	-	-	-	-	-	-
Min	-	-	-	-	-	-
Max	-	-	-	-	-	-

Medium	Rain water Coastal, Norway unfiltered ICP-MS	Rain water Inland, Norway unfiltered ICP-MS				
Median	-	-				
Min	-	-				
Max	-	-				

Concentrations in Air (ng/m^3) and Yearly Deposition (kg/km^2/yr)

Medium	Air World, remote	Air World, polluted	Bulk deposition West Germany	Throughfall deposition West Germany	Bulk deposition Kola, remote	Bulk deposition Kola, polluted
Median	-	-	-	-	-	-
Min	-	-	-	-	-	-
Max	-	-	-	-	-	-

*Estimated value

Concentrations in Plants (mg/kg)

Medium	Moss Norway conc. HNO$_3$ (1990)	Moss Germany conc. HNO$_3$ + H$_2$O$_2$ (1995)	Crustose lichen Germany conc. HNO$_3$	Lichen Northwest Territories oven-dried	Dandelion Europe oven-dried	Spruce bark Canada ashed
Median	-	-	-	-	-	-
Min	-	-	-	-	-	-
Max	-	-	-	-	-	-

*Mean   #Geometric mean

Concentrations in Human Fluids (mg/l)

Medium	Human blood Lombardy, Italy	Human serum Lombardy, Italy	Human urine Lombardy, Italy			
Mean	-	-	-			
Min	-	-	-			
Max	-	-	-			

HOLMIUM   Ho

Environmental Geochemistry

Biological impacts	Considered non-essential. Data to assess the toxicity of REE are scarce
Uses	Electric materials
Environmental pathways	Poorly understood. Probably mostly windblown dust, weathering of REE-bearing minerals. Generally, geogenic sources more important than anthropogenic ones
Environmental mobility	Oxidising conditions: very low Acid conditions: very low Reducing conditions: very low Neutral to alkaline conditions: very low Comments: -
Geochemical barriers	Mechanical
Natural association	REE-Li-Rb-Cs-Be-Nb-Ta-Zr-B-Th-U-F (pegmatites), REE-Th-P-Zr-Fe-Cu (monazite veins), REE-Th-Ba-Sr-P-F-N-C (carbonatites), REE-U-P-F (phosphorites), REE-Au-Ti-Sn-Sr-Th (placers)
Action levels	-
Remarks	Toxicity of REE decreases as atomic number increases. Inhaled REE probably cause pneumoconiosis. REE taken up orally (e.g., via drinking water) accumulate in the skeleton, teeth, lungs, liver and kidneys. Unusual magnetic properties. Little used
Suggested analytical method(s)	ICP-MS

World Yearly Production (t/y)	54,000 REE-minerals (in 1991)
Price of Purest Form, in Small Quantities ($/kg)	128,013 (99.99%)
Market Price ($/kg)	-

IODINE

Physico-Chemical Properties

Atomic number	53	Melting/boiling point (K)	386.85 / 457.95
Atomic mass	126.904,47(3)	Isotopes & isomers	1 stable + 41 unstable
Atomic radius (pm)	132	Acid/base of oxide	strong acid
Main oxidation state(s)	-1 (0,+1,+3,+4,+5,+7)	State (at 300 K, 1 atm.)	solid
Ionic radius (pm)	206 (-1), 58-109 (+5), 56-67 (+7)	Metallic character	pred. non-metal
		Element group(s)	halogen, heavy non-metal
Electronegativity (Pauling)	2.66		
Density (g/cm^3)	4.93	Affinity	atmophile, lithophile

Naturally occurring isotopes	Natural abundance (%)	Atomic mass	Half-life	Unstable isotopes	Longest half-life
^{127}I	100	126.904,47	stable	^{108}I to ^{142}I	17,000,000 y

Concentrations in Crust/Rocks (mg/kg)

Bulk continental crust	0.8 / 0.45 / -
Upper continental crust	1.4 / 0.15
Ultramafic rock	0.05
Ocean ridge basalt	-
Gabbro, basalt	0.11
Granite, granodiorite	0.17
Sandstone	0.01
Greywacke	-
Shale, schist	1.5
Limestone	1
Coal	5

Typical Minerals
lautarite (Ca(IO$_3$)$_2$), marshite (CuI), iodyrite (AgI), iodobromite (Ag(Cl,Br,I))

Possible Host Minerals
apatite, eudialyte, sodalite, hypersthene

Mass (kg) in

Continental crust	1.89x10^{+16}	Oceans	7.93x10^{+13}	Plants	5.52x10^{+9}

Concentrations in Media Sampled for the Kola Project (mg/kg)

Medium	Moss conc. HNO$_3$	Humus <2 mm	Humus <2 mm amm. acet.	Topsoil (0-5 cm) <2 mm aqua regia	B horizon <2 mm	B horizon <2 mm total (XRF)
Median	-	-	-	-	-	-
Min	-	-	-	-	-	-
Max	-	-	-	-	-	-

Medium	C horizon <2 mm aqua regia	C horizon <2 mm total (XRF)	C horizon <2 mm total (INAA)		Lake water unfiltered (mg/l)
Median	-	-	-		-
Min	-	-	-		-
Max	-	-	-		-

IODINE   I

Concentrations in Soils and Sediments (mg/kg)

Medium	Soil	Agricultural soil - Ap horizon	Agricultural soil - Top (0-25 cm)	Agricultural Soil - Bottom (50-75 cm)	Topsoil (0-15 cm)	Urban soil (0-2 cm)
	World	*Canada*	*Finland*	*Finland*	*England & Wales*	*Trondheim*
	<2 mm total	*<2 mm*	*<2 mm aqua regia*	*<2 mm aqua regia*	*<2 mm aqua regia*	*<2 mm aqua regia*
Median	2*	-	-	-	-	-
Min	-	-	-	-	-	-
Max	-	-	-	-	-	-

*Estimated mean

Medium	Forest soil - Humus	Forest soil - B horizon	Forest soil - C horizon	Till (C horizon)	Till (C horizon)	Laterite (25 ±15 cm)
	Norway	*Norway*	*Norway*	*Finland*	*Finland*	*Australia*
	<2 mm	*<2 mm*	*<2 mm*	*<0.063 mm*	*<0.063 mm*	*0.45-2 mm*
	7N HNO$_3$	*7N HNO$_3$*	*7N HNO$_3$*	*aqua regia*	*total*	*total*
Median	-	-	-	-	-	-
Min	-	-	-	-	-	-
Max	-	-	-	-	-	-

Medium	Stream sediment	Stream sediment	Stream sediment	Overbank sediment	Overbank sediment	Organic stream sediment
	Austria	*Southern Scotland*	*Harz, Germany*	*Norway*	*Norway*	*Finland*
	<0.18 mm	*<0.10 mm*	*<0.063 mm*	*<0.063 mm*	*<0.063 mm*	*<2 mm*
		total	*total*	*total (XRF)*	*7N HNO$_3$*	*Conc. HNO$_3$*
Median	-	-	-	-	-	-
Min	-	-	-	-	-	-
Max	-	-	-	-	-	-

#2nd percentile   ^98th percentile

Concentrations in Waters (mg/l)

Medium	Ocean water	Ocean water	Ocean water	Stream water	Stream water	Stream water
	World (1)	*World (2)*	*North Pacific*	*World*	*Nova Scotia, Canada*	*Finland*
					<0.45 µm	*<0.45 µm*
Median	0.006*	0.006*	0.058,004*	0.007*	-	-
Min	-	-	-	-	-	-
Max	-	-	-	-	-	-

*Estimated mean   #2nd percentile   ^98th percentile

Medium	Stream water	Stream water	Stream water	Lake water	Ground water	
	Romania	*Eastern India*	*Harz, Germany*	*Norway*	*Southern Norway*	
	unfiltered	*<0.2 µm*	*unfiltered*	*unfiltered*	*unfiltered*	
	ICP-MS	*ICP-MS*		*ICP-MS*	*ICP-MS*	
Median	0.004,1	0.025,9	-	-	0.002,025	
Min	0.000,1	0.014,2	-	-	0.000,31	
Max	0.198	0.110,5	-	-	0.3	

I   IODINE

Concentrations in Precipitation (mg/l)

Medium	Rain water Kola, remote <0.45 μm	Rain water Kola, polluted <0.45 μm	Snow melt-water Kola, remote <0.45 μm	Snow melt-water Kola, polluted <0.45 μm	Snow filter residue Kola, remote >0.45 μm	Snow filter residue Kola, polluted >0.45 μm
Median	-	-	-	-	-	-
Min	-	-	-	-	-	-
Max	-	-	-	-	-	-

Medium	Rain water Coastal, Norway unfiltered ICP-MS	Rain water Inland, Norway unfiltered ICP-MS				
Median	-	-				
Min	-	-				
Max	-	-				

Concentrations in Air (ng/m^3) and Yearly Deposition (kg/km^2/yr)

Medium	Air World, remote	Air World, polluted	Bulk deposition West Germany	Throughfall deposition West Germany	Bulk deposition Kola, remote	Bulk deposition Kola, polluted
Median	- / 1.1	13	-	-	-	-
Min	0.08	0.4	-	-	-	-
Max	4.4	60	-	-	-	-

*Estimated value

Concentrations in Plants (mg/kg)

Medium	Moss Norway conc. HNO$_3$ (1985)	Moss Germany conc. HNO$_3$ + H$_2$O$_2$ (1995)	Crustose lichen Germany conc. HNO$_3$	Lichen Northwest Territories oven-dried	Dandelion Europe oven-dried	Spruce bark Canada ashed
Median	2.7	-	-	-	-	-
Min	0.6	-	-	-	-	-
Max	47	-	-	-	-	-

*Mean   #Geometric mean

Concentrations in Human Fluids (mg/l)

Medium	Human blood Lombardy, Italy	Human serum Lombardy, Italy	Human urine Lombardy, Italy			
Mean	-	-	-			
Min	-	-	-			
Max	-	-	-			

IODINE — I

Environmental Geochemistry

Biological impacts	Essential for many organisms
Uses	Catalysts, stabilising agent, photography, medicine, pigment, often added to dietary salt
Environmental pathways	Natural sources most important, nuclear bomb testing and accidents
Environmental mobility	Oxidising conditions: very high — Acid conditions: very high Reducing conditions: very high — Neutral to alkaline conditions: very high Comments: -
Geochemical barriers	-
Natural association	Li-Na-K-B-P-W-F-Br-Cl-I-SO$_4$-CO$_3$ (brines, evaporites)
Action levels	-
Remarks	Deficiency of I causes goitre. Daily dietary intake suggested: 0.15-0.2 mg. Some mineral waters can contain up to several mg/l. Dangerous, radiogenic I-isotopes are released to the environment by nuclear bomb testing and nuclear accidents. I allergies are widespread. Rarely analysed for despite its importance to human health
Suggested analytical method(s)	Ion selective electrode, ICP-MS

World Yearly Production (t/y)	13,757 (in 1995)
Price of Purest Form, in Small Quantities ($/kg)	4663 (99.999,5%)
Market Price ($/kg)	12-16 (crude iodine crystal, min. 99.5%)

In   INDIUM

Physico-Chemical Properties

Atomic number	49	Melting/boiling point (K)	429.75 / 2345
Atomic mass	114.818(3)	Isotopes & isomers	1 stable + 65 unstable
Atomic radius (pm)	200	Acid/base of oxide	amphoteric
Main oxidation state(s)	0,+3 (+1,+2)	State (at 300 K, 1 atm.)	solid
Ionic radius (pm)	76-106 (+3)	Metallic character	pred. metal
Electronegativity (Pauling)	1.78	Element group(s)	heavy metal
Density (g/cm^3)	7.31	Affinity	chalcophile

Naturally occurring isotopes	Natural abundance (%)	Atomic mass	Half-life
^{113}In	4.29	112.904,06	stable
^{115}In	95.71	114.903,38	4.4x10^{+14} y

Unstable isotopes	Longest half-life
^{99}In to ^{133}In	4.4x10^{+14} y

Concentrations in Crust/Rocks (mg/kg)

Bulk continental crust	0.05 / 0.25 / 0.05
Upper continental crust	0.061 / 0.05
Ultramafic rock	0.02
Ocean ridge basalt	0.07
Gabbro, basalt	0.07
Granite, granodiorite	0.05
Sandstone	0.02
Greywacke	-
Shale, schist	0.07
Limestone	0.03
Coal	0.1

Typical Minerals
roquésite (CuInS$_2$), indite (FeInS$_4$), native In

Possible Host Minerals
tourmaline, biotite, amphiboles, pyroxenes, stannite, chalcopyrite, sphalerite

Mass (kg) in

Continental crust 1.18x10^{+15}	Oceans 1.32x10^{+8}	Plants 1.84x10^{+6}

Concentrations in Media Sampled for the Kola Project (mg/kg)

Medium	Moss conc. HNO$_3$	Humus <2 mm	Humus <2 mm amm. acet.	Topsoil (0-5 cm) <2 mm	B horizon <2 mm aqua regia	B horizon <2 mm total (XRF)
Median	-	-	-	-	-	-
Min	-	-	-	-	-	-
Max	-	-	-	-	-	-

Medium	C horizon <2 mm aqua regia	C horizon <2 mm total (XRF)	C horizon <2 mm total (INAA)		Lake water unfiltered (mg/l)
Median	-	-	-		-
Min	-	-	-		-
Max	-	-	-		-

INDIUM — In

Concentrations in Soils and Sediments (mg/kg)

Medium	Soil	Agricultural soil -	Agricultural soil - Top	Agricultural Soil - Bottom	Topsoil	Urban soil
	World	Ap horizon Canada	(0-25 cm) Finland	(50-75 cm) Finland	(0-15 cm) England & Wales	(0-2 cm) Trondheim
	<2 mm total	<2 mm	<2 mm aqua regia	<2 mm aqua regia	<2 mm aqua regia	<2 mm aqua regia
Median	0.07*	-	-	-	-	-
Min	-	-	-	-	-	-
Max	-	-	-	-	-	-

*Estimated mean

Medium	Forest soil - Humus	Forest soil - B horizon	Forest soil - C horizon	Till (C horizon)	Till (C horizon)	Laterite
	Norway	Norway	Norway	Finland	Finland	(25 ±15 cm) Australia
	<2 mm 7N HNO$_3$	<2 mm 7N HNO$_3$	<2 mm 7N HNO$_3$	<0.063 mm aqua regia	<0.063 mm total	0.45-2 mm total
Median	-	-	-	-	-	<0.05
Min	-	-	-	-	-	<0.05
Max	-	-	-	-	-	2

Medium	Stream sediment	Stream sediment	Stream sediment	Overbank sediment	Overbank sediment	Organic stream sediment
	Austria	Southern Scotland	Harz, Germany	Norway	Norway	Finland
	<0.18 mm	<0.10 mm total	<0.063 mm total	<0.063 mm total (XRF)	<0.063 mm 7N HNO$_3$	<2 mm Conc. HNO$_3$
Median	-	-	-	-	-	-
Min	-	-	-	-	-	-
Max	-	-	-	-	-	-

#2nd percentile   ^98th percentile

Concentrations in Waters (mg/l)

Medium	Ocean water	Ocean water	Ocean water	Stream water	Stream water	Stream water
	World (1)	World (2)	North Pacific	World	Nova Scotia, Canada	Finland
					<0.45 µm ICP-MS	<0.45 µm
Median	0.02* (?)	0.000,000,1*	0.000,000,01*	-	<0.000,01	-
Min	-	-	-	-	<0.000,01	-
Max	-	-	-	-	<0.000,01	-

*Estimated mean   #2nd percentile   ^98th percentile   (?) probably unit error in source

Medium	Stream water	Stream water	Stream water	Lake water	Ground water
	Romania	Eastern India	Harz, Germany	Norway	Southern Norway
	unfiltered ICP-MS	<0.2 µm ICP-MS	unfiltered	unfiltered ICP-MS	unfiltered ICP-MS
Median	0.000,002	<0.000,1	-	-	<0.000,001
Min	0.000,001	<0.000,1	-	-	<0.000,001
Max	0.000,12	0.000,16	-	-	0.000,02

INDIUM

Concentrations in Precipitation (mg/l)

Medium	Rain water	Rain water	Snow melt-water	Snow melt-water	Snow filter residue	Snow filter residue
	Kola, remote <0.45 µm	*Kola, polluted <0.45 µm*	*Kola, remote <0.45 µm*	*Kola, polluted <0.45 µm*	*Kola, remote >0.45 µm*	*Kola, polluted >0.45 µm*
Median	-	-	-	-	-	-
Min	-	-	-	-	-	-
Max	-	-	-	-	-	-

Medium	Rain water	Rain water				
	Coastal, Norway unfiltered ICP-MS	*Inland, Norway unfiltered ICP-MS*				
Median	-	-				
Min	-	-				
Max	-	-				

Concentrations in Air (ng/m^3) and Yearly Deposition (kg/km^2/yr)

Medium	Air	Air	Bulk deposition	Throughfall deposition	Bulk deposition	Bulk deposition
	World, remote	*World, polluted*	*West Germany*	*West Germany*	*Kola, remote*	*Kola, polluted*
Median	-	-	-	-	-	-
Min	-	-	-	-	-	-
Max	-	-	-	-	-	-

*Estimated value

Concentrations in Plants (mg/kg)

Medium	Moss	Moss	Crustose lichen	Lichen	Dandelion	Spruce bark
	Norway conc. HNO$_3$ (1990)	*Germany conc. HNO$_3$ + H$_2$O$_2$ (1995)*	*Germany conc. HNO$_3$*	*Northwest Territories oven-dried*	*Europe oven-dried*	*Canada ashed*
Median	-	0.002	-	-	-	-
Min	-	<0.000,1	-	-	-	-
Max	-	0.03	-	-	-	-

*Mean   #Geometric mean

Concentrations in Human Fluids (mg/l)

Medium	Human blood	Human serum	Human urine			
	Lombardy, Italy	*Lombardy, Italy*	*Lombardy, Italy*			
Mean	-	-	<0.000,15			
Min	-	-	-			
Max	-	-	-			

INDIUM   In

Environmental Geochemistry

Biological impacts	Considered non-essential. Baseline data to assess toxicity are lacking
Uses	Solder alloys, semi-conductors, solar cells, medicine
Environmental pathways	Zn smelters
Environmental mobility	Oxidising conditions: - Acid conditions: - Reducing conditions: - Neutral to alkaline conditions: - Comments: -
Geochemical barriers	-
Natural association	Ag-Zn-Cd-Cu-Ba-Sr-V-Cr-Mn-Fe-Ga-In-Ta-Ge-Sn-As-Sb-Bi-Se-Hg-Te-Pb (sulphide deposits)
Action levels	-
Remarks	Mostly a by-product of Zn smelting. In is a common trace constituent of sphalerite
Suggested analytical method(s)	INAA, ICP-MS

World Yearly Production (t/y)	230 (in 1995)
Price of Purest Form, in Small Quantities ($/kg)	4613 (99.999,5%)
Market Price ($/kg)	180

Ir   IRIDIUM

Physico-Chemical Properties

Atomic number	77	Melting/boiling point (K)	2719 / 4701
Atomic mass	192.217(3)	Isotopes & isomers	2 stable + 42 unstable
Atomic radius (pm)	187	Acid/base of oxide	base
Main oxidation state(s)	+4 (+2,+3,+6)	State (at 300 K, 1 atm.)	solid
Ionic radius (pm)	82 (+3), 76.5 (+4), 71 (+5)	Metallic character	pred. metal
Electronegativity (Pauling)	2.2	Element group(s)	heavy metal, noble metal, PGE
Density (g/cm^3)	22.42	Affinity	siderophile

Naturally occurring isotopes	Natural abundance (%)	Atomic mass	Half-life	Unstable isotopes	Longest half-life
^{191}Ir	37.3	190.960,59	stable	^{166}Ir to ^{198}Ir	241 y
^{193}Ir	62.7	192.962,92	stable		

Concentrations in Crust/Rocks (mg/kg)

Bulk continental crust	0.000,05 / 0.001 / 0.000,1
Upper continental crust	- / 0.000,02
Ultramafic rock	-
Ocean ridge basalt	-
Gabbro, basalt	-
Granite, granodiorite	-
Sandstone	-
Greywacke	0.000,05
Shale, schist	-
Limestone	-
Coal	<0.2

Typical Minerals
irarsite ((Ir,Ru,Rh,Pt)AsS),
hollingworthite ((Rh,Pd,Pt,Ir)AsS)

Possible Host Minerals
olivine, ilmenite, zircon, chromite, gadolinite

Mass (kg) in

Continental crust	1.18x10^{+12}	Oceans	-	Plants	1.84x10^{+5}

Concentrations in Media Sampled for the Kola Project (mg/kg)

Medium	Moss conc. HNO$_3$	Humus <2 mm 	Humus <2 mm amm. acet.	Topsoil (0-5 cm) <2 mm total (INAA)	B horizon <2 mm aqua regia	B horizon <2 mm total (XRF)
Median	-	-	-	<0.005	-	-
Min	-	-	-	<0.005	-	-
Max	-	-	-	<0.005	-	-

Medium	C horizon <2 mm aqua regia	C horizon <2 mm total (XRF)	C horizon <2 mm total (INAA)			Lake water unfiltered (mg/l)
Median	-	-	<0.005			-
Min	-	-	<0.005			-
Max	-	-	<0.005			-

IRIDIUM   Ir

Concentrations in Soils and Sediments (mg/kg)

Medium	Soil	Agricultural soil - Ap horizon	Agricultural soil - Top (0-25 cm)	Agricultural Soil - Bottom (50-75 cm)	Topsoil (0-15 cm)	Urban soil (0-2 cm)
	World	Canada	Finland	Finland	England & Wales	Trondheim
	<2 mm	<2 mm	<2 mm	<2 mm	<2 mm	<2 mm
	total	aqua regia	aqua regia	aqua regia	aqua regia	aqua regia
Median	-	-	-	-	-	-
Min	-	-	-	-	-	-
Max	-	-	-	-	-	-

*Estimated mean

Medium	Forest soil - Humus	Forest soil - B horizon	Forest soil - C horizon	Till (C horizon)	Till (C horizon)	Laterite (25 ±15 cm)
	Norway	Norway	Norway	Finland	Finland	Australia
	<2 mm	<2 mm	<2 mm	<0.063 mm	<0.063 mm	0.45-2 mm
	7N HNO$_3$	7N HNO$_3$	7N HNO$_3$	aqua regia	total	total
Median	-	-	-	-	-	-
Min	-	-	-	-	-	-
Max	-	-	-	-	-	-

Medium	Stream sediment	Stream sediment	Stream sediment	Overbank sediment	Overbank sediment	Organic stream sediment
	Austria	Southern Scotland	Harz, Germany	Norway	Norway	Finland
	<0.18 mm	<0.10 mm	<0.063 mm	<0.063 mm	<0.063 mm	<2 mm
		total	total	total (XRF)	7N HNO$_3$	Conc. HNO$_3$
Median	-	-	-	-	-	-
Min	-	-	-	-	-	-
Max	-	-	-	-	-	-

#2nd percentile   ^98th percentile

Concentrations in Waters (mg/l)

Medium	Ocean water	Ocean water	Ocean water	Stream water	Stream water	Stream water
	World (1)	World (2)	North Pacific	World	Nova Scotia, Canada	Finland
					<0.45 µm	<0.45 µm
Median	-	-	1.3x10^{-10}*	-	-	-
Min	-	-	-	-	-	-
Max	-	-	-	-	-	-

*Estimated mean   #2nd percentile   ^98th percentile

Medium	Stream water	Stream water	Stream water	Lake water	Ground water
	Romania	Eastern India	Harz, Germany	Norway	Southern Norway
	unfiltered	<0.2 µm	unfiltered	unfiltered	unfiltered
	ICP-MS	ICP-MS		ICP-MS	
Median	-	<0.000,1	-	<0.000,006	-
Min	-	<0.000,1	-	<0.000,006	-
Max	-	<0.000,1	-	0.000,011	-

Ir   IRIDIUM

Concentrations in Precipitation (mg/l)

Medium	Rain water Kola, remote <0.45 μm	Rain water Kola, polluted <0.45 μm	Snow melt- water Kola, remote <0.45 μm	Snow melt- water Kola, polluted <0.45 μm	Snow filter residue Kola, remote >0.45 μm	Snow filter residue Kola, polluted >0.45 μm
Median	-	-	-	-	-	-
Min	-	-	-	-	-	-
Max	-	-	-	-	-	-

Medium	Rain water Coastal, Norway unfiltered ICP-MS	Rain water Inland, Norway unfiltered ICP-MS				
Median	-	-				
Min	-	-				
Max	-	-				

Concentrations in Air (ng/m^3) and Yearly Deposition (kg/km^2/yr)

Medium	Air World, remote	Air World, polluted	Bulk deposition West Germany	Throughfall deposition West Germany	Bulk deposition Kola, remote	Bulk deposition Kola, polluted
Median	-	-	-	-	-	-
Min	-	-	-	-	-	-
Max	-	-	-	-	-	-

*Estimated value

Concentrations in Plants (mg/kg)

Medium	Moss Norway conc. HNO$_3$ (1990)	Moss Germany conc. HNO$_3$ + H$_2$O$_2$ (1995)	Crustose lichen Germany conc. HNO$_3$	Lichen Northwest Territories oven-dried	Dandelion Europe oven-dried	Spruce bark Canada ashed
Median	-	-	-	-	-	-
Min	-	-	-	-	-	-
Max	-	-	-	-	-	-

*Mean #Geometric mean

Concentrations in Human Fluids (mg/l)

Medium	Human blood Lombardy, Italy	Human serum Lombardy, Italy	Human urine Lombardy, Italy			
Mean	0.000,007,4	0.000,005	0.000,018			
Min	0.000,000,2	0.000,000,2	0.000,000,7			
Max	0.000,035	0.000,04	0.000,07			

IRIDIUM Ir

Environmental Geochemistry

Biological impacts	Considered non-essential
Uses	Special Pt alloys, ink pen tips
Environmental pathways	Poorly understood
Environmental mobility	Oxidising conditions: very low Acid conditions: very low Reducing conditions: very low Neutral to alkaline conditions: very low Comments: -
Geochemical barriers	Mechanical
Natural association	PGE-Ni-Cu-Co-As-Ag-Au-Te-Bi-Sn-Sb (massive Ni-Cu sulphide ores, e.g., Sudbury), PGE-Ag-Au-Cr-Fe-Cu-Ni-Co-S (sulphides, e.g., Merensky reef)
Action levels	-
Remarks	Very rare. Cosmic input (cosmic dust, meteorite impacts) significant. Ir anomalies in geological record suggested to be caused by meteorite impact 65 million years ago at Cretaceous/Tertiary boundary. Very rich ore deposits contain about 8 ppm PGE. PGE occur often in Ni ores. PGE usually occur together in the following proportions: Pt 25%, Ru 23%, Os 21%, Pd 18%, Ir 7%, Rh 6%. Some marine algae enrich PGE
Suggested analytical method(s)	ICP-MS

World Yearly Production (t/y)	300 all PGE (in 1995)
Price of Purest Form, in Small Quantities ($/kg)	167,125
Market Price ($/kg)	4340

K   POTASSIUM

Physico-Chemical Properties

Atomic number	19	Melting/boiling point (K)	336.53 / 1032
Atomic mass	39.098,3(1)	Isotopes & isomers	2 stable + 19 unstable
Atomic radius (pm)	277	Acid/base of oxide	strong base
Main oxidation state(s)	+1	State (at 300 K, 1 atm.)	solid
Ionic radius (pm)	151-173	Metallic character	metal
Electronegativity (Pauling)	0.82	Element group(s)	alkali metal, light metal
Density (g/cm^3)	0.862	Affinity	lithophile

Naturally occurring isotopes	Natural abundance (%)	Atomic mass	Half-life
^{39}K	93.258	38.963,71	stable
^{40}K	0.011,7	39.963,999	1.26x10^{+9} y
^{41}K	6.73	40.961,83	stable

Unstable isotopes	Longest half-life
^{35}K to ^{54}K	1.26x10^{+9} y

Concentrations in Crust/Rocks (mg/kg)

Bulk continental crust	21,400 / 20,900 / 9100
Upper continental crust	28,650 / 28,000
Ultramafic rock	5000
Ocean ridge basalt	2000
Gabbro, basalt	8000
Granite, granodiorite	33,000
Sandstone	11,000
Greywacke	16,604
Shale, schist	27,000
Limestone	3000
Coal	3500

Typical Minerals

sylvite (KCl), carnallite (KMgCl$_3$.6H$_2$O), kainite (MgSO$_4$.KCl.3H$_2$O) and major rock forming minerals, e.g., K-feldspar, mica

Possible Host Minerals

-

Mass (kg) in

Continental crust	5.05x10^{+20}	Oceans	5.02x10^{+17}	Plants	3.50x10^{+13}

Concentrations in Media Sampled for the Kola Project (mg/kg)

Medium	Moss conc. HNO$_3$	Humus <2 mm conc. HNO$_3$	Humus <2 mm amm. acet.	Topsoil (0-5 cm) <2 mm	B horizon <2 mm aqua regia	B horizon <2 mm total (XRF)
Median	4220	1000	676	-	630	12,453
Min	2260	300	102	-	<200	4234
Max	8590	5700	1480	-	10,400	48,650

Medium	C horizon <2 mm aqua regia	C horizon <2 mm total (XRF)	C horizon <2 mm total (INAA)		Lake water unfiltered (mg/l)
Median	1100	14,100	-		0.37
Min	<200	3600	-		0.08
Max	11,000	52,400	-		1.77

POTASSIUM K

Concentrations in Soils and Sediments (mg/kg)

Medium	Soil	Agricultural soil - Ap horizon	Agricultural soil - Top (0-25 cm)	Agricultural Soil - Bottom (50-75 cm)	Topsoil (0-15 cm)	Urban soil (0-2 cm)
	World	Canada	Finland	Finland	England & Wales	Trondheim
	<2 mm	<2 mm	<2 mm	<2 mm	<2 mm	<2 mm
	total		aqua regia	aqua regia	aqua regia	aqua regia
Median	14,000*	-	310	464	4626	2300
Min	-	-	<50	<50	60	400
Max	-	-	6030	9440	20,444	11,100

*Estimated mean

Medium	Forest soil - Humus	Forest soil - B horizon	Forest soil - C horizon	Till (C horizon)	Till (C horizon)	Laterite (25 ±15 cm)
	Norway	Norway	Norway	Finland	Finland	Australia
	<2 mm	<2 mm	<2 mm	<0.063 mm	<0.063 mm	0.45-2 mm
	7N HNO$_3$	7N HNO$_3$	7N HNO$_3$	aqua regia	total	total
Median	810	900	2000	1700	21,000	7347
Min	140	70	50	-	-	<80
Max	3900	17,500	21,400	-	-	51,057

Medium	Stream sediment	Stream sediment	Stream sediment	Overbank sediment	Overbank sediment	Organic stream sediment
	Austria	Southern Scotland	Harz, Germany	Norway	Norway	Finland
	<0.18 mm	<0.10 mm	<0.063 mm	<0.063 mm	<0.063 mm	<2 mm
	total (XRF)	total (DCES)	total (XRF)	total (XRF)	7N HNO$_3$ ICP-AES	Conc. HNO$_3$
Median	23,100	20,755	16,272	18,000	1400	1600
Min	-	<830	2657	1000	250	400#
Max	71,100	77,208	34,868	43,300	11,100	8500^

#2nd percentile   ^98th percentile

Concentrations in Waters (mg/l)

Medium	Ocean water	Ocean water	Ocean water	Stream water	Stream water	Stream water
	World (1)	World (2)	North Pacific	World	Nova Scotia, Canada	Finland
					<0.45 µm	<0.45 µm
					AAS	ICP-MS
Median	399*	380*	399*	2.3*	0.18	0.7
Min	-	-	-	-	<0.1	0.14#
Max	-	-	-	-	3.22	5.83^

*Estimated mean   #2nd percentile   ^98th percentile

Medium	Stream water	Stream water	Stream water	Lake water	Ground water
	Romania	Eastern India	Harz, Germany	Norway	Southern Norway
	unfiltered	<0.2 µm	unfiltered	unfiltered	unfiltered
	ICP-MS	ICP-MS	ICP-OES	ICP-MS	ICP-MS
Median	3.5	-	0.7	-	2.175
Min	0.1	-	0.1	-	<0.01
Max	90	-	5	-	24.27

K POTASSIUM

Concentrations in Precipitation (mg/l)

Medium	Rain water	Rain water	Snow melt-water	Snow melt-water	Snow filter residue	Snow filter residue
	Kola, remote <0.45 µm ICP-MS	Kola, polluted <0.45 µm ICP-MS	Kola, remote <0.45 µm ICP-MS	Kola, polluted <0.45 µm ICP-MS	Kola, remote >0.45 µm ICP-AES	Kola, polluted >0.45 µm ICP-AES
Median	0.05	0.09	0.02	0.085	0.01	0.055
Min	0.02	0.02	0.02	0.06	<0.01	0.03
Max	0.26	0.36	0.05	0.13	0.02	0.08

Medium	Rain water	Rain water				
	Coastal, Norway unfiltered ICP-MS	Inland, Norway unfiltered ICP-MS				
Median	-	-				
Min	-	-				
Max	-	-				

Concentrations in Air (ng/m^3) and Yearly Deposition (kg/km^2/yr)

Medium	Air	Air	Bulk deposition	Throughfall deposition	Bulk deposition	Bulk deposition
	World, remote	World, polluted	West Germany	West Germany	Kola, remote	Kola, polluted
Median	50 / 90	4300	-	-	-	-
Min	0.6	140	190	12,300	-	-
Max	170 (530)	40,000	1400	28,600	-	-

*Estimated value

Concentrations in Plants (mg/kg)

Medium	Moss	Moss	Crustose lichen	Lichen	Dandelion	Spruce bark
	Norway conc. HNO$_3$ (1990)	Germany conc. HNO$_3$ + H$_2$O$_2$ (1995)	Germany conc. HNO$_3$	Northwest Territories oven-dried total (INAA)	Europe oven-dried total (INAA)	Canada ashed total (INAA)
Median	-	9055	-	1660.7*	32,200#	29,400
Min	-	1380	-	-	-	10,600
Max	-	24,100	-	-	-	76,200

*Mean   #Geometric mean

Concentrations in Human Fluids (mg/l)

Medium	Human blood	Human serum	Human urine			
	Lombardy, Italy	Lombardy, Italy	Lombardy, Italy			
Mean	-	-	-			
Min	-	-	-			
Max	-	-	-			

POTASSIUM K

Environmental Geochemistry

Biological impacts	Essential for all organisms. Considered non-toxic, but concentrated K-salts kill plants
Uses	Fertilisers, alloys, chemical industry, gun powder (K-nitrate)
Environmental pathways	Fertilisers, windblown geogenic dust, sea spray. Generally, natural sources more important than anthropogenic ones
Environmental mobility	Oxidising conditions: low Acid conditions: low Reducing conditions: low Neutral to alkaline conditions: low Comments: -
Geochemical barriers	-
Natural association	Ba-Rb-K (in many minerals)
Action levels	Drinking water, MAC: 12 mg/l (Norway Shd)
Remarks	Important for functioning of nervous system. Enriched in plants relative to soil, making K fertilisation of agricultural soils a necessity. Many K compounds used as fertilisers, e.g., KCl, K_2SO_4 (for Cl-intolerant plants, e.g., potato, fruits), K_2CO_3, KNO_3
Suggested analytical method(s)	XRF, ICP-AES

World Yearly Production (t/y)	20,400,000 (in 1995, from 24,584,000 K_2O)
Price of Purest Form, in Small Quantities ($/kg)	107,000 (99.95%)
Market Price ($/kg)	0.17 (muriate of potash, bulk, 60% K_2O)

Kr KRYPTON

Physico-Chemical Properties

Atomic number	36	Melting/boiling point (K)	115.79 / 119.93
Atomic mass	83.80(1)	Isotopes & isomers	6 stable + 24 unstable
Atomic radius (pm)	103	Acid/base of oxide	-
Main oxidation state(s)	0	State (at 300 K, 1 atm.)	gas
Ionic radius (pm)	-	Metallic character	-
Electronegativity (Pauling)	-	Element group(s)	noble gas
Density (g/cm^3)	0.003,733	Affinity	atmophile

Naturally occurring isotopes	Natural abundance (%)	Atomic mass	Half-life
^{78}Kr	0.35	77.920,39	stable
^{80}Kr	2.25	79.916,38	stable
^{82}Kr	11.6	81.913,49	stable
^{83}Kr	11.5	82.914,14	stable
^{84}Kr	57	83.911,51	stable
^{86}Kr	17.3	85.910,62	stable

Unstable isotopes	Longest half-life
^{71}Kr to ^{97}Kr	210,000 y

Concentrations in Crust/Rocks (mg/kg)

Bulk continental crust	- / 0.000,1 / -
Upper continental crust	- / -
Ultramafic rock	-
Ocean ridge basalt	-
Gabbro, basalt	-
Granite, granodiorite	-
Sandstone	-
Greywacke	-
Shale, schist	-
Limestone	-
Coal	-

Typical Minerals
-

Possible Host Minerals
-

Mass (kg) in

Continental crust	2.36x10^{+12}	Oceans	2.64x10^{+11}	Plants	-

Concentrations in Media Sampled for the Kola Project (mg/kg)

Medium	Moss	Humus <2 mm amm. acet.	Humus <2 mm	Topsoil (0-5 cm) <2 mm	B horizon <2 mm aqua regia	B horizon <2 mm total (XRF)
	conc. HNO$_3$	<2 mm				
Median	-	-	-	-	-	-
Min	-	-	-	-	-	-
Max	-	-	-	-	-	-

Medium	C horizon <2 mm aqua regia	C horizon <2 mm total (XRF)	C horizon <2 mm total (INAA)			Lake water unfiltered (mg/l)
Median	-	-	-			-
Min	-	-	-			-
Max	-	-	-			-

KRYPTON   Kr

Concentrations in Soils and Sediments (mg/kg)

Medium	Soil	Agricultural soil - Ap horizon	Agricultural soil - Top (0-25 cm)	Agricultural Soil - Bottom (50-75 cm)	Topsoil (0-15 cm)	Urban soil (0-2 cm)
	World	Canada	Finland	Finland	England & Wales	Trondheim
	<2 mm	<2 mm	<2 mm	<2 mm	<2 mm	<2 mm
	total		aqua regia	aqua regia	aqua regia	aqua regia
Median	-	-	-	-	-	-
Min	-	-	-	-	-	-
Max	-	-	-	-	-	-

*Estimated mean

Medium	Forest soil - Humus	Forest soil - B horizon	Forest soil - C horizon	Till (C horizon)	Till (C horizon)	Laterite (25 ±15 cm)
	Norway	Norway	Norway	Finland	Finland	Australia
	<2 mm	<2 mm	<2 mm	<0.063 mm	<0.063 mm	0.45-2 mm
	7N HNO$_3$	7N HNO$_3$	7N HNO$_3$	aqua regia	total	total
Median	-	-	-	-	-	-
Min	-	-	-	-	-	-
Max	-	-	-	-	-	-

Medium	Stream sediment	Stream sediment	Stream sediment	Overbank sediment	Overbank sediment	Organic stream sediment
		Southern Scotland	Harz, Germany	Norway	Norway	Finland
	Austria					
	<0.18 mm	<0.10 mm	<0.063 mm	<0.063 mm	<0.063 mm	<2 mm
		total	total	total (XRF)	7N HNO$_3$	Conc. HNO$_3$
Median	-	-	-	-	-	-
Min	-	-	-	-	-	-
Max	-	-	-	-	-	-

#2nd percentile ^98th percentile

Concentrations in Waters (mg/l)

Medium	Ocean water World (1)	Ocean water World (2)	Ocean water North Pacific	Stream water World	Stream water Nova Scotia, Canada	Stream water Finland
					<0.45 µm	<0.45 µm
Median	0.000,21*	0.000,2*	0.000,31*	-	-	-
Min	-	-	-	-	-	-
Max	-	-	-	-	-	-

*Estimated mean   #2nd percentile   ^98th percentile

Medium	Stream water	Stream water	Stream water	Lake water	Ground water
	Romania	Eastern India	Harz, Germany	Norway	Southern Norway
	unfiltered	<0.2 µm	unfiltered	unfiltered	unfiltered
	ICP-MS	ICP-MS		ICP-MS	
Median	-	-	-	-	-
Min	-	-	-	-	-
Max	-	-	-	-	-

Kr KRYPTON

Concentrations in Precipitation (mg/l)

Medium	Rain water	Rain water	Snow melt-water	Snow melt-water	Snow filter residue	Snow filter residue
	Kola, remote <0.45 µm	Kola, polluted <0.45 µm	Kola, remote <0.45 µm	Kola, polluted <0.45 µm	Kola, remote >0.45 µm	Kola, polluted >0.45 µm
Median	-	-	-	-	-	-
Min	-	-	-	-	-	-
Max	-	-	-	-	-	-

Medium	Rain water	Rain water				
	Coastal, Norway unfiltered ICP-MS	Inland, Norway unfiltered ICP-MS				
Median	-	-				
Min	-	-				
Max	-	-				

Concentrations in Air (ng/m^3) and Yearly Deposition (kg/km^2/yr)

Medium	Air	Air	Bulk deposition	Throughfall deposition	Bulk deposition	Bulk deposition
	World, remote	World, polluted	West Germany	West Germany	Kola, remote	Kola, polluted
Median	-	-	-	-	-	-
Min	-	-	-	-	-	-
Max	-	-	-	-	-	-

*Estimated value

Concentrations in Plants (mg/kg)

Medium	Moss	Moss	Crustose lichen	Lichen	Dandelion	Spruce bark
	Norway conc. HNO$_3$ (1990)	Germany conc. HNO$_3$ + H$_2$O$_2$ (1995)	Germany conc. HNO$_3$	Northwest Territories oven-dried	Europe oven-dried	Canada ashed
Median	-	-	-	-	-	-
Min	-	-	-	-	-	-
Max	-	-	-	-	-	-

*Mean   #Geometric mean

Concentrations in Human Fluids (mg/l)

Medium	Human blood	Human serum	Human urine			
	Lombardy, Italy	Lombardy, Italy	Lombardy, Italy			
Mean	-	-	-			
Min	-	-	-			
Max	-	-	-			

KRYPTON   Kr

Environmental Geochemistry

Biological impacts	Considered non-essential
Uses	Electronics, light bulbs (fill gas)
Environmental pathways	Natural trace component of atmosphere
Environmental mobility	Oxidising conditions: very high Acid conditions: very high Reducing conditions: very high Neutral to alkaline conditions: very high Comments: Inert gas
Geochemical barriers	-
Natural association	-
Action levels	-
Remarks	Occurs in atmosphere at low concentrations (4.1 mg/m^3), together with the other noble gasses. ^{85}Kr is found in elevated concentrations in the atmosphere due to releases by the nuclear industry
Suggested analytical method(s)	-

World Yearly Production (t/y)	-
Price of Purest Form, in Small Quantities ($/kg)	146 $/l (99.995%)
Market Price ($/kg)	7000 $/m^3

La   LANTHANUM

Physico-Chemical Properties

Atomic number	57	Melting/boiling point (K)	1191 / 3737
Atomic mass	138.905,5(2)	Isotopes & isomers	1 stable + 31 unstable
Atomic radius (pm)	274	Acid/base of oxide	base
Main oxidation state(s)	+3	State (at 300 K, 1 atm.)	solid
Ionic radius (pm)	117-150	Metallic character	metal
Electronegativity (Pauling)	1.1	Element group(s)	heavy metal, REE
Density (g/cm^3)	6.145	Affinity	lithophile

Naturally occurring isotopes	Natural abundance (%)	Atomic mass	Half-life	Unstable isotopes	Longest half-life
^{138}La	0.090,2	137.907,11	1.06x10^{+11} y	^{120}La to ^{149}La	1.06x10^{+11} y
^{139}La	99.909,8	138.906,35	stable		

Concentrations in Crust/Rocks (mg/kg)

Bulk continental crust	30 / 39 / 16		
Upper continental crust	32.3 / 30		
Ultramafic rock	1		
Ocean ridge basalt	4		
Gabbro, basalt	6		
Granite, granodiorite	50		
Sandstone	20		
Greywacke	34		
Shale, schist	40		
Limestone	6		
Coal	10		

Typical Minerals
monazite ((Ce,La,Nd,Th)(PO$_4$,SiO$_4$)),
bastnaesite ((Ce,La)CO$_3$(F,OH)),
cerite ((Ce,La)$_9$(Mg,Fe)Si$_7$(O,OH,F)$_{28}$),
allanite ((Ca,Ce,La)$_2$FeAl$_2$OSi$_3$O$_{11}$(OH))

Possible Host Minerals
biotite, apatite, pyroxenes, feldspars, zircon

Mass (kg) in

Continental crust	7.08x10^{+17}	Oceans	3.97x10^{+9}	Plants	3.68x10^{+8}

Concentrations in Media Sampled for the Kola Project (mg/kg)

Medium	Moss conc. HNO$_3$	Humus <2 mm conc. HNO$_3$	Humus <2 mm amm. acet.	Topsoil (0-5 cm) <2 mm total (INAA)	B horizon <2 mm aqua regia	B horizon <2 mm total (XRF)
Median	<0.7	2.3	-	8.9	9.3	-
Min	<0.7	<0.7	-	0.5	1.7	-
Max	35.2	139	-	160	173	-

Medium	C horizon <2 mm aqua regia	C horizon <2 mm total (XRF)	C horizon <2 mm total (INAA)			Lake water unfiltered (mg/l)
Median	12.8	-	24			-
Min	3.5	-	6.1			-
Max	203	-	310			-

LANTHANUM — La

Concentrations in Soils and Sediments (mg/kg)

Medium	Soil	Agricultural soil - Ap horizon	Agricultural soil - Top (0-25 cm)	Agricultural Soil - Bottom (50-75 cm)	Topsoil (0-15 cm)	Urban soil (0-2 cm)
	World	Canada	Finland	Finland	England & Wales	Trondheim
	<2 mm	<2 mm	<2 mm	<2 mm	<2 mm	<2 mm
	total	total (INAA)	aqua regia	aqua regia	aqua regia	aqua regia
Median	35*	26	-	-	-	15.4
Min	-	1	-	-	-	1.4
Max	-	46	-	-	-	33.8

*Estimated mean

Medium	Forest soil - Humus	Forest soil - B horizon	Forest soil - C horizon	Till (C horizon)	Till (C horizon)	Laterite (25 ±15 cm)
	Norway	Norway	Norway	Finland	Finland	Australia
	<2 mm	<2 mm	<2 mm	<0.063 mm	<0.063 mm	0.45-2 mm
	7N HNO$_3$	7N HNO$_3$	7N HNO$_3$	aqua regia	total	total
Median	2	3.3	27.8	20.6	35.4	11
Min	<0.5	<1	<1	-	-	1.73
Max	197.6	292	697.3	-	-	70

Medium	Stream sediment	Stream sediment	Stream sediment	Overbank sediment	Overbank sediment	Organic stream sediment
	Austria	Southern Scotland	Harz, Germany	Norway	Norway	Finland
	<0.18 mm	<0.10 mm	<0.063 mm	<0.063 mm	<0.063 mm	<2 mm
	"total" (ICP-AES)	total (DCES)	total (XRF)	total (XRF)	7N HNO$_3$ ICP-AES	Conc. HNO$_3$
Median	51	41	50	-	37	21.2
Min	-	<15	<5	-	1	7.5#
Max	6732	920	1137	-	260	68.2^

#2nd percentile   ^98th percentile

Concentrations in Waters (mg/l)

Medium	Ocean water	Ocean water	Ocean water	Stream water	Stream water	Stream water
	World (1)	World (2)	North Pacific	World	Nova Scotia, Canada	Finland
					<0.45 µm	<0.45 µm
					ICP-MS	
Median	0.000,003,4*	0.000,003*	0.000,005,6*	0.000,03*	0.000,09	-
Min	-	-	-	-	<0.000,01	-
Max	-	-	-	-	0.000,84	-

*Estimated mean   #2nd percentile   ^98th percentile

Medium	Stream water	Stream water	Stream water	Lake water	Ground water
	Romania	Eastern India	Harz, Germany	Norway	Southern Norway
	unfiltered	<0.2 µm	unfiltered	unfiltered	unfiltered
	ICP-MS	ICP-MS	ICP-MS	ICP-MS	ICP-MS
Median	0.000,38	0.004,9	0.000,43	-	0.000,145
Min	0.000,01	0.000,5	0.000,01	-	<0.000,01
Max	0.114	0.054,1	0.014	-	0.112,479

La — LANTHANUM

Concentrations in Precipitation (mg/l)

Medium	Rain water	Rain water	Snow melt-water	Snow melt-water	Snow filter residue	Snow filter residue
	Kola, remote <0.45 μm	Kola, polluted <0.45 μm	Kola, remote <0.45 μm	Kola, polluted <0.45 μm	Kola, remote >0.45 μm ICP-AES	Kola, polluted >0.45 μm ICP-AES
Median	-	-	-	-	<0.000,07	0.000,37
Min	-	-	-	-	<0.000,07	0.000,16
Max	-	-	-	-	<0.000,07	0.001,38

Medium	Rain water	Rain water
	Coastal, Norway unfiltered ICP-MS	Inland, Norway unfiltered ICP-MS
Median	0.000,03	0.000,04
Min	<0.000,01	<0.000,01
Max	0.002,2	0.002,3

Concentrations in Air (ng/m^3) and Yearly Deposition (kg/km^2/yr)

Medium	Air	Air	Bulk deposition	Throughfall deposition	Bulk deposition	Bulk deposition
	World, remote	World, polluted	West Germany	West Germany	Kola, remote	Kola, polluted
Median	0.1 / 0.056	2	-	-	-	-
Min	0.000,5	<0.1	-	-	-	-
Max	<0.2 (0.5)	10	-	-	-	-

*Estimated value

Concentrations in Plants (mg/kg)

Medium	Moss	Moss	Crustose lichen	Lichen	Dandelion	Spruce bark
	Norway conc. HNO$_3$ (1990)	Germany conc. HNO$_3$ + H$_2$O$_2$ (1995)	Germany conc. HNO$_3$	Northwest Territories oven-dried	Europe oven-dried total (INAA)	Canada ashed total (INAA)
Median	0.44	0.377	7.6	-	0.43#	15
Min	<0.2	0.07	2	-	-	1.5
Max	4.5	9	87	-	-	67

*Mean   #Geometric mean

Concentrations in Human Fluids (mg/l)

Medium	Human blood	Human serum	Human urine
	Lombardy, Italy	Lombardy, Italy	Lombardy, Italy
Mean	0.001,42	<0.001	0.000,73
Min	0.000,13	<0.001	0.000,015
Max	0.004,75	<0.001	0.006

LANTHANUM — La

Environmental Geochemistry

Biological impacts	Considered non-essential. Generally low toxicity, but data to assess the health relevance of REE are scarce
Uses	Superconductors, lighters, catalysts, glass additives, glass polishing, fibre optics, ceramics (capacitors), batteries
Environmental pathways	Geogenic dust, mining and processing of alkaline rocks. Generally, geogenic sources more important than anthropogenic ones
Environmental mobility	Oxidising conditions: very low — Acid conditions: very low Reducing conditions: very low — Neutral to alkaline conditions: very low Comments: -
Geochemical barriers	Mechanical
Natural association	REE-Li-Rb-Cs-Be-Nb-Ta-Zr-B-Th-U-F (pegmatites), REE-Th-P-Zr-Fe-Cu (monazite veins), REE-Th-Ba-Sr-P-F-N-C (carbonatites), REE-U-P-F (phosphorites), REE-Au-Ti-Sn-Sr-Th (placers)
Action levels	-
Remarks	Toxicity of REE decreases as atomic number increases. Inhaled REE probably cause pneumoconiosis. REE taken up orally (e.g., via drinking water) accumulate in the skeleton, teeth, lungs, liver and kidneys. Some trees tend to accumulate La (e.g., hickory (*Carya*)). In 1991, 700-1000 t La-oxide were used in the glass industry, and 200-350 t in the ceramics industry
Suggested analytical method(s)	ICP-AES, ICP-MS (water)

World Yearly Production (t/y)	54,000 REE-minerals (in 1991)
Price of Purest Form, in Small Quantities ($/kg)	14,158 (99.99%)
Market Price ($/kg)	-

Li LITHIUM

Physico-Chemical Properties

Atomic number	3	Melting/boiling point (K)	453.65 / 1615
Atomic mass	6.941(2)	Isotopes & isomers	2 stable + 5 unstable
Atomic radius (pm)	205	Acid/base of oxide	strong base
Main oxidation state(s)	+1	State (at 300 K, 1 atm.)	solid
Ionic radius (pm)	73-106	Metallic character	metal
Electronegativity (Pauling)	0.98	Element group(s)	alkali metal, light metal
Density (g/cm^3)	0.534	Affinity	lithophile

Naturally occurring isotopes	Natural abundance (%)	Atomic mass	Half-life
^{6}Li	7.5	6.015,12	stable
^{7}Li	92.5	7.016	stable

Unstable isotopes	Longest half-life
^{5}Li to ^{11}Li	0.84 s

Concentrations in Crust/Rocks (mg/kg)

Bulk continental crust	18 / 20 / 13
Upper continental crust	22 / 20
Ultramafic rock	2
Ocean ridge basalt	10
Gabbro, basalt	10
Granite, granodiorite	30
Sandstone	10
Greywacke	-
Shale, schist	60
Limestone	5
Coal	30

Typical Minerals
spodumene (LiAlSi$_2$O$_6$),
lepidolite (K$_2$Li$_3$Al$_4$Si$_7$O$_{21}$(OH,F)$_3$),
petalite (LiAlSi$_4$O$_{10}$), amblygonite (LiAl(F,OH)PO$_4$),
montebrasite (LiAlPO$_4$(OH)), eucryptite (LiAlSiO$_4$)

Possible Host Minerals
micas (biotite), amphiboles

Mass (kg) in

Continental crust	4.25x10^{+17}	Oceans	2.38x10^{+14}	Plants	3.68x10^{+8}

Concentrations in Media Sampled for the Kola Project (mg/kg)

Medium	Moss conc. HNO$_3$	Humus <2 mm	Humus <2 mm amm. acet.	Topsoil (0-5 cm) <2 mm	B horizon <2 mm aqua regia	B horizon <2 mm total (XRF)
Median	-	-	<0.05	-	7.4	-
Min	-	-	<0.05	-	1.6	-
Max	-	-	0.17	-	69.5	-

Medium	C horizon <2 mm aqua regia	C horizon <2 mm total (XRF)	C horizon <2 mm total (INAA)			Lake water unfiltered (mg/l)
Median	7.2	-	-			<0.000,3
Min	1.7	-	-			<0.000,3
Max	70.9	-	-			0.001,67

LITHIUM   Li

Concentrations in Soils and Sediments (mg/kg)

Medium	Soil	Agricultural soil - Ap horizon	Agricultural soil - Top (0-25 cm)	Agricultural Soil - Bottom (50-75 cm)	Topsoil (0-15 cm)	Urban soil (0-2 cm)
	World	Canada	Finland	Finland	England & Wales	Trondheim
	<2 mm	<2 mm	<2 mm	<2 mm	<2 mm	<2 mm
	total		aqua regia	aqua regia	aqua regia	aqua regia
Median	20*	-	-	-	-	17.8
Min	-	-	-	-	-	0.9
Max	-	-	-	-	-	40.2

*Estimated mean

Medium	Forest soil - Humus	Forest soil - B horizon	Forest soil - C horizon	Till (C horizon)	Till (C horizon)	Laterite (25 ±15 cm)
	Norway	Norway	Norway	Finland	Finland	Australia
	<2 mm	<2 mm	<2 mm	<0.063 mm	<0.063 mm	0.45-2 mm
	7N HNO$_3$	7N HNO$_3$	7N HNO$_3$	aqua regia	total	total
Median	0.5	10.7	13.1	12.2	14.9	6.5
Min	<0.1	0.6	1.5	-	-	<0.5
Max	28.7	106	107.2	-	-	22

Medium	Stream sediment	Stream sediment	Stream sediment	Overbank sediment	Overbank sediment	Organic stream sediment
	Austria	Southern Scotland	Harz, Germany	Norway	Norway	Finland
	<0.18 mm	<0.10 mm	<0.063 mm	<0.063 mm	<0.063 mm	<2 mm
		total (DCES)	total (ICP-MS)	total (XRF)	7N HNO$_3$ ICP-AES	Conc. HNO$_3$
Median	-	48	31	-	7.8	9.5
Min	-	2	2	-	0.6	1.4#
Max	-	450	92	-	68.7	40.3^

#2nd percentile   ^98th percentile

Concentrations in Waters (mg/l)

Medium	Ocean water	Ocean water	Ocean water	Stream water	Stream water	Stream water
	World (1)	World (2)	North Pacific	World	Nova Scotia, Canada	Finland
					<0.45 µm	<0.45 µm
					ICP-MS	ICP-MS
Median	0.18*	0.18*	0.18*	0.003*	0.000,284	0.001,02
Min	-	-	-	-	<0.000,005	<0.000,3#
Max	-	-	-	-	0.005,629	0.011,9^

*Estimated mean   #2nd percentile   ^98th percentile

Medium	Stream water	Stream water	Stream water	Lake water	Ground water
	Romania	Eastern India	Harz, Germany	Norway	Southern Norway
	unfiltered	<0.2 µm	unfiltered	unfiltered	unfiltered
	ICP-MS	ICP-MS	ICP-MS	ICP-MS	ICP-MS
Median	0.011,2	0.003,3	0.002,3	0.000,17	0.003,616,5
Min	0.000,1	0.000,8	<0.000,1	<0.000,01	0.000,187
Max	0.148	0.019,1	0.073	0.134	0.059,641

Li   LITHIUM

Concentrations in Precipitation (mg/l)

Medium	Rain water	Rain water	Snow melt-water	Snow melt-water	Snow filter residue	Snow filter residue
	Kola, remote <0.45 µm ICP-MS	Kola, polluted <0.45 µm ICP-MS	Kola, remote <0.45 µm ICP-MS	Kola, polluted <0.45 µm ICP-MS	Kola, remote >0.45 µm ICP-AES	Kola, polluted >0.45 µm ICP-AES
Median	<0.000,1	0.000,12	<0.000,1	0.000,185	<0.000,07	0.000,145
Min	<0.000,1	<0.000,1	<0.000,1	<0.000,1	<0.000,07	0.000,07
Max	0.000,45	0.000,28	0.000,18	0.000,29	<0.000,07	0.000,74

Medium	Rain water	Rain water
	Coastal, Norway unfiltered ICP-MS	Inland, Norway unfiltered ICP-MS
Median	0.000,05	0.000,04
Min	<0.000,01	<0.000,01
Max	0.000,38	0.002,4

Concentrations in Air (ng/m^3) and Yearly Deposition (kg/km^2/yr)

Medium	Air	Air	Bulk deposition	Throughfall deposition	Bulk deposition	Bulk deposition
	World, remote	World, polluted	West Germany	West Germany	Kola, remote	Kola, polluted
Median	-	-	-	-	-	-
Min	-	-	-	-	-	-
Max	8.9	2.3	-	-	-	-

*Estimated value

Concentrations in Plants (mg/kg)

Medium	Moss	Moss	Crustose lichen	Lichen	Dandelion	Spruce bark
	Norway conc. HNO$_3$ (1990)	Germany conc. HNO$_3$ + H$_2$O$_2$ (1995)	Germany conc. HNO$_3$	Northwest Territories oven-dried	Europe oven-dried	Canada ashed aqua regia
Median	0.22	0.339	-	-	-	8
Min	0.03	<0.001	-	-	-	1
Max	1.7	25	-	-	-	43

*Mean   #Geometric mean

Concentrations in Human Fluids (mg/l)

Medium	Human blood	Human serum	Human urine			
	Lombardy, Italy	Lombardy, Italy	Lombardy, Italy			
Mean	-	-	-			
Min	-	-	-			
Max	-	-	-			

LITHIUM   Li

Environmental Geochemistry

Biological impacts	Considered essential. Toxicity considered low
Uses	Glass, ceramics, special alloys (with Pb, Al), batteries, electrodes, lubricants, greases, cooling agent in nuclear reactors, pharmaceutical industry, rock fuel, air purifier, H-bomb, flux in Al production (Li-carbonate)
Environmental pathways	Rock weathering, geogenic dust, Al production, (sea spray)
Environmental mobility	Oxidising conditions: low Acid conditions: low Reducing conditions: very low Neutral to alkaline conditions: low Comments: -
Geochemical barriers	Adsorption (Mn-oxides, clays)
Natural association	Li-Be-B-K-Rb-Cs-Nb-Ta-F-P-Sn-W-REE (granites, pegmatites), Li-B-F-Be-Sn-W-Mo (greisens), Li-Na-K-B-P-W-F-Br-Cl-I-SO_4-CO_3 (brines, saline evaporites)
Action levels	Drinking water, MAC: 0.03 mg/l (Russia MoH)
Remarks	Can interact with Mg in biological systems. Proven correlation between low Li levels in drinking water and high incidence of mental disorders
Suggested analytical method(s)	GF-AAS, ICP-AES, ICP-MS (water)

World Yearly Production (t/y)	9553 (in 1995)
Price of Purest Form, in Small Quantities ($/kg)	2688 (99.95%)
Market Price ($/kg)	86 (Li), 0.4 (spodumene conc., min. 7.25% Li_2O)

Lu — LUTETIUM

Physico-Chemical Properties

Atomic number	71	Melting/boiling point (K)	1936 / 3675
Atomic mass	174.967(1)	Isotopes & isomers	1 stable + 48 unstable
Atomic radius (pm)	225	Acid/base of oxide	weak base
Main oxidation state(s)	+3	State (at 300 K, 1 atm.)	solid
Ionic radius (pm)	100-117	Metallic character	metal
Electronegativity (Pauling)	1.27	Element group(s)	heavy metal, REE
Density (g/cm^3)	9.841	Affinity	lithophile

Naturally occurring isotopes	Natural abundance (%)	Atomic mass	Half-life
^{175}Lu	97.41	174.940,77	stable
^{176}Lu	2.59	175.942,68	3.8x10^{+10} y

Unstable isotopes	Longest half-life
^{150}Lu to ^{183}Lu	3.8x10^{+10} y

Concentrations in Crust/Rocks (mg/kg)

Bulk continental crust	0.35 / 0.8 / 0.3
Upper continental crust	0.27 / 0.32
Ultramafic rock	0.04
Ocean ridge basalt	0.6
Gabbro, basalt	0.5
Granite, granodiorite	0.7
Sandstone	0.3
Greywacke	0.37
Shale, schist	0.6
Limestone	0.1
Coal	0.2

Typical Minerals

monazite ((Ce,La,Nd,Th, Lu)(PO$_4$,SiO$_4$)),
bastnaesite ((Ce,La,Lu)CO$_3$(F,OH)),
cerite ((Ce,La,Lu)$_9$(Mg,Fe)Si$_7$(O,OH,F)$_{28}$),
allanite ((Ca,Ce,La,Lu)$_2$FeAl$_2$OSi$_3$O$_{11}$(OH))

Possible Host Minerals

biotite, apatite, pyroxenes, feldspars, zircon

Mass (kg) in

Continental crust	8.26x10^{+15}	Oceans	2.64x10^{+8}	Plants	5.52x10^{+6}

Concentrations in Media Sampled for the Kola Project (mg/kg)

Medium	Moss	Humus	Humus	Topsoil (0-5 cm)	B horizon	B horizon
	<2 mm	<2 mm	<2 mm	<2 mm	<2 mm	<2 mm
	conc. HNO$_3$		amm. acet.	total (INAA)	aqua regia	total (XRF)
Median	-	-	-	0.16	-	-
Min	-	-	-	<0.05	-	-
Max	-	-	-	1.1	-	-

Medium	C horizon	C horizon	C horizon		Lake water
	<2 mm	<2 mm	<2 mm		unfiltered
	aqua regia	total (XRF)	total (INAA)		(mg/l)
Median	-	-	0.3		-
Min	-	-	0.05		-
Max	-	-	2.67		-

LUTETIUM   Lu

Concentrations in Soils and Sediments (mg/kg)

Medium	Soil	Agricultural soil - Ap horizon	Agricultural soil - Top (0-25 cm)	Agricultural Soil - Bottom (50-75 cm)	Topsoil (0-15 cm)	Urban soil (0-2 cm)
	World	Canada	Finland	Finland	England & Wales	Trondheim
	<2 mm	<2 mm	<2 mm	<2 mm	<2 mm	<2 mm
	total	aqua regia	aqua regia	aqua regia	aqua regia	aqua regia
Median	0.5*	-	-	-	-	-
Min	-	-	-	-	-	-
Max	-	-	-	-	-	-

*Estimated mean

Medium	Forest soil - Humus	Forest soil - B horizon	Forest soil - C horizon	Till (C horizon)	Till (C horizon)	Laterite (25 ±15 cm)
	Norway	Norway	Norway	Finland	Finland	Australia
	<2 mm	<2 mm	<2 mm	<0.063 mm	<0.063 mm	0.45-2 mm
	7N HNO$_3$	7N HNO$_3$	7N HNO$_3$	aqua regia	total	total
Median	-	-	-	-	0.6	-
Min	-	-	-	-	-	-
Max	-	-	-	-	-	-

Medium	Stream sediment	Stream sediment	Stream sediment	Overbank sediment	Overbank sediment	Organic stream sediment
	Austria	Southern Scotland	Harz, Germany	Norway	Norway	Finland
	<0.18 mm	<0.10 mm	<0.063 mm	<0.063 mm	<0.063 mm	<2 mm
		total	total	total (XRF)	7N HNO$_3$	Conc. HNO$_3$
Median	-	-	-	-	-	-
Min	-	-	-	-	-	-
Max	-	-	-	-	-	-

#2nd percentile ^98th percentile

Concentrations in Waters (mg/l)

Medium	Ocean water	Ocean water	Ocean water	Stream water	Stream water	Stream water
	World (1)	World (2)	North Pacific	World	Nova Scotia, Canada	Finland
					<0.45 µm	<0.45 µm
					ICP-MS	
Median	0.000,000,15*	0.000,000,2*	0.000,000,23*	0.000,000,6*	<0.000,005	-
Min	-	-	-	-	<0.000,005	-
Max	-	-	-	-	0.000,016	-

*Estimated mean #2nd percentile ^98th percentile

Medium	Stream water	Stream water	Stream water	Lake water	Ground water
	Romania	Eastern India	Harz, Germany	Norway	Southern Norway
	unfiltered	<0.2 µm	unfiltered	unfiltered	unfiltered
	ICP-MS	ICP-MS	ICP-MS	ICP-MS	ICP-MS
Median	-	0.000,03	0.000,01	<0.000,002	0.000,003
Min	-	0.000,01	<0.000,001	<0.000,002	<0.000,001
Max	-	0.000,5	0.000,1	0.000,19	0.000,38

Lu — LUTETIUM

Concentrations in Precipitation (mg/l)

Medium	Rain water Kola, remote <0.45 µm	Rain water Kola, polluted <0.45 µm	Snow melt-water Kola, remote <0.45 µm	Snow melt-water Kola, polluted <0.45 µm	Snow filter residue Kola, remote >0.45 µm	Snow filter residue Kola, polluted >0.45 µm
Median	-	-	-	-	-	-
Min	-	-	-	-	-	-
Max	-	-	-	-	-	-

Medium	Rain water Coastal, Norway unfiltered ICP-MS	Rain water Inland, Norway unfiltered ICP-MS				
Median	-	-				
Min	-	-				
Max	-	-				

Concentrations in Air (ng/m^3) and Yearly Deposition (kg/km^2/yr)

Medium	Air World, remote	Air World, polluted	Bulk deposition West Germany	Throughfall deposition West Germany	Bulk deposition Kola, remote	Bulk deposition Kola, polluted
Median	-	-	-	-	-	-
Min	-	-	-	-	-	-
Max	-	-	-	-	-	-

*Estimated value

Concentrations in Plants (mg/kg)

Medium	Moss Norway conc. HNO$_3$ (1990)	Moss Germany conc. HNO$_3$ + H$_2$O$_2$ (1995)	Crustose lichen Germany conc. HNO$_3$	Lichen Northwest Territories oven-dried	Dandelion Europe oven-dried	Spruce bark Canada ashed total (INAA)
Median	-	-	-	-	-	0.21
Min	-	-	-	-	-	<0.05
Max	-	-	-	-	-	0.74

*Mean   #Geometric mean

Concentrations in Human Fluids (mg/l)

Medium	Human blood Lombardy, Italy	Human serum Lombardy, Italy	Human urine Lombardy, Italy			
Mean	0.000,2	<0.000,05	0.000,05			
Min	0.000,000,25	<0.000,05	0.000,001			
Max	0.000,8	<0.000,05	0.000,3			

LUTETIUM — Lu

Environmental Geochemistry

Biological impacts	Considered non-essential. Data to assess the toxicity of REE are scarce
Uses	Superconductors, catalyst
Environmental pathways	Poorly understood. Probably mostly windblown dust, weathering of REE-bearing minerals. Generally, geogenic sources more important than anthropogenic ones
Environmental mobility	Oxidising conditions: very low Acid conditions: very low Reducing conditions: very low Neutral to alkaline conditions: very low Comments: -
Geochemical barriers	Mechanical
Natural association	REE-Li-Rb-Cs-Be-Nb-Ta-Zr-B-Th-U-F (pegmatites), REE-Th-P-Zr-Fe-Cu (monazite veins), REE-Th-Ba-Sr-P-F-N-C (carbonatites), REE-U-P-F (phosphorites), REE-Au-Ti-Sn-Sr-Th (placers)
Action levels	-
Remarks	Toxicity of REE decreases as atomic number increases. Inhaled REE probably cause pneumoconiosis. REE taken up orally (e.g., via drinking water) accumulate in the skeleton, teeth, lungs, liver and kidneys. Some trees tend to accumulate Lu (e.g., hickory (*Carya*)). Most costly of all REE
Suggested analytical method(s)	ICP-MS

World Yearly Production (t/y)	54,000 REE-minerals (in 1991)
Price of Purest Form, in Small Quantities ($/kg)	506,500 (99.99%)
Market Price ($/kg)	-

Mg MAGNESIUM

Physico-Chemical Properties

Atomic number	12	Melting/boiling point (K)	923 / 1363
Atomic mass	24.305,0(6)	Isotopes & isomers	3 stable + 12 unstable
Atomic radius (pm)	172	Acid/base of oxide	base
Main oxidation state(s)	+2	State (at 300 K, 1 atm.)	solid
Ionic radius (pm)	71-103	Metallic character	metal
Electronegativity (Pauling)	1.31	Element group(s)	alkaline earth, light metal
Density (g/cm^3)	1.738	Affinity	lithophile

Naturally occurring isotopes	Natural abundance (%)	Atomic mass	Half-life	Unstable isotopes	Longest half-life
^{24}Mg	78.99	23.985,04	stable	^{20}Mg to ^{34}Mg	20.9 h
^{25}Mg	10	24.985,84	stable		
^{26}Mg	11.01	25.982,59	stable		

Concentrations in Crust/Rocks (mg/kg)

Bulk continental crust	22,000 / 23,300 / 32,000
Upper continental crust	13,510 / 13,300
Ultramafic rock	208,000
Ocean ridge basalt	46,000
Gabbro, basalt	46,000
Granite, granodiorite	5000
Sandstone	7000
Greywacke	13,869
Shale, schist	16,000
Limestone	4000
Coal	2600

Typical Minerals

oxides and hydroxides, e.g., peridase (MgO), spinel (HgAl$_2$O$_4$), brucite Mg (OH)$_2$;
silicates, e.g., forsterite (Mg$_2$SiO$_4$), pyrope (Mg$_2$Al$_2$(SiO$_4$)$_3$), enstatite (Mg$_3$Si$_2$O$_6$), tremolite (Ca$_2$Mg$_5$(Si$_8$O$_{22}$)(OH)$_2$), phlogopite (kMg$_3$Si$_3$AlO$_{10}$ (OH)$_2$)
carbonates, e.g., magnesite (MgCO$_3$), dolomite ((Ca,Mg)CO$_3$);
sulphates, e.g., kieserite (MgSO$_4 \cdot$H$_2$O);
phosphates and arsenates, e.g., farringtonite (Mg$_3$(PO$_4$)$_2$), hornesite (Mg$_3$(AsO$_4$)$_2 \cdot$8 H$_2$O

Possible Host Minerals

--

Mass (kg) in

Continental crust 5.19x10^{+20}	Oceans 1.71x10^{+18}	Plants 3.68x10^{+8}

Concentrations in Media Sampled for the Kola Project (mg/kg)

Medium	Moss <2 mm conc. HNO$_3$	Humus <2 mm conc. HNO$_3$	Humus <2 mm amm. acet.	Topsoil (0-5 cm) <2 mm	B horizon <2 mm aqua regia	B horizon <2 mm total (XRF)
Median	1090	750	450	-	3030	10,434
Min	518	240	58.3	-	340	724
Max	2380	5830	2120	-	23,800	36,970

Medium	C horizon <2 mm aqua regia	C horizon <2 mm total (XRF)	C horizon <2 mm total (INAA)			Lake water unfiltered (mg/l)
Median	3700	11,500	-			0.885
Min	370	1200	-			0.02
Max	70,500	73,200	-			2.21

MAGNESIUM   Mg

Concentrations in Soils and Sediments (mg/kg)

Medium	Soil	Agricultural soil - Ap horizon	Agricultural soil - Top (0-25 cm)	Agricultural Soil - Bottom (50-75 cm)	Topsoil (0-15 cm)	Urban soil (0-2 cm)
	World	Canada	Finland	Finland	England & Wales	Trondheim
	<2 mm total	<2 mm	<2 mm aqua regia	<2 mm aqua regia	<2 mm aqua regia	<2 mm aqua regia
Median	9000*	-	2220	2760	3005	12,900
Min	-	-	149	120	41	1530
Max	-	-	11,000	18,800	62,690	30,400

*Estimated mean

Medium	Forest soil - Humus	Forest soil - B horizon	Forest soil - C horizon	Till (C horizon)	Till (C horizon)	Laterite (25 ±15 cm)
	Norway	Norway	Norway	Finland	Finland	Australia
	<2 mm 7N HNO₃	<2 mm 7N HNO₃	<2 mm 7N HNO₃	<0.063 mm aqua regia	<0.063 mm total	0.45-2 mm total
Median	1100	3300	6200	4200	10,000	482
Min	190	290	220	-	-	<60
Max	10,900	37,000	51,000	-	-	37,085

Medium	Stream sediment	Stream sediment Southern Scotland	Stream sediment Harz, Germany	Overbank sediment	Overbank sediment	Organic stream sediment
	Austria			Norway	Norway	Finland
	<0.18 mm "total" (ICP-AES)	<0.10 mm total (DCES)	<0.063 mm total (XRF)	<0.063 mm total (XRF)	<0.063 mm 7N HNO₃ ICP-AES	<2 mm Conc. HNO₃
Median	11,200	13,262	4583	14,400	4600	3800
Min	-	<603	663	1900	140	1100#
Max	208,400	188,074	32,743	97,700	74,200	10,800^

#2nd percentile   ^98th percentile

Concentrations in Waters (mg/l)

Medium	Ocean water	Ocean water	Ocean water	Stream water	Stream water	Stream water
	World (1)	World (2)	North Pacific	World	Nova Scotia, Canada	Finland
					<0.45 µm ICP-AES	<0.45 µm ICP-MS
Median	1290*	1290*	1280*	4.1*	0.75	1.39
Min	-	-	-	-	<0.1	0.53#
Max	-	-	-	-	9.1	11^

*Estimated mean   #2nd percentile   ^98th percentile

Medium	Stream water	Stream water	Stream water	Lake water	Ground water
	Romania	Eastern India	Harz, Germany	Norway	Southern Norway
	unfiltered ICP-MS	<0.2 µm ICP-MS	unfiltered ICP-MS	unfiltered ICP-MS	unfiltered ICP-AES
Median	15	-	2.065	-	4.25
Min	0.3	-	0.1	-	<0.2
Max	116	-	11.6	-	33

Mg   MAGNESIUM

Concentrations in Precipitation (mg/l)

Medium	Rain water	Rain water	Snow melt-water	Snow melt-water	Snow filter residue	Snow filter residue
	Kola, remote <0.45 µm ICP-AES	Kola, polluted <0.45 µm ICP-AES	Kola, remote <0.45 µm ICP-AES	Kola, polluted <0.45 µm ICP-AES	Kola, remote >0.45 µm ICP-AES	Kola, polluted >0.45 µm ICP-AES
Median	0.02	0.04	<0.05	0.105	<0.01	0.105
Min	<0.01	<0.01	<0.05	<0.05	<0.01	0.04
Max	0.04	0.1	0.05	0.13	<0.01	0.54

Medium	Rain water	Rain water				
	Coastal, Norway unfiltered ICP-MS	Inland, Norway unfiltered ICP-MS				
Median	0.096	0.035				
Min	0.006,3	0.003,7				
Max	1.8	1.9				

Concentrations in Air (ng/m^3) and Yearly Deposition (kg/km^2/yr)

Medium	Air	Air	Bulk deposition	Throughfall deposition	Bulk deposition	Bulk deposition
	World, remote	World, polluted	West Germany	West Germany	Kola, remote	Kola, polluted
Median	50 / 5.4	1700	-	-	24.7*	137*
Min	0.006,7	60	90	110	-	-
Max	480 (800)	11,000	270	750	-	-

*Estimated value

Concentrations in Plants (mg/kg)

Medium	Moss	Moss	Crustose lichen	Lichen	Dandelion	Spruce bark
	Norway conc. HNO$_3$ (1990)	Germany conc. HNO$_3$ + H$_2$O$_2$ (1995)	Germany conc. HNO$_3$	Northwest Territories oven-dried total (INAA)	Europe oven-dried	Canada ashed aqua regia
Median	1200	1350	-	637.6*	-	15,925
Min	470	227	-	-	-	5100
Max	4600	13,900	-	-	-	69,180

*Mean   #Geometric mean

Concentrations in Human Fluids (mg/l)

Medium	Human blood	Human serum	Human urine			
	Lombardy, Italy	Lombardy, Italy	Lombardy, Italy			
Mean	-	-	-			
Min	-	-	-			
Max	-	-	-			

MAGNESIUM   Mg

Environmental Geochemistry

Biological impacts	Essential for all organisms. Practically non-toxic under normal circumstances. Mg deficiencies are much more widespread than Mg toxicity problems, and are reported, e.g., in humans and cattle
Uses	Many alloys (e.g., with Al), fertilisers, fire resistant materials (e.g., refractory brick), reduction agent, anti-corrosion agent, special cements
Environmental pathways	Rock weathering, windblown dust, sea spray, fertilisers, liming (e.g., dolomite). Generally, natural sources more important than anthropogenic ones
Environmental mobility	Oxidising conditions: high Acid conditions: high Reducing conditions: high Neutral to alkaline conditions: high Comments: -
Geochemical barriers	-
Natural association	Fe-Mg-Ca (substitution in many minerals)
Action levels	Drinking water, MAC: 20 mg/l (Norway Shd)
Remarks	Mostly produced by extraction from sea water. Mg atom important in photosynthesis. Mg deficiency in soil (e.g., forest soils under acid deposition) leads to chlorosis in plants, even on Mg rich substrates (high Mg mobility). Mg deficiency usually occurs when exchangeable Mg drops below 5% of the total cation exchange capacity (CEC). Nuts concentrate Mg. Mg-based fertilisers: $MgSO_4$, $MgCl_2$, $(Ca,Mg)CO_3$ (dolomite)
Suggested analytical method(s)	XRF, ICP-AES

World Yearly Production (t/y)	311,000 (in 1995, for use as light metal from magnesite)
Price of Purest Form, in Small Quantities ($/kg)	2245 (99.8%)
Market Price ($/kg)	2.65

Mn MANGANESE

Physico-Chemical Properties

Atomic number	25
Atomic mass	54.938,049(9)
Atomic radius (pm)	179
Main oxidation state(s)	+2 (+1,+3,+4,+5,+6,+7)
Ionic radius (pm)	80-110 (+2), 72-78.5 (+3), 53-67 (+4), 47 (+5), 39.5 (+6)
Electronegativity (Pauling)	1.15
Density (g/cm^3)	7.44
Melting/boiling point (K)	1519 / 2334
Isotopes & isomers	1 stable + 23 unstable
Acid/base of oxide	strong acid
State (at 300 K, 1 atm.)	solid
Metallic character	pred. metal
Element group(s)	heavy metal
Affinity	lithophile

Naturally occurring isotopes	Natural abundance (%)	Atomic mass	Half-life
^{55}Mn	100	54.938,05	stable

Unstable isotopes	Longest half-life
^{46}Mn to ^{66}Mn	3,700,000 y

Concentrations in Crust/Rocks (mg/kg)

Bulk continental crust	716 / 950 / 1400
Upper continental crust	527 / 600
Ultramafic rock	1200
Ocean ridge basalt	1400
Gabbro, basalt	1500
Granite, granodiorite	400
Sandstone	100
Greywacke	775
Shale, schist	850
Limestone	700
Coal	40

Typical Minerals
pyrolusite (MnO$_2$), manganite (MnO(OH)), hausmannite (Mn$_3$O$_4$), rhodocrosite (MnCO$_3$), psilomelane (BaMn$_9$O$_{18}$.2H$_2$O), cryptomelane (KMn$_8$O$_{16}$)

Possible Host Minerals
garnets, olivine, pyroxenes, amphiboles, micas, calcite, dolomite

Mass (kg) in

Continental crust 1.69x10^{+19}	Oceans 2.64x10^{+11}	Plants 3.68x10^{+11}

Concentrations in Media Sampled for the Kola Project (mg/kg)

Medium	Moss conc. HNO$_3$	Humus <2 mm conc. HNO$_3$	Humus <2 mm amm. acet.	Topsoil (0-5 cm) <2 mm	B horizon <2 mm aqua regia	B horizon <2 mm total (XRF)
Median	433	126	68.9	-	103	465
Min	28.5	11.1	2.75	-	23.8	<80
Max	1170	5470	844	-	1450	2091

Medium	C horizon <2 mm aqua regia	C horizon <2 mm total (XRF)	C horizon <2 mm total (INAA)			Lake water unfiltered (mg/l)
Median	128	540	-			0.003,6
Min	33.8	150	-			0.000,28
Max	2140	3560	-			0.061,9

MANGANESE   Mn

Concentrations in Soils and Sediments (mg/kg)

Medium	Soil	Agricultural soil - Ap horizon	Agricultural soil - Top (0-25 cm)	Agricultural Soil - Bottom (50-75 cm)	Topsoil (0-15 cm)	Urban soil (0-2 cm)
	World	Canada	Finland	Finland	England & Wales	Trondheim
	<2 mm	<2 mm	<2 mm	<2 mm	<2 mm	<2 mm
	total	total (AAS)	aqua regia	aqua regia	aqua regia	aqua regia
Median	530*	396	128	110	577	442
Min	-	80	4.5	<2	3	43
Max	-	2400	1690	850	42,603	4410

*Estimated mean

Medium	Forest soil - Humus	Forest soil - B horizon	Forest soil - C horizon	Till (C horizon)	Till (C horizon)	Laterite (25 ±15 cm)
	Norway	Norway	Norway	Finland	Finland	Australia
	<2 mm	<2 mm	<2 mm	<0.063 mm	<0.063 mm	0.45-2 mm
	7N HNO$_3$	7N HNO$_3$	7N HNO$_3$	aqua regia	total	total
Median	60	160	300	170	500	155
Min	<10	20	40	-	-	<80
Max	10,000	1900	6100	-	-	1162

Medium	Stream sediment	Stream sediment	Stream sediment	Overbank sediment	Overbank sediment	Organic stream sediment
	Austria	Southern Scotland	Harz, Germany	Norway	Norway	Finland
	<0.18 mm	<0.10 mm	<0.063 mm	<0.063 mm	<0.063 mm	<2 mm
	"total" (ICP-AES)	total (DCES)	total (XRF)	total (XRF)	7N HNO$_3$	Conc. HNO$_3$
Median	850	1960	1394	770	-	546
Min	-	<20	77	80	-	105#
Max	16,000	236	51,743	12,930	-	9573^

#2nd percentile ^98th percentile

Concentrations in Waters (mg/l)

Medium	Ocean water	Ocean water	Ocean water	Stream water	Stream water	Stream water
	World (1)	World (2)	North Pacific	World	Nova Scotia, Canada	Finland
					<0.45 μm	<0.45 μm
					ICP-MS	ICP-MS
Median	0.000,2*	0.000,2*	0.000,02*	0.004*	0.005,3	0.029
Min	-	-	-	-	<0.000,1	0.000,9#
Max	-	-	-	-	0.255,4	0.216^

*Estimated mean   #2nd percentile   ^98th percentile

Medium	Stream water	Stream water	Stream water	Lake water	Ground water	
	Romania	Eastern India	Harz, Germany	Norway	Southern Norway	
	unfiltered	<0.2 μm	unfiltered	unfiltered	unfiltered	
	ICP-MS	ICP-MS	ICP-MS	ICP-MS	ICP-MS	
Median	0.077	0.094,4	0.022	0.003,43	0.007,5	
Min	0.000,1	0.019,3	<0.001	<0.000,2	0.000,1	
Max	2.75	0.459,4	1.51	0.327	2.975	

Mn   MANGANESE

Concentrations in Precipitation (mg/l)

Medium	Rain water	Rain water	Snow melt-water	Snow melt-water	Snow filter residue	Snow filter residue
	Kola, remote <0.45 μm ICP-MS	Kola, polluted <0.45 μm ICP-MS	Kola, remote <0.45 μm ICP-MS	Kola, polluted <0.45 μm ICP-MS	Kola, remote >0.45 μm ICP-AES	Kola, polluted >0.45 μm ICP-AES
Median	0.002,1	0.002	0.000,45	0.003,36	0.000,1	0.004,87
Min	0.000,45	0.001	0.000,35	0.002,51	0.000,8	0.002,18
Max	0.006,11	0.004,4	0.003,77	0.005,72	0.000,41	0.031,9

Medium	Rain water	Rain water				
	Coastal, Norway unfiltered ICP-MS	Inland, Norway unfiltered ICP-MS				
Median	0.000,37	0.001,9				
Min	0.000,02	0.000,1				
Max	0.034	0.072				

Concentrations in Air (ng/m^3) and Yearly Deposition (kg/km^2/yr)

Medium	Air	Air	Bulk deposition	Throughfall deposition	Bulk deposition	Bulk deposition
	World, remote	World, polluted	West Germany	West Germany	Kola, remote	Kola, polluted
Median	2	90	-	-	-	-
Min	0.01	6	3	32	-	-
Max	3	900	72	530	-	-

*Estimated value

Concentrations in Plants (mg/kg)

Medium	Moss	Moss	Crustose lichen	Lichen	Dandelion	Spruce bark
	Norway conc. HNO$_3$ (1990)	Germany conc. HNO$_3$ + H$_2$O$_2$ (1995)	Germany conc. HNO$_3$	Northwest Territories oven-dried total (INAA)	Europe oven-dried conc. HNO$_3$ (AAS)	Canada ashed aqua regia
Median	300	337	190	47.7*	54#	11,683
Min	32	19	40	-	-	1641
Max	3200	2300	1850	-	-	10,137

*Mean   #Geometric mean

Concentrations in Human Fluids (mg/l)

Medium	Human blood	Human serum	Human urine			
	Lombardy, Italy	Lombardy, Italy	Lombardy, Italy			
Mean	0.008,8	0.000,6	0.001,02			
Min	0.005	0.000,3	0.000,1			
Max	0.012,4	0.001,35	0.003			

MANGANESE   Mn

Environmental Geochemistry

Biological impacts	Essential for all organisms. Non-toxic. Mn deficiencies in humans and animals are more widespread than Mn toxicity problems
Uses	Steel, alloys (e.g., with Al, Mg, Cu), batteries, catalyst, fertilisers, pigment, wood preservative, fungicide, antiknock agent (MMT) in gasoline (as a replacement for Pb in unleaded fuel)
Environmental pathways	Rock weathering, windblown dust, agriculture, traffic, mining and smelting, steel production. Generally, geogenic sources more important than anthropogenic ones
Environmental mobility	Oxidising conditions: very low Acid conditions: low Reducing conditions: low Neutral to alkaline conditions: low Comments: -
Geochemical barriers	pH, precipitation of Mn-Fe oxides, hydroxides, oxy-hydroxides
Natural association	Fe-Mn (substitution in many minerals, seafloor polymetallic nodules)
Action levels	Drinking water, MAC: 0.05 mg/l (Norway Shd), 0.05 mg/l (Canada CWQG); recommended: 0.5 mg/l (WHO); MAC: 0.1 mg/l (Russia MoH)
Remarks	Various plants accumulate Mn (e.g., nuts, legumes, heather, tea). Mn deficiency causes growth disturbances in plants. Very high doses of Mn required for toxic effects, but occur on acid soils. Composted bark used in gardening can lead to Mn toxicity. Deficiency in humans and animals causes failure in reproduction and impaired growth. Mn interacts with Fe uptake in animals at very high doses. Concentrations of 0.1 mg/l in drinking water may give rise to consumer complaints due to staining of laundry and sanitary ware
Suggested analytical method(s)	XRF, ICP-AES, ICP-MS (water)

World Yearly Production (t/y)   8,200,000 (in 1995)
Price of Purest Form, in Small Quantities ($/kg)   1046 (99.985%)
Market Price ($/kg)   1.5

Mo   MOLYBDENUM

Physico-Chemical Properties

Atomic number	42	Melting/boiling point (K)	2896 / 4912
Atomic mass	95.94(1)	Isotopes & isomers	7 stable + 23 unstable
Atomic radius (pm)	201	Acid/base of oxide	strong acid
Main oxidation state(s)	+2,+6 (+3,+4,+5)	State (at 300 K, 1 atm.)	solid
Ionic radius (pm)	83 (+2), 79 (+4), 60 (+5)	Metallic character	pred. non-metal
Electronegativity (Pauling)	2.16	Element group(s)	heavy non-metal
Density (g/cm^3)	10.22	Affinity	chalcophile, siderophile

Naturally occurring isotopes	Natural abundance (%)	Atomic mass	Half-life
^{92}Mo	14.84	91.906,81	stable
^{94}Mo	9.25	93.905,09	stable
^{95}Mo	15.92	94.905,84	stable
^{96}Mo	16.68	95.904,68	stable
^{97}Mo	9.55	96.906,02	stable
^{98}Mo	24.13	97.905,41	stable
^{100}Mo	9.63	99.907,48	stable

Unstable isotopes	Longest half-life
^{84}Mo to ^{110}Mo	3500 y

Concentrations in Crust/Rocks (mg/kg)

Bulk continental crust	1.1 / 1.2 / 1
Upper continental crust	1.4 / 1.5
Ultramafic rock	0.3
Ocean ridge basalt	1.1
Gabbro, basalt	1.2
Granite, granodiorite	1.5
Sandstone	0.3
Greywacke	-
Shale, schist	2
Limestone	0.3
Coal	3

Typical Minerals
molybdenite (MoS$_2$), wulfenite (PbMoO$_4$), powellite (Ca(Mo,W)O$_4$)

Possible Host Minerals
scheelite, wolframite, does not substitute into silicates

Mass (kg) in

Continental crust	2.60x10^{+16}	Oceans	1.32x10^{+13}	Plants	9.20x10^{+8}

Concentrations in Media Sampled for the Kola Project (mg/kg)

Medium	Moss	Humus <2 mm	Humus <2 mm	Topsoil (0-5 cm) <2 mm	B horizon <2 mm	B horizon <2 mm
	conc. HNO$_3$	conc. HNO$_3$	amm. acet.	total (INAA)	aqua regia	total (XRF)
Median	0.08	0.258	<0.05	<1	0.3	-
Min	0.016	0.086	<0.05	<1	<0.2	-
Max	1.08	5.45	0.12	20	5.8	-

Medium	C horizon <2 mm	C horizon <2 mm	C horizon <2 mm	Lake water unfiltered
	aqua regia	total (XRF)	total (INAA)	(mg/l)
Median	<0.2	-	<1	0.000,08
Min	<0.2	-	<1	<0.000,03
Max	3.4	-	19	0.001,18

MOLYBDENUM — Mo

Concentrations in Soils and Sediments (mg/kg)

Medium	Soil	Agricultural soil - Ap horizon	Agricultural soil - Top (0-25 cm)	Agricultural Soil - Bottom (50-75 cm)	Topsoil (0-15 cm)	Urban soil (0-2 cm)
	World	Canada	Finland	Finland	England & Wales	Trondheim
	<2 mm total	<2 mm total (AAS)	<2 mm aqua regia	<2 mm aqua regia	<2 mm aqua regia	<2 mm aqua regia
Median	1.2*	2	0.224	<0.2	-	<1
Min	-	<2	<0.2	<0.2	-	<1
Max	-	32	12	7.97	-	7.2

*Estimated mean

Medium	Forest soil - Humus	Forest soil - B horizon	Forest soil - C horizon	Till (C horizon)	Till (C horizon)	Laterite (25 ±15 cm)
	Norway	Norway	Norway	Finland	Finland	Australia
	<2 mm $7N\ HNO_3$	<2 mm $7N\ HNO_3$	<2 mm $7N\ HNO_3$	<0.063 mm aqua regia	<0.063 mm total	0.45-2 mm total
Median	0.6	<1	<1	0.2	1	1
Min	<0.5	<1	<1	-	-	<0.2
Max	9.5	54.6	32	-	-	5.5

Medium	Stream sediment	Stream sediment	Stream sediment	Overbank sediment	Overbank sediment	Organic stream sediment
	Austria	Southern Scotland	Harz, Germany	Norway	Norway	Finland
	<0.18 mm "total" (ICP-AES)	<0.10 mm total (DCES)	<0.063 mm total (ICP-MS)	<0.063 mm total (XRF)	<0.063 mm $7N\ HNO_3$ ICP-AES	<2 mm Conc. HNO_3
Median	0.7	<5	1.5	-	2.1	0.91
Min	-	<5	0.19	-	1	0.24#
Max	160	137	8.5	-	27	7.59^

#2nd percentile   ^98th percentile

Concentrations in Waters (mg/l)

Medium	Ocean water	Ocean water	Ocean water	Stream water	Stream water	Stream water
	World (1)	World (2)	North Pacific	World	Nova Scotia, Canada	Finland
					<0.45 µm ICP-MS	<0.45 µm ICP-MS
Median	0.01*	0.01*	0.01*	0.000,5*	<0.000,05	0.000,15
Min	-	-	-	-	<0.000,05	<0.000,03#
Max	-	-	-	-	0.012,95	0.001,22^

*Estimated mean   #2nd percentile   ^98th percentile

Medium	Stream water	Stream water	Stream water	Lake water	Ground water
	Romania	Eastern India	Harz, Germany	Norway	Southern Norway
	unfiltered ICP-MS	<0.2 µm ICP-MS	unfiltered ICP-MS	unfiltered ICP-MS	unfiltered ICP-MS
Median	0.000,51	0.000,2	0.000,05	<0.000,04	0.001,63
Min	0.000,03	0.000,1	<0.000,01	<0.000,04	<0.000,05
Max	0.017	0.052,2	0.001,1	0.006,95	0.289,38

Mo   MOLYBDENUM

Concentrations in Precipitation (mg/l)

Medium	Rain water	Rain water	Snow melt-water	Snow melt-water	Snow filter residue	Snow filter residue
	Kola, remote <0.45 μm ICP-MS	Kola, polluted <0.45 μm ICP-MS	Kola, remote <0.45 μm ICP-MS	Kola, polluted <0.45 μm ICP-MS	Kola, remote >0.45 μm ICP-AES	Kola, polluted >0.45 μm ICP-AES
Median	<0.000,03	0.001,6	<0.000,05	0.000,23	<0.000,07	0.001,49
Min	<0.000,03	0.000,46	<0.000,05	0.000,09	<0.000,07	0.000,93
Max	<0.000,03	0.007,36	<0.000,05	0.000,35	0.000,09	0.004,36

Medium	Rain water	Rain water				
	Coastal, Norway unfiltered ICP-MS	Inland, Norway unfiltered ICP-MS				
Median	0.000,04	0.000,05				
Min	<0.000,01	<0.000,01				
Max	0.002,6	0.000,57				

Concentrations in Air (ng/m^3) and Yearly Deposition (kg/km^2/yr)

Medium	Air	Air	Bulk deposition	Throughfall deposition	Bulk deposition	Bulk deposition
	World, remote	World, polluted	West Germany	West Germany	Kola, remote	Kola, polluted
Median	-	-	-	-	0.043,1*	0.923*
Min	-	<0.2	-	-	-	-
Max	<0.2	10	-	-	-	-

*Estimated value

Concentrations in Plants (mg/kg)

Medium	Moss	Moss	Crustose lichen	Lichen	Dandelion	Spruce bark
	Norway conc. HNO$_3$ (1990)	Germany conc. HNO$_3$ + H$_2$O$_2$ (1995)	Germany conc. HNO$_3$	Northwest Territories oven-dried	Europe oven-dried	Canada ashed total (INAA)
Median	0.15	0.262	2	-	-	3
Min	<0.01	0.05	1.1	-	-	<2
Max	1.2	5.5	9.1	-	-	38

*Mean   #Geometric mean

Concentrations in Human Fluids (mg/l)

Medium	Human blood	Human serum	Human urine			
	Lombardy, Italy	Lombardy, Italy	Lombardy, Italy			
Mean	-	-	-			
Min	-	-	-			
Max	-	-	-			

MOLYBDENUM   Mo

Environmental Geochemistry

Biological impacts	Essential for all organisms, except some bacteria. Considered toxic. Probably non-carcinogenic. More toxic to cows and sheep than to humans
Uses	Alloys, catalysts, anti-corrosion agents, flame retardant, lubricant, pigments, fertilisers
Environmental pathways	U mining, Mo mining and smelting, oil refining, oil and coal combustion, phosphate fertilisers, sewage sludge, phosphate detergents, geogenic dust, weathering
Environmental mobility	Oxidising conditions: high Acid conditions: high Reducing conditions: very low Neutral to alkaline conditions: very high Comments: -
Geochemical barriers	Sulphide, reducing condition, adsorption, presence of Pb, Fe and carbonate ions
Natural association	Mo-W-Re-Cu-Sn-Be-B-F-P-Zn-Bi-Fe (pegmatites), Mo-Bi-W-F-Be (greisens), Mo-Cu-Re-Ag-Au-Zn (porphyry Cu), Mo-U-Se-V-Cu (sandstone type U deposits)
Action levels	Drinking water, recommended: 0.07 mg/l (WHO); MAC: 0.25 mg/l (Russia MoH). Ground water, background: 0.005 mg/l; remediate: 0.3 mg/l (Netherlands VROM). Soil, background: 10 mg/kg; remediate: 200 mg/kg (Netherlands VROM). Agricultural soil, maximum tolerable concentration: 5 mg/kg (Germany)
Remarks	Mo often produced as by-product of Cu and W mining. Small amounts in soil important for optimum plant growth. Mo deficiency in acid soil frequent. Vegetables grown on alkaline soils can contain very high Mo levels. Mo interacts with Cu and S (molybdenosis is a secondary Cu deficiency). Too high Mo concentrations in soils reported to cause decreased animal production in sheep and cow farming. Some Mo-compounds used in fertilisers: Na_2MoO_4, $(NH_4)_6Mo_7O_{24}.4H_2O$, MoO_3
Suggested analytical method(s)	ICP-MS, GF-AAS

World Yearly Production (t/y)	119,000 (in 1995)
Price of Purest Form, in Small Quantities ($/kg)	4570 (99.999%)
Market Price ($/kg)	10 (Mo-oxide, 56%)

N NITROGEN

Physico-Chemical Properties

Atomic number	7	Melting/boiling point (K)	63 / 77.36
Atomic mass	14.006,74(7)	Isotopes & isomers	2 stable + 10 unstable
Atomic radius (pm)	75	Acid/base of oxide	strong acid
Main oxidation state(s)	-3,+3 (+2,+4,+5)	State (at 300 K, 1 atm.)	gas
Ionic radius (pm)	132 (-3), 30 (+3), 4.4-27 (+5)	Metallic character	non-metal
		Element group(s)	light non-metal
Electronegativity (Pauling)	3.04	Affinity	atmophile, biophile
Density (g/cm^3)	0.001,250,6		

Naturally occurring isotopes	Natural abundance (%)	Atomic mass	Half-life	Unstable isotopes	Longest half-life
^{14}N	99.634	14.003,074	stable	^{12}N to ^{23}N	9.97 min
^{15}N	0.366	15.000,11	stable		

Concentrations in Crust/Rocks (mg/kg)

Bulk continental crust	60 / 19 / -	
Upper continental crust	83 / -	
Ultramafic rock	-	
Ocean ridge basalt	-	
Gabbro, basalt	-	
Granite, granodiorite	-	
Sandstone	-	
Greywacke	-	
Shale, schist	-	
Limestone	-	
Coal	12,000	

Typical Minerals
nitre (KNO$_3$), soda-nitre (NaNO$_3$), and many other nitrates, osbornite (TiN)

Possible Host Minerals
cordierite, micas, K-feldspar

Mass (kg) in

Continental crust	1.42x10^{+18}	Oceans	1.50x10^{+16}	Plants	4.60x10^{+13}

Concentrations in Media Sampled for the Kola Project (mg/kg)

Medium	Moss	Humus	Humus	Topsoil (0-5 cm)	B horizon	B horizon
	conc. HNO$_3$	<2 mm CHN-anal.	<2 mm amm. acet.	<2 mm	<2 mm aqua regia	<2 mm total (XRF)
Median	-	13,000	-	-	-	-
Min	-	5000	-	-	-	-
Max	-	20,000	-	-	-	-

Medium	C horizon	C horizon	C horizon			Lake water
	<2 mm aqua regia	<2 mm total (XRF)	<2 mm total (INAA)			unfiltered (mg/l)
Median	-	-	-			-
Min	-	-	-			-
Max	-	-	-			-

NITROGEN N

Concentrations in Soils and Sediments (mg/kg)

Medium	Soil	Agricultural soil - Ap horizon	Agricultural soil - Top (0-25 cm)	Agricultural Soil - Bottom (50-75 cm)	Topsoil (0-15 cm)	Urban soil (0-2 cm)
	World	Canada	Finland	Finland	England & Wales	Trondheim
	<2 mm	<2 mm	<2 mm	<2 mm	<2 mm	<2 mm
	total		aqua regia	aqua regia	aqua regia	aqua regia
Median	-	-	-	-	-	-
Min	-	-	-	-	-	-
Max	-	-	-	-	-	-

*Estimated mean

Medium	Forest soil - Humus	Forest soil - B horizon	Forest soil - C horizon	Till (C horizon)	Till (C horizon)	Laterite (25 ±15 cm)
	Norway	Norway	Norway	Finland	Finland	Australia
	<2 mm	<2 mm	<2 mm	<0.063 mm	<0.063 mm	0.45-2 mm
	7N HNO$_3$	7N HNO$_3$	7N HNO$_3$	aqua regia	total	total
Median	-	-	-	-	-	-
Min	-	-	-	-	-	-
Max	-	-	-	-	-	-

Medium	Stream sediment	Stream sediment	Stream sediment	Overbank sediment	Overbank sediment	Organic stream sediment
	Austria	Southern Scotland	Harz, Germany	Norway	Norway	Finland
	<0.18 mm	<0.10 mm	<0.063 mm	<0.063 mm	<0.063 mm	<2 mm
	total	total	total	total (XRF)	7N HNO$_3$	C, H, N-analyser
Median	-	-	-	-	-	5000
Min	-	-	-	-	-	1000#
Max	-	-	-	-	-	17,000^

#2nd percentile ^98th percentile

Concentrations in Waters (mg/l)

Medium	Ocean water	Ocean water	Ocean water	Stream water	Stream water	Stream water
	World (1)	World (2)	North Pacific	World	Nova Scotia, Canada	Finland
					<0.45 µm	<0.45 µm
Median	0.5*	11.5*	8.3*	-	-	-
Min	-	-	-	-	-	-
Max	-	-	-	-	-	-

*Estimated mean #2nd percentile ^98th percentile

Medium	Stream water	Stream water	Stream water	Lake water	Ground water	
	Romania	Eastern India	Harz, Germany	Norway	Southern Norway	
	unfiltered	<0.2 µm	unfiltered	unfiltered	unfiltered	
	ICP-MS	ICP-MS		ICP-MS		
Median	-	-	-	-	-	
Min	-	-	-	-	-	
Max	-	-	-	-	-	

N NITROGEN

Concentrations in Precipitation (mg/l)

Medium	Rain water	Rain water	Snow melt-water	Snow melt-water	Snow filter residue	Snow filter residue
	Kola, remote <0.45 µm	Kola, polluted <0.45 µm	Kola, remote <0.45 µm	Kola, polluted <0.45 µm	Kola, remote >0.45 µm	Kola, polluted >0.45 µm
Median	-	-	-	-	-	-
Min	-	-	-	-	-	-
Max	-	-	-	-	-	-

Medium	Rain water	Rain water				
	Coastal, Norway unfiltered ICP-MS	Inland, Norway unfiltered ICP-MS				
Median	-	-				
Min	-	-				
Max	-	-				

Concentrations in Air (ng/m^3) and Yearly Deposition (kg/km^2/yr)

Medium	Air	Air	Bulk deposition	Throughfall deposition	Bulk deposition	Bulk deposition
	World, remote	World, polluted	West Germany	West Germany	Kola, remote	Kola, polluted
Median	-	-	-	-	-	-
Min	-	-	-	-	-	-
Max	-	-	-	-	-	-

*Estimated value

Concentrations in Plants (mg/kg)

Medium	Moss	Moss	Crustose lichen	Lichen	Dandelion	Spruce bark
	Norway conc. HNO$_3$ (1990)	Germany conc. HNO$_3$ + H$_2$O$_2$ (1995)	Germany conc. HNO$_3$	Northwest Territories oven-dried	Europe oven-dried	Canada ashed
Median	-	-	-	-	-	-
Min	-	-	-	-	-	-
Max	-	-	-	-	-	-

*Mean #Geometric mean

Concentrations in Human Fluids (mg/l)

Medium	Human blood	Human serum	Human urine			
	Lombardy, Italy	Lombardy, Italy	Lombardy, Italy			
Mean	-	-	-			
Min	-	-	-			
Max	-	-	-			

NITROGEN N

Environmental Geochemistry

Biological impacts	Essential for all organisms. N_2 non-toxic, but many N-compounds are toxic (e.g., CN^-)
Uses	General chemical industry, fertilisers, welding, deep-freezing, coolant, inert gas in research laboratories
Environmental pathways	Natural major component of atmosphere, fertilisers
Environmental mobility	Oxidising conditions: - Acid conditions: - Reducing conditions: - Neutral to alkaline conditions: - Comments: Highly dependent on compound, e.g., nitrates are very soluble and do not bind to soils; they have a high potential to migrate to ground water
Geochemical barriers	-
Natural association	C-H-O-(N)-(S) (organic matter), N-O (nitrate, nitrite)
Action levels	Drinking water, MAC: 10 mg NO_3^-/l, 0.05 mg NO_2^-/l, 0.5 mg NH_4^+/l, 1 mg N/l as other than NO_3^- or NO_2^- (Norway Shd), 10 mg N/l as NO_3^-, 1 mg N/l as NO_2^-, or 10 mg N/l as sum NO_3^- + NO_2^- (US EPA), 10 mg N/l as NO_3^-, 1 mg N/l as NO_2^- mg/l (Canada CWQG); recommended: 50 mg NO_3^-/l, 3 mg NO_2^-/l (WHO)
Remarks	N is one of the most abundant elements in the Universe (cosmic abundance similar to that of C and O). Major constituent of Earth atmosphere (enriched relative to crust in the proportion 3:1). Of all the N-O groups, only the NO_3^- ion exists in minerals. Ammonium (NH_4^+) can replace K in the lattice of silicate minerals (e.g., K-feldspar) and take part in ion exchange processes in clay minerals. Average N content of granites and granodiorites is 21 mg/kg, and that of gabbros is 5 mg/kg. Isotopic fractionation of N in nature used for research. N is a component of amino acids. Ammonia (NH_3) is an important compound of the atmosphere of many planets (Jupiter, Saturn). Concentrations of 1.5 mg NH_3/l in drinking water may give rise to consumer complaints due to odour and taste. The US EPA suggests that drinking water exceeding action levels can lead to serious illness and sometimes death in the short term, and to dieresis, increased starchy deposits and hemorrhaging of the spleen in the long term
Suggested analytical method(s)	-

World Yearly Production (t/y)	120,000,000 (in 1979)
Price of Purest Form, in Small Quantities ($/kg)	-
Market Price ($/kg)	0.25 ($NaNO_3$, 98%)

Na SODIUM

Physico-Chemical Properties

Atomic number	11	Melting/boiling point (K)	370.95 / 1156
Atomic mass	22.989,770(2)	Isotopes & isomers	1 stable + 17 unstable
Atomic radius (pm)	223	Acid/base of oxide	strong base
Main oxidation state(s)	+1	State (at 300 K, 1 atm.)	solid
Ionic radius (pm)	113-153	Metallic character	metal
Electronegativity (Pauling)	0.93	Element group(s)	alkali metal, light metal
Density (g/cm^3)	0.971	Affinity	lithophile

Naturally occurring isotopes	Natural abundance (%)	Atomic mass	Half-life
^{23}Na	100	22.989,77	stable

Unstable isotopes	Longest half-life
^{19}Na to ^{35}Na	2,605,000 y

Concentrations in Crust/Rocks (mg/kg)

Bulk continental crust	23,600 / 23,600 / 23,000
Upper continental crust	25,670 / 28,900
Ultramafic rock	6000
Ocean ridge basalt	20,000
Gabbro, basalt	20,000
Granite, granodiorite	25,000
Sandstone	17,000
Greywacke	22,257
Shale, schist	13,000
Limestone	6000
Coal	900

Typical Minerals

albite (NaAlSi$_3$O$_8$), halite (NaCl), cryolite (Na$_3$AlF$_6$), soda (Na$_2$CO$_3$) and many major rock forming minerals (e.g., plagioclases, micas, amphiboles, pyroxenes)

Possible Host Minerals

-

Mass (kg) in

Continental crust	5.57x10^{+20}	Oceans	1.42x10^{+19}	Plants	2.76x10^{+11}

Concentrations in Media Sampled for the Kola Project (mg/kg)

| Medium | Moss | Humus <2 mm | Humus <2 mm | Topsoil (0-5 cm) <2 mm | B horizon <2 mm | B horizon <2 mm |
	conc. HNO$_3$	conc. HNO$_3$	amm. acet.	total (INAA)	aqua regia	total (XRF)
Median	71.6	60	35.6	17,200	130	21,738
Min	<10	5	8.8	400	20	371
Max	918	2350	450	33,300	4550	32,792

Medium	C horizon <2 mm aqua regia	C horizon <2 mm total (XRF)	C horizon <2 mm total (INAA)		Lake water unfiltered (mg/l)
Median	140	24,500	27,600		2.2
Min	20	800	1900		0.8
Max	19,400	48,700	53,100		12.9

SODIUM Na

Concentrations in Soils and Sediments (mg/kg)

Medium	Soil	Agricultural soil - Ap horizon	Agricultural soil - Top (0-25 cm)	Agricultural Soil - Bottom (50-75 cm)	Topsoil (0-15 cm)	Urban soil (0-2 cm)
	World	Canada	Finland	Finland	England & Wales	Trondheim
	<2 mm	<2 mm	<2 mm	<2 mm	<2 mm	<2 mm
	total	total (INAA)	aqua regia	aqua regia	aqua regia	aqua regia
Median	10,000*	10,000	61.1	75.5	242	2300
Min	-	400	<50	<50	31	700
Max	-	24,700	401	572	25,152	7000

*Estimated mean

Medium	Forest soil - Humus	Forest soil - B horizon	Forest soil - C horizon	Till (C horizon)	Till (C horizon)	Laterite (25 ±15 cm)
	Norway	Norway	Norway	Finland	Finland	Australia
	<2 mm	<2 mm	<2 mm	<0.063 mm	<0.063 mm	0.45-2 mm
	7N HNO$_3$	7N HNO$_3$	7N HNO$_3$	aqua regia	total	total
Median	130	150	290	200	23,000	519
Min	10	10	<10	-	-	<75
Max	1700	4900	3700	-	-	15,580

Medium	Stream sediment	Stream sediment	Stream sediment	Overbank sediment	Overbank sediment	Organic stream sediment
	Austria	Southern Scotland	Harz, Germany	Norway	Norway	Finland
	<0.18 mm	<0.10 mm	<0.063 mm	<0.063 mm	<0.063 mm	<2 mm
	"total" (ICP-AES)	total	total (XRF)	total (XRF)	7N HNO$_3$ ICP-AES	Conc. HNO$_3$
Median	14,800	-	6825	17,400	300	270
Min	-	-	445	1400	200	110#
Max	50,400	-	13,354	34,000	3900	676^

#2nd percentile ^98th percentile

Concentrations in Waters (mg/l)

Medium	Ocean water	Ocean water	Ocean water	Stream water	Stream water	Stream water
	World (1)	World (2)	North Pacific	World	Nova Scotia, Canada	Finland
					<0.45 µm	<0.45 µm
					AAS	ICP-AES
Median	10,800*	10,770*	10,780*	6.1*	3.35	2.1
Min	-	-	-	-	<0.1	1.2#
Max	-	-	-	-	325	26.1^

*Estimated mean #2nd percentile ^98th percentile

Medium	Stream water	Stream water	Stream water	Lake water	Ground water
	Romania	Eastern India	Harz, Germany	Norway	Southern Norway
	unfiltered	<0.2 µm	unfiltered	unfiltered	unfiltered
	ICP-MS	ICP-MS	ICP-OES	ICP-MS	ICP-AES
Median	19	-	3.4	-	17.3
Min	0.1	-	0.4	-	1.2
Max	1510	-	29	-	508

Na SODIUM

Concentrations in Precipitation (mg/l)

Medium	Rain water	Rain water	Snow melt-water	Snow melt-water	Snow filter residue	Snow filter residue
	Kola, remote <0.45 μm ICP-AES	Kola, polluted <0.45 μm ICP-AES	Kola, remote <0.45 μm ICP-AES	Kola, polluted <0.45 μm ICP-AES	Kola, remote >0.45 μm ICP-AES	Kola, polluted >0.45 μm ICP-AES
Median	<0.1	0.3	0.2	0.5	<0.01	0.02
Min	<0.1	<0.1	0.1	0.2	<0.01	<0.01
Max	0.13	0.72	0.2	0.7	<0.01	0.1

Medium	Rain water	Rain water				
	Coastal, Norway unfiltered ICP-MS	Inland, Norway unfiltered ICP-MS				
Median	0.6	0.14				
Min	<0.002	0.004				
Max	8.1	0.65				

Concentrations in Air (ng/m^3) and Yearly Deposition (kg/km^2/yr)

Medium	Air	Air	Bulk deposition	Throughfall deposition	Bulk deposition	Bulk deposition
	World, remote	World, polluted	West Germany	West Germany	Kola, remote	Kola, polluted
Median	280 / 280	1200	-	-	-	-
Min	3	120	100	200	-	-
Max	3300 (5600)	5500	1200	27,300	-	-

*Estimated value

Concentrations in Plants (mg/kg)

Medium	Moss	Moss	Crustose lichen	Lichen	Dandelion	Spruce bark
	Norway conc. HNO$_3$ (1990)	Germany conc. HNO$_3$ + H$_2$O$_2$ (1995)	Germany conc. HNO$_3$	Northwest Territories oven-dried total (INAA)	Europe oven-dried total (INAA)	Canada ashed total (INAA)
Median	120	254	-	233.1*	774# (?)	11,700
Min	25	0.67	-	-	-	3850
Max	870	2940	-	-	-	42,600

*Mean   #Geometric mean

Concentrations in Human Fluids (mg/l)

Medium	Human blood	Human serum	Human urine			
	Lombardy, Italy	Lombardy, Italy	Lombardy, Italy			
Mean	-	-	-			
Min	-	-	-			
Max	-	-	-			

SODIUM — Na

Environmental Geochemistry

Biological impacts	Essential for many organisms. Toxic to plants and animals at high levels
Uses	Detergents, fertilisers, antiknock agent in gasoline (as a replacement for Pb in unleaded fuel), cooling agent in nuclear reactors, chemical industry
Environmental pathways	Sea spray, road salting, fertilisers, waste waters, geogenic sources (mineral weathering, dust)
Environmental mobility	Oxidising conditions: high — Acid conditions: high Reducing conditions: high — Neutral to alkaline conditions: high Comments: -
Geochemical barriers	-
Natural association	Cl-Na-Mg-Br-S (sea spray and brines)
Action levels	Drinking water, MAC: 150 mg/l (Norway Shd), 200 mg/l (Russia MoH)
Remarks	High levels of Na (e.g., >1000 mg/l in drinking water) can cause high blood pressure, arteriosclerosis in humans with regular intake. $NaClO_3$ used as herbicide. $C_{18}H_{29}NaO_3S$: detergent (yearly production ca. 500,000 tons). Na compounds used in fertilisers: $NaNO_3$, Na_2MoO_4. Concentrations of 200 mg/l in drinking water may give rise to consumer complaints due to taste
Suggested analytical method(s)	XRF, ICP-AES

World Yearly Production (t/y)	53,000,000 (in 1979)
Price of Purest Form, in Small Quantities ($/kg)	91,750 (99.95%)
Market Price ($/kg)	0.12 (soda ash, Na_2CO_3)

Nb NIOBIUM

Physico-Chemical Properties

Atomic number	41	Melting/boiling point (K)	2750 / 5017
Atomic mass	92.906,38(2)	Isotopes & isomers	1 stable + 41 unstable
Atomic radius (pm)	208	Acid/base of oxide	weak acid
Main oxidation state(s)	+5 (+2,+3,+4?)	State (at 300 K, 1 atm.)	solid
Ionic radius (pm)	86 (+3), 82-93 (+4), 62-88 (+5)	Metallic character	pred. non-metal
		Element group(s)	heavy non-metal
Electronegativity (Pauling)	1.6	Affinity	lithophile
Density (g/cm^3)	8.57		

Naturally occurring isotopes	Natural abundance (%)	Atomic mass	Half-life
^{93}Nb	100	92.906,38	stable

Unstable isotopes	Longest half-life
^{82}Nb to ^{107}Nb	3.7x10^{+7} y

Concentrations in Crust/Rocks (mg/kg)

Bulk continental crust	19 / 20 / 11
Upper continental crust	26 / 25
Ultramafic rock	1
Ocean ridge basalt	4
Gabbro, basalt	10
Granite, granodiorite	18
Sandstone	10
Greywacke	8.4
Shale, schist	17
Limestone	0.1
Coal	2

Typical Minerals
pyrochlore ((Na,Ca)$_2$(Nb,Ta)$_2$O$_6$(OH,F)), columbite-tantalite ((Fe,Mn)(Nb-Ta)$_2$O$_6$))

Possible Host Minerals
biotite, rutile, ilmenite, sphene, cassiterite, zircon

Mass (kg) in

Continental crust 4.48x10^{+17}	Oceans 1.32x10^{+10}	Plants 9.20x10^{+7}

Concentrations in Media Sampled for the Kola Project (mg/kg)

Medium	Moss conc. HNO$_3$	Humus <2 mm	Humus <2 mm amm. acet.	Topsoil (0-5 cm) <2 mm	B horizon <2 mm aqua regia	B horizon <2 mm total (XRF)
Median	-	-	-	-	-	-
Min	-	-	-	-	-	-
Max	-	-	-	-	-	-

Medium	C horizon <2 mm aqua regia	C horizon <2 mm total (XRF)	C horizon <2 mm total (INAA)		Lake water unfiltered (mg/l)
Median	-	-	-		-
Min	-	-	-		-
Max	-	-	-		-

NIOBIUM   Nb

Concentrations in Soils and Sediments (mg/kg)

Medium	Soil	Agricultural soil - Ap horizon	Agricultural soil - Top (0-25 cm)	Agricultural Soil - Bottom (50-75 cm)	Topsoil (0-15 cm)	Urban soil (0-2 cm)
	World	Canada	Finland	Finland	England & Wales	Trondheim
	<2 mm total	<2 mm	<2 mm aqua regia	<2 mm aqua regia	<2 mm aqua regia	<2 mm aqua regia
Median	12*	-	-	-	-	-
Min	-	-	-	-	-	-
Max	-	-	-	-	-	-

*Estimated mean

Medium	Forest soil - Humus	Forest soil - B horizon	Forest soil - C horizon	Till (C horizon)	Till (C horizon)	Laterite (25 ±15 cm)
	Norway	Norway	Norway	Finland	Finland	Australia
	<2 mm $7N\ HNO_3$	<2 mm $7N\ HNO_3$	<2 mm $7N\ HNO_3$	<0.063 mm aqua regia	<0.063 mm total	0.45-2 mm total
Median	-	-	-	-	-	5.6
Min	-	-	-	-	-	0.5
Max	-	-	-	-	-	18

Medium	Stream sediment	Stream sediment	Stream sediment	Overbank sediment	Overbank sediment	Organic stream sediment
	Austria	Southern Scotland	Harz, Germany	Norway	Norway	Finland
	<0.18 mm total (XRF)	<0.10 mm total	<0.063 mm total (XRF)	<0.063 mm total (XRF)	<0.063 mm $7N\ HNO_3$	<2 mm Conc. HNO_3
Median	20.7	-	24	37	-	-
Min	-	-	<5	7	-	-
Max	243	-	81	123	-	-

#2nd percentile   ^98th percentile

Concentrations in Waters (mg/l)

Medium	Ocean water	Ocean water	Ocean water	Stream water	Stream water	Stream water
	World (1)	World (2)	North Pacific	World	Nova Scotia, Canada	Finland
					<0.45 µm	<0.45 µm
Median	0.000,01*	0.000,01*	<0.000,005*	<0.001*	-	-
Min	-	-	-	-	-	-
Max	-	-	-	-	-	-

*Estimated mean   #2nd percentile   ^98th percentile

Medium	Stream water	Stream water	Stream water	Lake water	Ground water
	Romania	Eastern India	Harz, Germany	Norway	Southern Norway
	unfiltered ICP-MS	<0.2 µm ICP-MS	unfiltered ICP-MS	unfiltered ICP-MS	unfiltered ICP-MS
Median	0.000,034	-	0.000,03	<0.000,04	0.000,008
Min	0.000,002	-	<0.000,01	<0.000,04	<0.000,002
Max	0.000,36	-	0.000,47	0.000,254	0.001,75

Nb NIOBIUM

Concentrations in Precipitation (mg/l)

Medium	Rain water	Rain water	Snow melt-water	Snow melt-water	Snow filter residue	Snow filter residue
	Kola, remote <0.45 μm	Kola, polluted <0.45 μm	Kola, remote <0.45 μm	Kola, polluted <0.45 μm	Kola, remote >0.45 μm	Kola, polluted >0.45 μm
Median	-	-	-	-	-	-
Min	-	-	-	-	-	-
Max	-	-	-	-	-	-

Medium	Rain water	Rain water				
	Coastal, Norway unfiltered ICP-MS	Inland, Norway unfiltered ICP-MS				
Median	-	-				
Min	-	-				
Max	-	-				

Concentrations in Air (ng/m^3) and Yearly Deposition (kg/km^2/yr)

Medium	Air	Air	Bulk deposition	Throughfall deposition	Bulk deposition	Bulk deposition
	World, remote	World, polluted	West Germany	West Germany	Kola, remote	Kola, polluted
Median	-	-	-	-	-	-
Min	-	-	-	-	-	-
Max	-	-	-	-	-	-

*Estimated value

Concentrations in Plants (mg/kg)

Medium	Moss	Moss	Crustose lichen	Lichen	Dandelion	Spruce bark
	Norway conc. HNO$_3$ (1990)	Germany conc. HNO$_3$ + H$_2$O$_2$ (1995)	Germany conc. HNO$_3$	Northwest Territories oven-dried	Europe oven-dried	Canada ashed total (INAA)
Median	-	0.102	-	-	-	14
Min	-	0.02	-	-	-	<5
Max	-	1.1	-	-	-	65

*Mean   #Geometric mean

Concentrations in Human Fluids (mg/l)

Medium	Human blood	Human serum	Human urine			
	Lombardy, Italy	Lombardy, Italy	Lombardy, Italy			
Mean	-	-	-			
Min	-	-	-			
Max	-	-	-			

NIOBIUM   Nb

Environmental Geochemistry

Biological impacts	Considered non-essential for humans and plants. Little is known about toxicity of Nb
Uses	Alloys (e.g., carbon and stainless steel, Ti, Al), welding, nuclear industry
Environmental pathways	Geogenic sources likely to be more important than anthropogenic ones
Environmental mobility	Oxidising conditions: very low Acid conditions: very low Reducing conditions: very low Neutral to alkaline conditions: very low Comments: -
Geochemical barriers	Mechanical
Natural association	Nb-Ta-Sn-W-Li-Be-Ti-Rb-Cs-U-Th-B-Zr-Hf-P-F-REE (granites and syenitic pegmatites), Nb-Ta-Na-K-Ba-Sr-Ti-Zr-U-Th-Cu-Zn-P-S-F-REE (carbonatites), Nb-Ta-Ti-Ga-Be-Al (bauxite developed on alkaline rocks)
Action levels	Drinking water, MAC: 0.01 mg/l (Russia MoH)
Remarks	Some organisms tend to accumulate Nb (e.g., ascidian (sea squirt)). Nb occurs invariably with Ta (Nb/Ta = 10)
Suggested analytical method(s)	ICP-MS

World Yearly Production (t/y)	18,300 Nb-Ta (in 1995)
Price of Purest Form, in Small Quantities ($/kg)	2908 (99.8%)
Market Price ($/kg)	7 (65% columbite concentrate)

Nd   NEODYMIUM

Physico-Chemical Properties

Atomic number	60	Melting/boiling point (K)	1294 / 3347
Atomic mass	144.24(3)	Isotopes & isomers	6 stable + 28 unstable
Atomic radius (pm)	264	Acid/base of oxide	base
Main oxidation state(s)	+2,+3,+5	State (at 300 K, 1 atm.)	solid
Ionic radius (pm)	143-149 (+2), 112-141 (+3)	Metallic character	metal
		Element group(s)	heavy metal, REE
Electronegativity (Pauling)	1.14	Affinity	lithophile
Density (g/cm^3)	7.008		

Naturally occurring isotopes	Natural abundance (%)	Atomic mass	Half-life
^{142}Nd	27.13	141.907,72	stable
^{143}Nd	12.18	142.909,81	stable
^{144}Nd	23.8	143.910,08	2.1x10^{+15} y
^{145}Nd	8.3	144.912,57	stable
^{146}Nd	17.19	145.913,11	stable
^{148}Nd	5.76	147.916,89	stable
^{150}Nd	5.64	149.920,89	stable

Unstable isotopes	Longest half-life
^{127}Nd to ^{156}Nd	2.1x10^{+15} y

Concentrations in Crust/Rocks (mg/kg)

Bulk continental crust	27 / 41.5 / 16
Upper continental crust	25.9 / 26
Ultramafic rock	-
Ocean ridge basalt	-
Gabbro, basalt	-
Granite, granodiorite	-
Sandstone	-
Greywacke	25
Shale, schist	-
Limestone	-
Coal	10

Typical Minerals
monazite ((Ce,La,Nd,Th)(PO$_4$,SiO$_4$)),
bastnaesite ((Ce,La,Nd)CO$_3$(F,OH)),
cerite ((Ce,La,Nd)$_9$(Mg,Fe)Si$_7$(O,OH,F)$_{28}$),
allanite ((Ca,Ce,La,Nd)$_2$FeAl$_2$OSi$_3$O$_{11}$(OH))

Possible Host Minerals
biotite, apatite, pyroxenes, feldspars, zircon

Mass (kg) in

Continental crust	6.37x10^{+17}	Oceans	3.97x10^{+9}	Plants	3.68x10^{+8}

Concentrations in Media Sampled for the Kola Project (mg/kg)

Medium	Moss	Humus <2 mm	Humus <2 mm	Topsoil (0-5 cm) <2 mm	B horizon <2 mm	B horizon <2 mm
	conc. HNO$_3$		amm. acet.	total (INAA)	aqua regia	total (XRF)
Median	-	-	-	7	-	-
Min	-	-	-	<5	-	-
Max	-	-	-	110	-	-

Medium	C horizon <2 mm	C horizon <2 mm	C horizon <2 mm		Lake water unfiltered (mg/l)
	aqua regia	total (XRF)	total (INAA)		
Median	-	-	18		-
Min	-	-	<5		-
Max	-	-	220		-

NEODYMIUM   Nd

Concentrations in Soils and Sediments (mg/kg)

Medium	Soil World <2 mm total	Agricultural soil - Ap horizon Canada <2 mm	Agricultural soil - Top (0-25 cm) Finland <2 mm aqua regia	Agricultural Soil - Bottom (50-75 cm) Finland <2 mm aqua regia	Topsoil (0-15 cm) England & Wales <2 mm aqua regia	Urban soil (0-2 cm) Trondheim <2 mm aqua regia
Median	-	-	-	-	-	-
Min	-	-	-	-	-	-
Max	-	-	-	-	-	-

*Estimated mean

Medium	Forest soil - Humus Norway <2 mm 7N HNO$_3$	Forest soil - B horizon Norway <2 mm 7N HNO$_3$	Forest soil - C horizon Norway <2 mm 7N HNO$_3$	Till (C horizon) Finland <0.063 mm aqua regia	Till (C horizon) Finland <0.063 mm total	Laterite (25 ±15 cm) Australia 0.45-2 mm total
Median	-	-	-	-	-	-
Min	-	-	-	-	-	-
Max	-	-	-	-	-	-

Medium	Stream sediment Austria <0.18 mm 	Stream sediment Southern Scotland <0.10 mm total	Stream sediment Harz, Germany <0.063 mm total	Overbank sediment Norway <0.063 mm total (XRF)	Overbank sediment Norway <0.063 mm 7N HNO$_3$	Organic stream sediment Finland <2 mm Conc. HNO$_3$
Median	-	-	-	-	-	-
Min	-	-	-	-	-	-
Max	-	-	-	-	-	-

#2nd percentile ^98th percentile

Concentrations in Waters (mg/l)

Medium	Ocean water World (1)	Ocean water World (2)	Ocean water North Pacific	Stream water World	Stream water Nova Scotia, Canada <0.45 μm ICP-MS	Stream water Finland <0.45 μm
Median	0.000,002,8*	0.000,003*	0.000,003,3*	-	0.000,146	-
Min	-	-	-	-	<0.000,005	-
Max	-	-	-	-	0.000,745	-

*Estimated mean   #2nd percentile   ^98th percentile

Medium	Stream water Romania unfiltered ICP-MS	Stream water Eastern India <0.2 μm ICP-MS	Stream water Harz, Germany unfiltered ICP-MS	Lake water Norway unfiltered ICP-MS	Ground water Southern Norway unfiltered ICP-MS	
Median	-	0.004,8	0.000,44	0.000,152	0.000,155	
Min	-	0.000,5	<0.000,001	<0.000,01	<0.000,005	
Max	-	0.059,8	0.005,4	0.002,5	0.049,082	

Nd — NEODYMIUM

Concentrations in Precipitation (mg/l)

Medium	Rain water Kola, remote <0.45 µm	Rain water Kola, polluted <0.45 µm	Snow melt-water Kola, remote <0.45 µm	Snow melt-water Kola, polluted <0.45 µm	Snow filter residue Kola, remote >0.45 µm	Snow filter residue Kola, polluted >0.45 µm
Median	-	-	-	-	-	-
Min	-	-	-	-	-	-
Max	-	-	-	-	-	-

Medium	Rain water Coastal, Norway unfiltered ICP-MS	Rain water Inland, Norway unfiltered ICP-MS
Median	-	-
Min	-	-
Max	-	-

Concentrations in Air (ng/m^3) and Yearly Deposition (kg/km^2/yr)

Medium	Air World, remote	Air World, polluted	Bulk deposition West Germany	Throughfall deposition West Germany	Bulk deposition Kola, remote	Bulk deposition Kola, polluted
Median	-	-	-	-	-	-
Min	-	-	-	-	-	-
Max	-	-	-	-	-	-

*Estimated value

Concentrations in Plants (mg/kg)

Medium	Moss Norway conc. HNO$_3$ (1990)	Moss Germany conc. HNO$_3$ + H$_2$O$_2$ (1995)	Crustose lichen Germany conc. HNO$_3$	Lichen Northwest Territories oven-dried	Dandelion Europe oven-dried	Spruce bark Canada ashed
Median	-	-	7	-	-	-
Min	-	-	1.6	-	-	-
Max	-	-	75	-	-	-

*Mean   #Geometric mean

Concentrations in Human Fluids (mg/l)

Medium	Human blood Lombardy, Italy	Human serum Lombardy, Italy	Human urine Lombardy, Italy
Mean	0.001,39	-	0.003,84
Min	0.000,075	-	0.000,2
Max	0.003,75	-	0.012

NEODYMIUM Nd

Environmental Geochemistry

Biological impacts	Considered non-essential. Data to assess the toxicity of REE are scarce
Uses	Magnets, lighters, lasers, glass additives, magneto-optical materials, electrical ceramics, high-temperature glazes
Environmental pathways	Poorly understood. Probably mostly windblown dust, weathering of REE-bearing minerals. Generally, geogenic sources more important than anthropogenic ones
Environmental mobility	Oxidising conditions: very low Acid conditions: very low Reducing conditions: very low Neutral to alkaline conditions: very low Comments: -
Geochemical barriers	Mechanical
Natural association	REE-Li-Rb-Cs-Be-Nb-Ta-Zr-B-Th-U-F (pegmatites), REE-Th-P-Zr-Fe-Cu (monazite veins), REE-Th-Ba-Sr-P-F-N-C (carbonatites), REE-U-P-F (phosphorites), REE-Au-Ti-Sn-Sr-Th (placers)
Action levels	-
Remarks	Toxicity of REE decreases as atomic number increases. Inhaled REE probably cause pneumoconiosis. REE taken up orally (e.g., via drinking water) accumulate in the skeleton, teeth, lungs, liver and kidneys. Some trees tend to accumulate Nd (e.g., hickory (*Carya*)). In 1991, 350 t Nd-oxide were used in the glass industry, and 100 t in the ceramics industry
Suggested analytical method(s)	ICP-MS

World Yearly Production (t/y)	54,000 REE-minerals (in 1991)
Price of Purest Form, in Small Quantities ($/kg)	71,375 (99.99%)
Market Price ($/kg)	-

Ne — NEON

Physico-Chemical Properties

Atomic number	10	Melting/boiling point (K)	24.56 / 27.07
Atomic mass	20.179,7(6)	Isotopes & isomers	3 stable + 11 unstable
Atomic radius (pm)	51	Acid/base of oxide	-
Main oxidation state(s)	0	State (at 300 K, 1 atm.)	gas
Ionic radius (pm)	-	Metallic character	-
Electronegativity (Pauling)	-	Element group(s)	noble gas
Density (g/cm^3)	0.000,899,9	Affinity	atmophile

Naturally occurring isotopes	Natural abundance (%)	Atomic mass	Half-life
^{20}Ne	90.48	19.992,44	stable
^{21}Ne	0.27	20.993,85	stable
^{22}Ne	9.25	21.991,39	stable

Unstable isotopes	Longest half-life
^{16}Ne to ^{29}Ne	3.38 min

Concentrations in Crust/Rocks (mg/kg)

Bulk continental crust	- / 0.005 / -
Upper continental crust	- / -
Ultramafic rock	-
Ocean ridge basalt	-
Gabbro, basalt	-
Granite, granodiorite	-
Sandstone	-
Greywacke	-
Shale, schist	-
Limestone	-
Coal	-

Typical Minerals
-

Possible Host Minerals
-

Mass (kg) in

Continental crust	1.18x10^{+14}	Oceans	1.59x10^{+11}	Plants	-

Concentrations in Media Sampled for the Kola Project (mg/kg)

Medium	Moss conc. HNO$_3$	Humus <2 mm	Humus <2 mm amm. acet.	Topsoil (0-5 cm) <2 mm	B horizon <2 mm aqua regia	B horizon <2 mm total (XRF)
Median	-	-	-	-	-	-
Min	-	-	-	-	-	-
Max	-	-	-	-	-	-

Medium	C horizon <2 mm aqua regia	C horizon <2 mm total (XRF)	C horizon <2 mm total (INAA)		Lake water unfiltered (mg/l)
Median	-	-	-		-
Min	-	-	-		-
Max	-	-	-		-

NEON   Ne

Concentrations in Soils and Sediments (mg/kg)

Medium	Soil	Agricultural soil - Ap horizon	Agricultural soil - Top (0-25 cm)	Agricultural Soil - Bottom (50-75 cm)	Topsoil (0-15 cm)	Urban soil (0-2 cm)
	World	Canada	Finland	Finland	England & Wales	Trondheim
	<2 mm total	<2 mm	<2 mm aqua regia	<2 mm aqua regia	<2 mm aqua regia	<2 mm aqua regia
Median	-	-	-	-	-	-
Min	-	-	-	-	-	-
Max	-	-	-	-	-	-

*Estimated mean

Medium	Forest soil - Humus	Forest soil - B horizon	Forest soil - C horizon	Till (C horizon)	Till (C horizon)	Laterite (25 ±15 cm)
	Norway	Norway	Norway	Finland	Finland	Australia
	<2 mm 7N HNO$_3$	<2 mm 7N HNO$_3$	<2 mm 7N HNO$_3$	<0.063 mm aqua regia	<0.063 mm total	0.45-2 mm total
Median	-	-	-	-	-	-
Min	-	-	-	-	-	-
Max	-	-	-	-	-	-

Medium	Stream sediment	Stream sediment	Stream sediment	Overbank sediment	Overbank sediment	Organic stream sediment
	Austria	Southern Scotland	Harz, Germany	Norway	Norway	Finland
	<0.18 mm total	<0.10 mm total	<0.063 mm total	<0.063 mm total (XRF)	<0.063 mm 7N HNO$_3$	<2 mm Conc. HNO$_3$
Median	-	-	-	-	-	-
Min	-	-	-	-	-	-
Max	-	-	-	-	-	-

#2nd percentile   ^98th percentile

Concentrations in Waters (mg/l)

Medium	Ocean water	Ocean water	Ocean water	Stream water	Stream water	Stream water
	World (1)	World (2)	North Pacific	World	Nova Scotia, Canada	Finland
					<0.45 μm	<0.45 μm
Median	0.000,12*	0.000,12*	0.000,16*	-	-	-
Min	-	-	-	-	-	-
Max	-	-	-	-	-	-

*Estimated mean   #2nd percentile   ^98th percentile

Medium	Stream water	Stream water	Stream water	Lake water	Ground water	
	Romania	Eastern India	Harz, Germany	Norway	Southern Norway	
	unfiltered ICP-MS	<0.2 μm ICP-MS	unfiltered	unfiltered ICP-MS	unfiltered	
Median	-	-	-	-	-	
Min	-	-	-	-	-	
Max	-	-	-	-	-	

Ne — NEON

Concentrations in Precipitation (mg/l)

Medium	Rain water Kola, remote <0.45 µm	Rain water Kola, polluted <0.45 µm	Snow melt- water Kola, remote <0.45 µm	Snow melt- water Kola, polluted <0.45 µm	Snow filter residue Kola, remote >0.45 µm	Snow filter residue Kola, polluted >0.45 µm
Median	-	-	-	-	-	-
Min	-	-	-	-	-	-
Max	-	-	-	-	-	-

Medium	Rain water Coastal, Norway unfiltered ICP-MS	Rain water Inland, Norway unfiltered ICP-MS				
Median	-	-				
Min	-	-				
Max	-	-				

Concentrations in Air (ng/m^3) and Yearly Deposition (kg/km^2/yr)

Medium	Air World, remote	Air World, polluted	Bulk deposition West Germany	Throughfall deposition West Germany	Bulk deposition Kola, remote	Bulk deposition Kola, polluted
Median	-	-	-	-	-	-
Min	-	-	-	-	-	-
Max	-	-	-	-	-	-

*Estimated value

Concentrations in Plants (mg/kg)

Medium	Moss Norway conc. HNO$_3$ (1990)	Moss Germany conc. HNO$_3$ + H$_2$O$_2$ (1995)	Crustose lichen Germany conc. HNO$_3$	Lichen Northwest Territories oven-dried	Dandelion Europe oven-dried	Spruce bark Canada ashed
Median	-	-	-	-	-	-
Min	-	-	-	-	-	-
Max	-	-	-	-	-	-

*Mean #Geometric mean

Concentrations in Human Fluids (mg/l)

Medium	Human blood Lombardy, Italy	Human serum Lombardy, Italy	Human urine Lombardy, Italy			
Mean	-	-	-			
Min	-	-	-			
Max	-	-	-			

NEON   Ne

Environmental Geochemistry

Biological impacts	Considered non-essential
Uses	Cooling agent, fluorescent lights (fill gas)
Environmental pathways	Natural trace component of atmosphere
Environmental mobility	Oxidising conditions: very high Acid conditions: very high Reducing conditions: very high Neutral to alkaline conditions: very high Comments: Inert gas
Geochemical barriers	-
Natural association	-
Action levels	-
Remarks	Occurs in atmosphere at low concentrations (16 mg/m^3), together with the other noble gasses
Suggested analytical method(s)	-

World Yearly Production (t/y)	-
Price of Purest Form, in Small Quantities ($/kg)	146 (99.999%)
Market Price ($/kg)	350 $/m^3

Ni   NICKEL

Physico-Chemical Properties

Atomic number	28	Melting/boiling point (K)	1728 / 3186
Atomic mass	58.693,4(2)	Isotopes & isomers	5 stable + 19 unstable
Atomic radius (pm)	162	Acid/base of oxide	base
Main oxidation state(s)	0,+2 (-1,+1,+3,+4)	State (at 300 K, 1 atm.)	solid
Ionic radius (pm)	63-83 (+2), 70-74 (+3), 62 (+4)	Metallic character	metal
		Element group(s)	heavy metal
Electronegativity (Pauling)	1.91	Affinity	chalcophile, siderophile
Density (g/cm^3)	8.902		

Naturally occurring isotopes	Natural abundance (%)	Atomic mass	Half-life
^{58}Ni	68.077	57.935,348	stable
^{60}Ni	26.233	59.930,79	stable
^{61}Ni	1.14	60.931,06	stable
^{62}Ni	3.634	61.928,348	stable
^{64}Ni	0.926	63.927,969	stable

Unstable isotopes	Longest half-life
^{51}Ni to ^{74}Ni	7.6x10^{+4} y

Concentrations in Crust/Rocks (mg/kg)

Bulk continental crust	56 / 84 / 105
Upper continental crust	18.6 / 20
Ultramafic rock	2000
Ocean ridge basalt	140
Gabbro, basalt	130
Granite, granodiorite	5
Sandstone	2
Greywacke	24
Shale, schist	70
Limestone	5
Coal	20

Typical Minerals
nickeline (NiAs), gersdorffite (NiAsS), pentlandite ((Fe,Ni)$_9$S$_8$), Ni-pyrrhotite (Fe$_{1-x}$S, containing up to 5% Ni), kullerudite (NiSe$_2$), ullmannite (NiSbS), polydymite (Ni$_3$S$_4$), garnierite ((Ni,Mg)$_3$Si$_2$O$_5$(OH)$_4$)

Possible Host Minerals
olivine, pyroxenes, amphiboles, micas, garnets, pyrite, chalcopyrite

Mass (kg) in

Continental crust	1.32x10^{+18}	Oceans	2.25x10^{+12}	Plants	2.67x10^{+9}

Concentrations in Media Sampled for the Kola Project (mg/kg)

Medium	Moss conc. HNO$_3$	Humus <2 mm conc. HNO$_3$	Humus <2 mm amm. acet.	Topsoil (0-5 cm) <2 mm total (INAA)	B horizon <2 mm aqua regia	B horizon <2 mm total (XRF)
Median	5.39	9.18	0.82	<20	16.1	-
Min	0.96	1.5	<0.05	<20	1.4	-
Max	396	2880	429	1300	179	-

Medium	C horizon <2 mm aqua regia	C horizon <2 mm total (XRF)	C horizon <2 mm total (INAA)		Lake water unfiltered (mg/l)
Median	18.6	-	<20		0.000,69
Min	1.2	-	<20		0.000,01
Max	228	-	280		0.304

NICKEL Ni

Concentrations in Soils and Sediments (mg/kg)

Medium	Soil	Agricultural soil - Ap horizon	Agricultural soil - Top (0-25 cm)	Agricultural Soil - Bottom (50-75 cm)	Topsoil (0-15 cm)	Urban soil (0-2 cm)
	World	Canada	Finland	Finland	England & Wales	Trondheim
	<2 mm	<2 mm	<2 mm	<2 mm	<2 mm	<2 mm
	total	total (AAS)	aqua regia	aqua regia	aqua regia	aqua regia
Median	20*	19	8.11	10.5	22.6	45
Min	-	3	<2	<2	0.8	6
Max	-	115	60.1	59.7	440	231

*Estimated mean

Medium	Forest soil - Humus	Forest soil - B horizon	Forest soil - C horizon	Till (C horizon)	Till (C horizon)	Laterite (25 ±15 cm)
	Norway	Norway	Norway	Finland	Finland	Australia
	<2 mm	<2 mm	<2 mm	<0.063 mm	<0.063 mm	0.45-2 mm
	7N HNO$_3$	7N HNO$_3$	7N HNO$_3$	aqua regia	total	total
Median	3.2	8.5	19.1	16.7	24.1	12
Min	<1	<2	<2	-	-	<2
Max	487.2	148	435	-	-	890

Medium	Stream sediment	Stream sediment	Stream sediment	Overbank sediment	Overbank sediment	Organic stream sediment
	Austria	Southern Scotland	Harz, Germany	Norway	Norway	Finland
	<0.18 mm	<0.10 mm	<0.063 mm	<0.063 mm	<0.063 mm	<2 mm
	"total" (ICP-AES)	total (DCES)	total (XRF)	total (XRF)	7N HNO$_3$ ICP-AES	Conc. HNO$_3$
Median	30	59	28	41	44	13.9
Min	-	<6	<7	16	2	4.5#
Max	2471	244	508	1509	1100	52.4^

#2nd percentile   ^98th percentile

Concentrations in Waters (mg/l)

Medium	Ocean water	Ocean water	Ocean water	Stream water	Stream water	Stream water
	World (1)	World (2)	North Pacific	World	Nova Scotia, Canada	Finland
					<0.45 μm	<0.45 μm
					ICP-MS	ICP-MS
Median	0.000,56*	0.001,7*	0.000,48*	0.000,3*	<0.000,2	0.000,52
Min	-	-	-	-	<0.000,2	0.000,09#
Max	-	-	-	-	0.001,3	0.010,4^

*Estimated mean   #2nd percentile   ^98th percentile

Medium	Stream water	Stream water	Stream water	Lake water	Ground water
	Romania	Eastern India	Harz, Germany	Norway	Southern Norway
	unfiltered	<0.2 μm	unfiltered	unfiltered	unfiltered
	ICP-MS	ICP-MS	ICP-MS	ICP-MS	ICP-MS
Median	0.003,4	0.007,6	0.002,9	0.000,328	0.000,74
Min	0.000,2	0.003,1	<0.000,1	<0.000,1	<0.000,02
Max	0.198	0.041,4	0.039	0.004,82	0.014,3

Ni NICKEL

Concentrations in Precipitation (mg/l)

Medium	Rain water	Rain water	Snow melt-water	Snow melt-water	Snow filter residue	Snow filter residue
	Kola, remote <0.45 µm ICP-MS	Kola, polluted <0.45 µm ICP-MS	Kola, remote <0.45 µm ICP-MS	Kola, polluted <0.45 µm ICP-MS	Kola, remote >0.45 µm ICP-AES	Kola, polluted >0.45 µm ICP-AES
Median	0.000,09	0.057	0.000,17	0.258	<0.000,1	0.595
Min	<0.000,06	0.024	0.000,16	0.209	<0.000,1	0.38
Max	0.000,21	0.132	0.000,37	0.708	0.000,2	2.23

Medium	Rain water	Rain water
	Coastal, Norway unfiltered ICP-MS	Inland, Norway unfiltered ICP-MS
Median	0.000,3	0.000,66
Min	<0.000,01	0.000,02
Max	0.005,5	0.007

Concentrations in Air (ng/m^3) and Yearly Deposition (kg/km^2/yr)

Medium	Air	Air	Bulk deposition	Throughfall deposition	Bulk deposition	Bulk deposition
	World, remote	World, polluted	West Germany	West Germany	Kola, remote	Kola, polluted
Median	- / 0.9	60	-	-	0.22*	845*
Min	0.08	<1	0.2	1.37	-	-
Max	3.2	120	5.1	1.8	-	-

*Estimated value

Concentrations in Plants (mg/kg)

Medium	Moss	Moss	Crustose lichen	Lichen	Dandelion	Spruce bark
	Norway conc. HNO$_3$ (1990)	Germany conc. HNO$_3$ + H$_2$O$_2$ (1995)	Germany conc. HNO$_3$	Northwest Territories oven-dried total (XRF)	Europe oven-dried conc. HNO$_3$ (AAS)	Canada ashed aqua regia
Median	1.6	1.643	-	2.5*	2.87#	40
Min	0.5	0.41	-	-	-	8
Max	320	16	-	-	-	122

*Mean #Geometric mean

Concentrations in Human Fluids (mg/l)

Medium	Human blood	Human serum	Human urine
	Lombardy, Italy	Lombardy, Italy	Lombardy, Italy
Mean	0.002,3	0.001,2	0.000,9
Min	0.000,6	0.000,24	0.000,1
Max	0.003,8	0.003,7	0.003,9

NICKEL Ni

Environmental Geochemistry

Biological impacts	Essential for some organisms. Ni^{2+} compounds relatively non-toxic. Some other compounds extremely toxic and/or carcinogenic. Known to cause allergies
Uses	More than 3000 known alloys (e.g., with Fe, Zn, Mn, Co, Ti, Mo), electroplating, batteries, pigments, catalysts (e.g., in margarine production), magnetic tapes
Environmental pathways	Cu-Ni smelters, steel works, chemical industry, petroleum refining, waste disposal and incineration, sewage sludge, fertilisers, traffic, fuel combustion, weathering, geogenic dust, volcanoes
Environmental mobility	Oxidising conditions: medium Acid conditions: high Reducing conditions: very low Neutral to alkaline conditions: very low Comments: -
Geochemical barriers	Sulphide, adsorption, pH
Natural association	Ni-Co-Fe-Cu-Au-Ag-PGE-Se-Te-As-S (massive sulphide deposits, e.g., Sudbury, Norilsk), Ni-Co-Fe-Cu-S (veins in sulphide lenses), U-Cu-Ag-Co-Ni-As-V-Se-Au-Mo (unconformity U deposits), Ni-Co-Fe-Mn-Cr (residual laterite deposits), Mn-Ni-Cu-Co (deep-sea Mn nodules)
Action levels	Drinking water, MAC: 0.05 mg/l (Norway Shd), 0.1 mg/l (US EPA); recommended: 0.02 mg/l (WHO); MAC: 0.1 mg/l (Russia MoH). Ground water, background: 0.015 mg/l; remediate: 0.075 mg/l (Netherlands VROM). Soil, background: 35 mg/kg; remediate: 210 mg/kg (Netherlands VROM)
Remarks	Ni deficiency results in growth retardation in animals. High levels in soils can result in plant growth problems (chlorosis, plant death). Acid rain can mobilise Ni in soils. Phosphate fertilisers increase the Ni availability. Liming and K-fertilisers reduce Ni availability. Most Ni compounds are relatively soluble at pH<6.5, but insoluble at pH>6.7. More than 70 plants known to accumulate Ni (e.g., kale, nuts and cocoa). Other plants do not take up Ni (e.g., corn, celery, potato, cabbage, carrot, spinach). The US EPA suggests that drinking water exceeding action levels does not lead to any health problems in the short term, and can lead to decreased body weight, heart and liver damage, and skin irritation in the long term
Suggested analytical method(s)	ICP-AES, XRF, ICP-MS (water)

World Yearly Production (t/y)	1,013,000 (in 1995)
Price of Purest Form, in Small Quantities ($/kg)	10,010 (99.99%)
Market Price ($/kg)	6.4

Np NEPTUNIUM

Physico-Chemical Properties

Atomic number	93	Melting/boiling point (K)	917 / -
Atomic mass	[237]	Isotopes & isomers	0 stable + 20 unstable
Atomic radius (pm)	-	Acid/base of oxide	amphoteric
Main oxidation state(s)	+5 (+3,+4,+6)	State (at 300 K, 1 atm.)	solid
Ionic radius (pm)	124 (+2), 115 (+3), 101-112 (+4), 89 (+5), 86 (+6), 85 (+7)	Metallic character	metal
		Element group(s)	heavy metal
Electronegativity (Pauling)	1.36	Affinity	-
Density (g/cm^3)	20.25		

Naturally occurring isotopes	Natural abundance (%)	Atomic mass	Half-life
-	-	-	-

Unstable isotopes	Longest half-life
^{226}Np to ^{242}Np	2,140,000 y

Concentrations in Crust/Rocks (mg/kg)

Bulk continental crust	- / - / -
Upper continental crust	- / -
Ultramafic rock	-
Ocean ridge basalt	-
Gabbro, basalt	-
Granite, granodiorite	-
Sandstone	-
Greywacke	-
Shale, schist	-
Limestone	-
Coal	-

Typical Minerals
-

Possible Host Minerals
uraninite/pitchblende, carnotite

Mass (kg) in

Continental crust	-	Oceans	-	Plants	-

Concentrations in Media Sampled for the Kola Project (mg/kg)

Medium	Moss	Humus	Humus	Topsoil (0-5 cm)	B horizon	B horizon
	<2 mm	<2 mm	<2 mm	<2 mm	<2 mm	<2 mm
	conc. HNO$_3$		amm. acet.		aqua regia	total (XRF)
Median	-	-	-	-	-	-
Min	-	-	-	-	-	-
Max	-	-	-	-	-	-

Medium	C horizon	C horizon	C horizon			Lake water
	<2 mm	<2 mm	<2 mm			unfiltered
	aqua regia	total (XRF)	total (INAA)			(mg/l)
Median	-	-	-			-
Min	-	-	-			-
Max	-	-	-			-

NEPTUNIUM Np

Environmental Geochemistry

Biological impacts	Considered non-essential. Radiotoxic
Uses	Neutron detectors
Environmental pathways	Nuclear reactor accidents
Environmental mobility	Oxidising conditions: - Acid conditions: - Reducing conditions: - Neutral to alkaline conditions: - Comments: -
Geochemical barriers	-
Natural association	U-Pu-Np
Action levels	-
Remarks	Biological half-life in the human body is about 100 years. Trace quantities of Np are present in nature as a result of transmutation reactions in U ores produced by the neutrons present. Np forms tri- and tetra-halides and oxides of various compositions. ^{237}Np is available at a price of 660,000 $/kg plus packaging (closed marked)
Suggested analytical method(s)	-

World Yearly Production (t/y)	0.00x
Price of Purest Form, in Small Quantities ($/kg)	-
Market Price ($/kg)	-

O — OXYGEN

Physico-Chemical Properties

Atomic number	8	Melting/boiling point (K)	54.36 / 90.2
Atomic mass	15.999,4(3)	Isotopes & isomers	3 stable + 10 unstable
Atomic radius (pm)	65	Acid/base of oxide	-
Main oxidation state(s)	-2	State (at 300 K, 1 atm.)	gas
Ionic radius (pm)	121-128	Metallic character	non-metal
Electronegativity (Pauling)	3.44	Element group(s)	light non-metal
Density (g/cm^3)	0.001,429	Affinity	atmophile, biophile, lithophile

Naturally occurring isotopes	Natural abundance (%)	Atomic mass	Half-life	Unstable isotopes	Longest half-life
^{16}O	99.762	15.994,915	stable	^{12}O to ^{24}O	122.2 s
^{17}O	0.038	16.999,132	stable		
^{18}O	0.2	17.999,16	stable		

Concentrations in Crust/Rocks (mg/kg)

Bulk continental crust	472,000 / 461,000 / -
Upper continental crust	- / -
Ultramafic rock	-
Ocean ridge basalt	-
Gabbro, basalt	-
Granite, granodiorite	-
Sandstone	-
Greywacke	-
Shale, schist	-
Limestone	-
Coal	133,000

Typical Minerals

simple oxides, e.g. quartz (SiO$_2$), ice (H$_2$O), rutile (TiO$_2$), zincite (ZnO), corundum (Al$_2$O$_3$);
complex oxides, e.g., zircon (ZrSiO$_4$), spinel (MgAl$_2$O$_4$), brookite (Fe$_2$TiO$_5$);
compounds, e.g., culcite (CaCO$_3$), barite (BaSO$_4$), scheelite (CaWO$_4$);
hydroxides and hydrates, e.g., goethite (FeO(OH)), diaspore (AlO(OH)), gypsum (CaSO$_4 \cdot$ 2H$_2$O)

Possible Host Minerals

.

Mass (kg) in

Continental crust	1.11x10^{+22}	Oceans	1.14x10^{+21}	Plants	7.82x10^{+14}

Concentrations in Media Sampled for the Kola Project (mg/kg)

| Medium | Moss | Humus | Humus | Topsoil (0-5 cm) | B horizon | B horizon |
| | <2 mm | <2 mm | <2 mm | <2 mm | <2 mm | <2 mm |
	conc. HNO$_3$		amm. acet.		aqua regia	total (XRF)
Median	-	-	-	-	-	-
Min	-	-	-	-	-	-
Max	-	-	-	-	-	-

| Medium | C horizon | C horizon | C horizon | | | Lake water |
| | <2 mm | <2 mm | <2 mm | | | unfiltered |
	aqua regia	total (XRF)	total (INAA)			(mg/l)
Median	-	-	-			-
Min	-	-	-			-
Max	-	-	-			-

OXYGEN — O

Concentrations in Soils and Sediments (mg/kg)

Medium	Soil	Agricultural soil - Ap horizon	Agricultural soil - Top (0-25 cm)	Agricultural Soil - Bottom (50-75 cm)	Topsoil (0-15 cm)	Urban soil (0-2 cm)
	World	Canada	Finland	Finland	England & Wales	Trondheim
	<2 mm	<2 mm	<2 mm	<2 mm	<2 mm	<2 mm
	total	aqua regia	aqua regia	aqua regia	aqua regia	aqua regia
Median	-	-	-	-	-	-
Min	-	-	-	-	-	-
Max	-	-	-	-	-	-

*Estimated mean

Medium	Forest soil - Humus	Forest soil - B horizon	Forest soil - C horizon	Till (C horizon)	Till (C horizon)	Laterite (25 ±15 cm)
	Norway	Norway	Norway	Finland	Finland	Australia
	<2 mm	<2 mm	<2 mm	<0.063 mm	<0.063 mm	0.45-2 mm
	7N HNO$_3$	7N HNO$_3$	7N HNO$_3$	aqua regia	total	total
Median	-	-	-	-	-	-
Min	-	-	-	-	-	-
Max	-	-	-	-	-	-

Medium	Stream sediment	Stream sediment	Stream sediment	Overbank sediment	Overbank sediment	Organic stream sediment
	Austria	Southern Scotland	Harz, Germany	Norway	Norway	Finland
	<0.18 mm	<0.10 mm	<0.063 mm	<0.063 mm	<0.063 mm	<2 mm
		total	total	total (XRF)	7N HNO$_3$	Conc. HNO$_3$
Median	-	-	-	-	-	-
Min	-	-	-	-	-	-
Max	-	-	-	-	-	-

#2nd percentile ^98th percentile

Concentrations in Waters (mg/l)

Medium	Ocean water	Ocean water	Ocean water	Stream water	Stream water	Stream water
	World (1)	World (2)	North Pacific	World	Nova Scotia, Canada	Finland
					<0.45 µm	<0.45 µm
Median	857,000*	6*	2.8*	-	-	-
Min	-	(dissolved	(dissolved	-	-	-
Max	-	oxygen?)	oxygen?)	-	-	-

*Estimated mean   #2nd percentile   ^98th percentile

Medium	Stream water	Stream water	Stream water	Lake water	Ground water	
	Romania	Eastern India	Harz, Germany	Norway	Southern Norway	
	unfiltered	<0.2 µm	unfiltered	unfiltered	unfiltered	
	ICP-MS	ICP-MS		ICP-MS		
Median	-	-	-	-	-	
Min	-	-	-	-	-	
Max	-	-	-	-	-	

O OXYGEN

Concentrations in Precipitation (mg/l)

Medium	Rain water	Rain water	Snow melt-water	Snow melt-water	Snow filter residue	Snow filter residue
	Kola, remote <0.45 µm	Kola, polluted <0.45 µm	Kola, remote <0.45 µm	Kola, polluted <0.45 µm	Kola, remote >0.45 µm	Kola, polluted >0.45 µm
Median	-	-	-	-	-	-
Min	-	-	-	-	-	-
Max	-	-	-	-	-	-

Medium	Rain water	Rain water				
	Coastal, Norway unfiltered ICP-MS	Inland, Norway unfiltered ICP-MS				
Median	-	-				
Min	-	-				
Max	-	-				

Concentrations in Air (ng/m^3) and Yearly Deposition (kg/km^2/yr)

Medium	Air	Air	Bulk deposition	Throughfall deposition	Bulk deposition	Bulk deposition
	World, remote	World, polluted	West Germany	West Germany	Kola, remote	Kola, polluted
Median	-	-	-	-	-	-
Min	-	-	-	-	-	-
Max	-	-	-	-	-	-

*Estimated value

Concentrations in Plants (mg/kg)

Medium	Moss	Moss	Crustose lichen	Lichen	Dandelion	Spruce bark
	Norway conc. HNO$_3$ (1990)	Germany conc. HNO$_3$ + H$_2$O$_2$ (1995)	Germany conc. HNO$_3$	Northwest Territories oven-dried	Europe oven-dried	Canada ashed
Median	-	-	-	-	-	-
Min	-	-	-	-	-	-
Max	-	-	-	-	-	-

*Mean #Geometric mean

Concentrations in Human Fluids (mg/l)

Medium	Human blood	Human serum	Human urine			
	Lombardy, Italy	Lombardy, Italy	Lombardy, Italy			
Mean	-	-	-			
Min	-	-	-			
Max	-	-	-			

OXYGEN — O

Environmental Geochemistry

Biological impacts	Essential for most organisms. Almost never toxic, but O_3 (ozone) is toxic. O is part of many toxic compounds
Uses	Many industrial processes require oxygen, welding, sewage treatment, water treatment
Environmental pathways	Major natural component of atmosphere
Environmental mobility	Oxidising conditions: - Acid conditions: - Reducing conditions: - Neutral to alkaline conditions: - Comments: Highly dependent on compound
Geochemical barriers	-
Natural association	C-H-O-(N)-(S) (organic matter), H-O (water)
Action levels	-
Remarks	Most abundant element in the Earth's crust. Occurs in over half of the known minerals. Major constituent of Earth atmosphere. Too high concentrations lead to muscular cramping. Isotopic fractionation important in nature
Suggested analytical method(s)	-

World Yearly Production (t/y)	-
Price of Purest Form, in Small Quantities ($/kg)	146 $/l (99.99%)
Market Price ($/kg)	1.75 $/m^3

Os   OSMIUM

Physico-Chemical Properties

Atomic number	76	Melting/boiling point (K)	3306 / 5285
Atomic mass	190.23(3)	Isotopes & isomers	6 stable + 35 unstable
Atomic radius (pm)	192	Acid/base of oxide	weak acid
Main oxidation state(s)	+3,+4,+6,+8 (0,+1,+2,+5,+7)	State (at 300 K, 1 atm.)	solid
		Metallic character	pred. metal
Ionic radius (pm)	77 (+4), 71.5 (+5), 63-68.5 (+6), 66.5 (+7), 53 (+8)	Element group(s)	heavy metal, noble metal, PGE
		Affinity	siderophile
Electronegativity (Pauling)	2.2		
Density (g/cm^3)	22.57		

Naturally occurring isotopes	Natural abundance (%)	Atomic mass	Half-life
^{184}Os	0.02	183.952,49	stable
^{186}Os	1.58	185.953,84	2x10^{+15} y
^{187}Os	1.6	186.955,75	stable
^{188}Os	13.2	187.955,84	stable
^{189}Os	16.1	188.958,15	stable
^{190}Os	26.4	189.958,45	stable
^{192}Os	41	191.961,48	stable

Unstable isotopes	Longest half-life
^{162}Os to ^{196}Os	2x10^{+15} y

Concentrations in Crust/Rocks (mg/kg)

Bulk continental crust	0.000,05 / 0.001,5 / 0.000,05
Upper continental crust	- / 0.000,05
Ultramafic rock	-
Ocean ridge basalt	-
Gabbro, basalt	-
Granite, granodiorite	-
Sandstone	-
Greywacke	-
Shale, schist	-
Limestone	-
Coal	-

Typical Minerals
laurite ((Ru,Os)S$_2$)

Possible Host Minerals
olivine, ilmenite, zircon, chromite, gadolinite

Mass (kg) in

Continental crust	1.18x10^{+12}	Oceans	-	Plants	2.76x10^{+4}

Concentrations in Media Sampled for the Kola Project (mg/kg)

Medium	Moss <2 mm conc. HNO$_3$	Humus <2 mm	Humus <2 mm amm. acet.	Topsoil (0-5 cm) <2 mm	B horizon <2 mm aqua regia	B horizon <2 mm total (XRF)
Median	-	-	-	-	-	-
Min	-	-	-	-	-	-
Max	-	-	-	-	-	-

Medium	C horizon <2 mm aqua regia	C horizon <2 mm total (XRF)	C horizon <2 mm total (INAA)		Lake water unfiltered (mg/l)
Median	-	-	-		-
Min	-	-	-		-
Max	-	-	-		-

OSMIUM   Os

Concentrations in Soils and Sediments (mg/kg)

Medium	Soil	Agricultural soil - Ap horizon	Agricultural soil - Top (0-25 cm)	Agricultural Soil - Bottom (50-75 cm)	Topsoil (0-15 cm)	Urban soil (0-2 cm)
	World	Canada	Finland	Finland	England & Wales	Trondheim
	<2 mm total	<2 mm	<2 mm aqua regia	<2 mm aqua regia	<2 mm aqua regia	<2 mm aqua regia
Median	-	-	-	-	-	-
Min	-	-	-	-	-	-
Max	-	-	-	-	-	-

*Estimated mean

Medium	Forest soil - Humus	Forest soil - B horizon	Forest soil - C horizon	Till (C horizon)	Till (C horizon)	Laterite (25 ±15 cm)
	Norway	Norway	Norway	Finland	Finland	Australia
	<2 mm $7N\ HNO_3$	<2 mm $7N\ HNO_3$	<2 mm $7N\ HNO_3$	<0.063 mm aqua regia	<0.063 mm total	0.45-2 mm total
Median	-	-	-	-	-	-
Min	-	-	-	-	-	-
Max	-	-	-	-	-	-

Medium	Stream sediment	Stream sediment	Stream sediment	Overbank sediment	Overbank sediment	Organic stream sediment
	Austria	Southern Scotland	Harz, Germany	Norway	Norway	Finland
	<0.18 mm	<0.10 mm total	<0.063 mm total	<0.063 mm total (XRF)	<0.063 mm $7N\ HNO_3$	<2 mm Conc. HNO_3
Median	-	-	-	-	-	-
Min	-	-	-	-	-	-
Max	-	-	-	-	-	-

#2nd percentile   ^98th percentile

Concentrations in Waters (mg/l)

Medium	Ocean water	Ocean water	Ocean water	Stream water	Stream water	Stream water
	World (1)	World (2)	North Pacific	World	Nova Scotia, Canada	Finland
					<0.45 µm	<0.45 µm
Median	-	-	0.000,000,002*	-	-	-
Min	-	-	-	-	-	-
Max	-	-	-	-	-	-

*Estimated mean   #2nd percentile   ^98th percentile

Medium	Stream water	Stream water	Stream water	Lake water	Ground water
	Romania	Eastern India	Harz, Germany	Norway	Southern Norway
	unfiltered ICP-MS	<0.2 µm ICP-MS	unfiltered	unfiltered ICP-MS	unfiltered
Median	-	<0.000,01	-	<0.000,5	-
Min	-	<0.000,01	-	<0.000,5	-
Max	-	<0.000,01	-	0.000,948	-

Os   OSMIUM

Concentrations in Precipitation (mg/l)

Medium	Rain water Kola, remote <0.45 μm	Rain water Kola, polluted <0.45 μm	Snow melt-water Kola, remote <0.45 μm	Snow melt-water Kola, polluted <0.45 μm	Snow filter residue Kola, remote >0.45 μm	Snow filter residue Kola, polluted >0.45 μm
Median	-	-	-	-	-	-
Min	-	-	-	-	-	-
Max	-	-	-	-	-	-

Medium	Rain water Coastal, Norway unfiltered ICP-MS	Rain water Inland, Norway unfiltered ICP-MS				
Median	-	-				
Min	-	-				
Max	-	-				

Concentrations in Air (ng/m^3) and Yearly Deposition (kg/km^2/yr)

Medium	Air World, remote	Air World, polluted	Bulk deposition West Germany	Throughfall deposition West Germany	Bulk deposition Kola, remote	Bulk deposition Kola, polluted
Median	-	-	-	-	-	-
Min	-	-	-	-	-	-
Max	-	-	-	-	-	-

*Estimated value

Concentrations in Plants (mg/kg)

Medium	Moss Norway conc. HNO$_3$ (1990)	Moss Germany conc. HNO$_3$ + H$_2$O$_2$ (1995)	Crustose lichen Germany conc. HNO$_3$	Lichen Northwest Territories oven-dried	Dandelion Europe oven-dried	Spruce bark Canada ashed
Median	-	-	-	-	-	-
Min	-	-	-	-	-	-
Max	-	-	-	-	-	-

*Mean   #Geometric mean

Concentrations in Human Fluids (mg/l)

Medium	Human blood Lombardy, Italy	Human serum Lombardy, Italy	Human urine Lombardy, Italy			
Mean	-	-	-			
Min	-	-	-			
Max	-	-	-			

OSMIUM   Os

Environmental Geochemistry

Biological impacts	Considered non-essential. OsO_4 is very toxic
Uses	Hard alloys
Environmental pathways	Geogenic sources likely to be more important than anthropogenic ones
Environmental mobility	Oxidising conditions: very low Acid conditions: very low Reducing conditions: very low Neutral to alkaline conditions: very low Comments: -
Geochemical barriers	Mechanical
Natural association	PGE-Ni-Cu-Co-As-Ag-Au-Te-Bi-Sn-Sb (massive Ni-Cu sulphide ores, e.g., Sudbury), PGE-Ag-Au-Cr-Fe-Cu-Ni-Co-S (sulphides, e.g., Merensky reef)
Action levels	-
Remarks	Os-isotopes are the subject of scientific research. Very rich ore deposits contain about 8 ppm PGE. PGE occur often in Ni ores. PGE usually occur together in the following proportions: Pt 25%, Ru 23%, Os 21%, Pd 18%, Ir 7%, Rh 6%. Cosmic input via meteorites, cosmic dust, etc. significant. Some marine algae enrich PGE
Suggested analytical method(s)	ICP-MS, INAA

World Yearly Production (t/y)	300 all PGE (in 1995)
Price of Purest Form, in Small Quantities ($/kg)	100,000
Market Price ($/kg)	14,470

P   PHOSPHORUS

Physico-Chemical Properties

Atomic number	15	Melting/boiling point (K)	317.3 / 553.65
Atomic mass	30.973,761(2)	Isotopes & isomers	1 stable + 16 unstable
Atomic radius (pm)	123	Acid/base of oxide	weak acid
Main oxidation state(s)	+5 (-3,+3,+4)	State (at 300 K, 1 atm.)	solid
Ionic radius (pm)	58 (+3), 31-52 (+5)	Metallic character	non-metal
Electronegativity (Pauling)	2.19	Element group(s)	light non-metal
Density (g/cm^3)	1.82 (white), 2.35 (red), 2.7 (black)	Affinity	biophile, lithophile, siderophile

Naturally occurring isotopes	Natural abundance (%)	Atomic mass	Half-life	Unstable isotopes	Longest half-life
^{31}P	100	30.973,762	stable	^{26}P to ^{42}P	26.3 d

Concentrations in Crust/Rocks (mg/kg)

Bulk continental crust	757 / 1050 / -
Upper continental crust	665 / 700
Ultramafic rock	220
Ocean ridge basalt	800
Gabbro, basalt	1200
Granite, granodiorite	750
Sandstone	30
Greywacke	567
Shale, schist	800
Limestone	350
Coal	150

Typical Minerals
apatite ($Ca_5(PO_4,CO_3)_3(F,OH,Cl)$),
monazite ((Ce,La,Nd,Th,Sm)(PO_4,SiO_4)),
xenotime (YPO_4), and many other phosphates

Possible Host Minerals
olivine, garnets, pyroxenes, amphiboles, micas, feldspars

Mass (kg) in

Continental crust	1.79x10^{+19}	Oceans	7.93x10^{+13}	Plants	3.68x10^{+13}

Concentrations in Media Sampled for the Kola Project (mg/kg)

Medium	Moss conc. HNO_3	Humus <2 mm conc. HNO_3	Humus <2 mm amm. acet.	Topsoil (0-5 cm) <2 mm	B horizon <2 mm aqua regia	B horizon <2 mm total (XRF)
Median	1265	930	132	-	405	436
Min	511	192	25.6	-	64	87
Max	3800	9280	419	-	3260	3186

Medium	C horizon <2 mm aqua regia	C horizon <2 mm total (XRF)	C horizon <2 mm total (INAA)			Lake water unfiltered (mg/l)
Median	392	390	-			-
Min	59	40	-			-
Max	7170	5890	-			-

PHOSPHORUS P

Concentrations in Soils and Sediments (mg/kg)

Medium	Soil	Agricultural soil - Ap horizon	Agricultural soil - Top (0-25 cm)	Agricultural Soil - Bottom (50-75 cm)	Topsoil (0-15 cm)	Urban soil (0-2 cm)
	World	Canada	Finland	Finland	England & Wales	Trondheim
	<2 mm total	<2 mm	<2 mm aqua regia	<2 mm aqua regia	<2 mm aqua regia	<2 mm aqua regia
Median	750*	-	786	526	765	795
Min	-	-	<100	103	41	50
Max	-	-	2070	1660	6273	2480

*Estimated mean

Medium	Forest soil - Humus	Forest soil - B horizon	Forest soil - C horizon	Till (C horizon)	Till (C horizon)	Laterite (25 ±15 cm)
	Norway	Norway	Norway	Finland	Finland	Australia
	<2 mm $7N\ HNO_3$	<2 mm $7N\ HNO_3$	<2 mm $7N\ HNO_3$	<0.063 mm aqua regia	<0.063 mm total	0.45-2 mm total
Median	750	570	985	620	650	218
Min	80	50	10	-	-	<40
Max	2800	11,700	4900	-	-	2007

Medium	Stream sediment	Stream sediment	Stream sediment	Overbank sediment	Overbank sediment	Organic stream sediment
	Austria	Southern Scotland	Harz, Germany	Norway	Norway	Finland
	<0.18 mm "total" (ICP-AES)	<0.10 mm total	<0.063 mm total (XRF)	<0.063 mm total (XRF)	<0.063 mm $7N\ HNO_3$ ICP-AES	<2 mm Conc. HNO_3
Median	1050	-	655	1300	500	870
Min	-	-	175	100	10	390#
Max	13,240	-	3753	5600	4100	2113^

#2nd percentile ^98th percentile

Concentrations in Waters (mg/l)

Medium	Ocean water	Ocean water	Ocean water	Stream water	Stream water	Stream water
	World (1)	World (2)	North Pacific	World	Nova Scotia, Canada	Finland
					<0.45 μm	<0.45 μm
Median	0.06*	0.06*	0.062*	0.02*	-	-
Min	-	-	-	-	-	-
Max	-	-	-	-	-	-

*Estimated mean #2nd percentile ^98th percentile

Medium	Stream water	Stream water	Stream water	Lake water	Ground water
	Romania	Eastern India	Harz, Germany	Norway	Southern Norway
	unfiltered ICP-MS	<0.2 μm ICP-MS	unfiltered	unfiltered ICP-MS	unfiltered ICP-MS
Median	-	0.087	-	-	0.005,365
Min	-	0.048	-	-	<0.000,03
Max	-	1.286	-	-	0.74

P PHOSPHORUS

Concentrations in Precipitation (mg/l)

Medium	Rain water	Rain water	Snow melt-water	Snow melt-water	Snow filter residue	Snow filter residue
	Kola, remote <0.45 µm ICP-AES	Kola, polluted <0.45 µm ICP-AES	Kola, remote <0.45 µm ICP-AES	Kola, polluted <0.45 µm ICP-AES	Kola, remote >0.45 µm ICP-AES	Kola, polluted >0.45 µm ICP-AES
Median	<0.05	<0.05	<0.1	<0.1	0.000,8	0.028,4
Min	<0.05	<0.05	<0.1	<0.1	0.000,5	0.014,9
Max	<0.05	<0.05	<0.1	<0.1	0.002,8	0.084,7

Medium	Rain water	Rain water				
	Coastal, Norway unfiltered ICP-MS	Inland, Norway unfiltered ICP-MS				
Median	-	-				
Min	-	-				
Max	-	-				

Concentrations in Air (ng/m^3) and Yearly Deposition (kg/km^2/yr)

Medium	Air	Air	Bulk deposition	Throughfall deposition	Bulk deposition	Bulk deposition
	World, remote	World, polluted	West Germany	West Germany	Kola, remote	Kola, polluted
Median	-	-	-	-	-	-
Min	-	180	-	-	-	-
Max	-	1100	-	-	-	-

*Estimated value

Concentrations in Plants (mg/kg)

Medium	Moss	Moss	Crustose lichen	Lichen	Dandelion	Spruce bark
	Norway conc. HNO$_3$ (1990)	Germany conc. HNO$_3$ + H$_2$O$_2$ (1995)	Germany conc. HNO$_3$	Northwest Territories oven-dried	Europe oven-dried	Canada ashed aqua regia
Median	-	-	-	-	-	10,130
Min	-	-	-	-	-	1200
Max	-	-	-	-	-	43,180

*Mean   #Geometric mean

Concentrations in Human Fluids (mg/l)

Medium	Human blood	Human serum	Human urine			
	Lombardy, Italy	Lombardy, Italy	Lombardy, Italy			
Mean	-	-	-			
Min	-	-	-			
Max	-	-	-			

PHOSPHORUS P

Environmental Geochemistry

Biological impacts	Essential for all organisms. Toxic at high doses (60-100 mg P lethal to humans)
Uses	Detergents, fertilisers, chemical industry, military (explosives and warfare chemicals), pyrotechnics, semi-conductors, insecticide, herbicide, fungicide, matches
Environmental pathways	Agriculture, waste water, geogenic dust
Environmental mobility	Oxidising conditions: low Acid conditions: low Reducing conditions: low Neutral to alkaline conditions: low Comments: Some important compounds are very mobile, e.g., the dissolved phosphate anion
Geochemical barriers	-
Natural association	Phosphates of many elements occur in nature
Action levels	Drinking water (Norway), MAC: 5 mg/l P_2O_5 (=2.2 mg/l P)
Remarks	Uptake of P by plants can result in P deficiency in soil; phosphate (PO_4) is an important fertiliser, improving yield. Too much P in lake water leads to eutrophication. PH_3 and P_4 very toxic to fish. Many P-compounds toxic. Many organic P-compounds exist
Suggested analytical method(s)	XRF, ICP-AES

World Yearly Production (t/y)	18,200,000 (in 1995, from 41,664,000 P_2O_5)
Price of Purest Form, in Small Quantities ($/kg)	26,331 (99.999,95%)
Market Price ($/kg)	0.05 (phosphate, 75%)

Pa — PROTACTINIUM

Physico-Chemical Properties

Atomic number	91	Melting/boiling point (K)	1845 / -
Atomic mass	231.035,88(2)	Isotopes & isomers	0 stable + 24 unstable
Atomic radius (pm)	-	Acid/base of oxide	weak base
Main oxidation state(s)	+5 (+3,+4)	State (at 300 K, 1 atm.)	solid
Ionic radius (pm)	118 (+3), 104-115 (+4), 92-109 (+5)	Metallic character	metal
Electronegativity (Pauling)	1.5	Element group(s)	actinide, heavy metal
Density (g/cm^3)	15.37	Affinity	-

Naturally occurring isotopes	Natural abundance (%)	Atomic mass	Half-life	Unstable isotopes	Longest half-life
-	-	-	-	^{215}Pa to ^{238}Pa	32,500 y

Concentrations in Crust/Rocks (mg/kg)

Bulk continental crust	- / 0.000,001,4 / -
Upper continental crust	- / -
Ultramafic rock	-
Ocean ridge basalt	-
Gabbro, basalt	-
Granite, granodiorite	-
Sandstone	-
Greywacke	-
Shale, schist	-
Limestone	-
Coal	-

Typical Minerals
-

Possible Host Minerals
uraninite/pitchblende

Mass (kg) in

Continental crust	3.30x10^{+10}	Oceans	66,100	Plants	?

Concentrations in Media Sampled for the Kola Project (mg/kg)

Medium	Moss conc. HNO_3	Humus <2 mm	Humus <2 mm amm. acet.	Topsoil (0-5 cm) <2 mm	B horizon <2 mm aqua regia	B horizon <2 mm total (XRF)
Median	-	-	-	-	-	-
Min	-	-	-	-	-	-
Max	-	-	-	-	-	-

Medium	C horizon <2 mm aqua regia	C horizon <2 mm total (XRF)	C horizon <2 mm total (INAA)	Lake water unfiltered (mg/l)
Median	-	-	-	-
Min	-	-	-	-
Max	-	-	-	-

PROTACTINIUM — Pa

Concentrations in Soils and Sediments (mg/kg)

Medium	Soil	Agricultural soil - Ap horizon	Agricultural soil - Top (0-25 cm)	Agricultural Soil - Bottom (50-75 cm)	Topsoil (0-15 cm)	Urban soil (0-2 cm)
	World	Canada	Finland	Finland	England & Wales	Trondheim
	<2 mm	<2 mm	<2 mm	<2 mm	<2 mm	<2 mm
	total		aqua regia	aqua regia	aqua regia	aqua regia
Median	-	-	-	-	-	-
Min	-	-	-	-	-	-
Max	-	-	-	-	-	-

*Estimated mean

Medium	Forest soil - Humus	Forest soil - B horizon	Forest soil - C horizon	Till (C horizon)	Till (C horizon)	Laterite (25 ±15 cm)
	Norway	Norway	Norway	Finland	Finland	Australia
	<2 mm	<2 mm	<2 mm	<0.063 mm	<0.063 mm	0.45-2 mm
	7N HNO$_3$	7N HNO$_3$	7N HNO$_3$	aqua regia	total	total
Median	-	-	-	-	-	-
Min	-	-	-	-	-	-
Max	-	-	-	-	-	-

Medium	Stream sediment	Stream sediment	Stream sediment	Overbank sediment	Overbank sediment	Organic stream sediment
	Austria	Southern Scotland	Harz, Germany	Norway	Norway	Finland
	<0.18 mm	<0.10 mm	<0.063 mm	<0.063 mm	<0.063 mm	<2 mm
		total	total	total (XRF)	7N HNO$_3$	Conc. HNO$_3$
Median	-	-	-	-	-	-
Min	-	-	-	-	-	-
Max	-	-	-	-	-	-

#2nd percentile   ^98th percentile

Concentrations in Waters (mg/l)

Medium	Ocean water	Ocean water	Ocean water	Stream water	Stream water	Stream water
	World (1)	World (2)	North Pacific	World	Nova Scotia, Canada	Finland
					<0.45 μm	<0.45 μm
Median	5×10^{-11}*	5×10^{-11}*	-	-	-	-
Min	-	-	-	-	-	-
Max	-	-	-	-	-	-

*Estimated mean   #2nd percentile   ^98th percentile

Medium	Stream water	Stream water	Stream water	Lake water	Ground water
	Romania	Eastern India	Harz, Germany	Norway	Southern Norway
	unfiltered	<0.2 μm	unfiltered	unfiltered	unfiltered
	ICP-MS	ICP-MS		ICP-MS	
Median	-	-	-	-	-
Min	-	-	-	-	-
Max	-	-	-	-	-

Pa PROTACTINIUM

Concentrations in Precipitation (mg/l)

Medium	Rain water Kola, remote <0.45 µm	Rain water Kola, polluted <0.45 µm	Snow melt- water Kola, remote <0.45 µm	Snow melt- water Kola, polluted <0.45 µm	Snow filter residue Kola, remote >0.45 µm	Snow filter residue Kola, polluted >0.45 µm
Median	-	-	-	-	-	-
Min	-	-	-	-	-	-
Max	-	-	-	-	-	-

Medium	Rain water Coastal, Norway unfiltered ICP-MS	Rain water Inland, Norway unfiltered ICP-MS				
Median	-	-				
Min	-	-				
Max	-	-				

Concentrations in Air (ng/m^3) and Yearly Deposition (kg/km^2/yr)

Medium	Air World, remote	Air World, polluted	Bulk deposition West Germany	Throughfall deposition West Germany	Bulk deposition Kola, remote	Bulk deposition Kola, polluted
Median	-	-	-	-	-	-
Min	-	-	-	-	-	-
Max	-	-	-	-	-	-

*Estimated value

Concentrations in Plants (mg/kg)

Medium	Moss Norway conc. HNO$_3$ (1990)	Moss Germany conc. HNO$_3$ + H$_2$O$_2$ (1995)	Crustose lichen Germany conc. HNO$_3$	Lichen Northwest Territories oven-dried	Dandelion Europe oven-dried	Spruce bark Canada ashed
Median	-	-	-	-	-	-
Min	-	-	-	-	-	-
Max	-	-	-	-	-	-

*Mean   #Geometric mean

Concentrations in Human Fluids (mg/l)

Medium	Human blood Lombardy, Italy	Human serum Lombardy, Italy	Human urine Lombardy, Italy			
Mean	-	-	-			
Min	-	-	-			
Max	-	-	-			

PROTACTINIUM — Pa

Environmental Geochemistry

Biological impacts	Considered non-essential
Uses	Not known
Environmental pathways	Poorly understood
Environmental mobility	Oxidising conditions: - Acid conditions: - Reducing conditions: - Neutral to alkaline conditions: - Comments: -
Geochemical barriers	-
Natural association	-
Action levels	
Remarks	^{231}Pa occurs in nature as a product of U decay (1 atom ^{231}Pa per $1\times10^{+8}$ atoms U). One of the most expensive naturally occurring elements (closed market). In 1959-1961, 125 g of pure (99.9%) Pa were produced at a cost of US$ 500,000
Suggested analytical method(s)	-

World Yearly Production (t/y)	-
Price of Purest Form, in Small Quantities ($/kg)	4,000,000 (see Remarks)
Market Price ($/kg)	-

Pb LEAD

Physico-Chemical Properties

Atomic number	82	Melting/boiling point (K)	600.61 / 2022
Atomic mass	207.2(1)	Isotopes & isomers	4 stable + 40 unstable
Atomic radius (pm)	181	Acid/base of oxide	amphoteric
Main oxidation state(s)	+2 (+4)	State (at 300 K, 1 atm.)	solid
Ionic radius (pm)	112-163 (+2), 79-108 (+4)	Metallic character	pred. metal
Electronegativity (Pauling)	2.33	Element group(s)	heavy metal
Density (g/cm^3)	11.35	Affinity	chalcophile

Naturally occurring isotopes	Natural abundance (%)	Atomic mass	Half-life
^{204}Pb	1.4	203.973,03	stable
^{206}Pb	24.1	205.974,45	stable
^{207}Pb	22.1	206.975,88	stable
^{208}Pb	52.4	207.976,64	stable

Unstable isotopes	Longest half-life
^{182}Pb to ^{214}Pb	15,100,000 y

Concentrations in Crust/Rocks (mg/kg)

Bulk continental crust	14.8 / 14 / 8
Upper continental crust	17 / 20
Ultramafic rock	0.05
Ocean ridge basalt	1
Gabbro, basalt	4
Granite, granodiorite	20
Sandstone	10
Greywacke	14.2
Shale, schist	22
Limestone	5
Coal	20

Typical Minerals
galena (PbS), anglesite (PbSO$_4$), cerussite (PbCO$_3$), minium (Pb$_3$O$_4$)

Possible Host Minerals
K-feldspar, plagioclase, micas, zircon, magnetite

Mass (kg) in

Continental crust	3.49x10^{+17}	Oceans	6.61x10^{+8}	Plants	1.84x10^{+9}

Concentrations in Media Sampled for the Kola Project (mg/kg)

Medium	Moss	Humus	Humus	Topsoil (0-5 cm)	B horizon	B horizon
	<2 mm	<2 mm	<2 mm	<2 mm	<2 mm	<2 mm
	conc. HNO$_3$	conc. HNO$_3$	amm. acet.		aqua regia	total (XRF)
Median	2.98	18.8	4.8	-	3.05	-
Min	0.84	4.07	0.8	-	0.81	-
Max	29.4	1110	381	-	27.7	-

Medium	C horizon	C horizon	C horizon	Lake water
	<2 mm	<2 mm	<2 mm	unfiltered
	aqua regia	total (XRF)	total (INAA)	(mg/l)
Median	1.6	-	-	0.000,07
Min	0.3	-	-	<0.000,03
Max	45.3	-	-	0.004,76

LEAD   Pb

Concentrations in Soils and Sediments (mg/kg)

Medium	Soil	Agricultural soil - Ap horizon	Agricultural soil - Top (0-25 cm)	Agricultural Soil - Bottom (50-75 cm)	Topsoil (0-15 cm)	Urban soil (0-2 cm)
	World	Canada	Finland	Finland	England & Wales	Trondheim
	<2 mm total	<2 mm total (AAS)	<2 mm aqua regia	<2 mm aqua regia	<2 mm aqua regia	<2 mm aqua regia
Median	17*	14	7.45	<5	40	35
Min	-	3	<5	<5	3	9
Max	-	192	43.2	25.5	16,338	976

*Estimated mean

Medium	Forest soil - Humus	Forest soil - B horizon	Forest soil - C horizon	Till (C horizon)	Till (C horizon)	Laterite (25 ±15 cm)
	Norway	Norway	Norway	Finland	Finland	Australia
	<2 mm 7N HNO$_3$	<2 mm 7N HNO$_3$	<2 mm 7N HNO$_3$	<0.063 mm aqua regia	<0.063 mm total	0.45-2 mm total
Median	32.8	13.4	13	3.2	-	14
Min	2.6	<5	<5	-	-	0.3
Max	488.5	591	196.2	-	-	83

Medium	Stream sediment	Stream sediment	Stream sediment	Overbank sediment	Overbank sediment	Organic stream sediment
	Austria	Southern Scotland	Harz, Germany	Norway	Norway	Finland
	<0.18 mm "total" (ICP-AES)	<0.10 mm total (DCES)	<0.063 mm total (XRF)	<0.063 mm total (XRF)	<0.063 mm 7N HNO$_3$ AAS	<2 mm Conc. HNO$_3$
Median	26	44	195	67	20	8.3
Min	-	<12	24	2	0.1	2.6#
Max	9000	147	9870	219	157	28.5^

#2nd percentile   ^98th percentile

Concentrations in Waters (mg/l)

Medium	Ocean water	Ocean water	Ocean water	Stream water	Stream water	Stream water
	World (1)	World (2)	North Pacific	World	Nova Scotia, Canada	Finland
					<0.45 µm ICP-MS	<0.45 µm ICP-MS
Median	0.000,03*	0.000,000,5*	0.000,002,7*	0.003*	<0.000,1	0.000,23
Min	-	-	-	-	<0.000,1	0.000,06#
Max	-	-	-	-	0.045,9	0.001,13^

*Estimated mean   #2nd percentile   ^98th percentile

Medium	Stream water	Stream water	Stream water	Lake water	Ground water
	Romania	Eastern India	Harz, Germany	Norway	Southern Norway
	unfiltered ICP-MS	<0.2 µm ICP-MS	unfiltered ICP-MS	unfiltered ICP-MS	unfiltered ICP-MS
Median	0.000,73	0.003,04	0.002,1	0.000,18	0.000,3
Min	0.000,04	0.000,75	<0.000,01	<0.000,03	0.000,033
Max	0.081	0.033,54	0.58	0.014,8	0.044,4

Pb   LEAD

Concentrations in Precipitation (mg/l)

Medium	Rain water	Rain water	Snow melt-water	Snow melt-water	Snow filter residue	Snow filter residue
	Kola, remote <0.45 μm ICP-MS	Kola, polluted <0.45 μm ICP-MS	Kola, remote <0.45 μm ICP-MS	Kola, polluted <0.45 μm ICP-MS	Kola, remote >0.45 μm ICP-AES	Kola, polluted >0.45 μm ICP-AES
Median	0.000,56	0.006,3	0.000,6	0.003,38	<0.000,3	0.005,05
Min	0.000,1	0.002,1	0.000,5	0.000,92	<0.000,3	0.003,1
Max	0.001,4	0.040,5	0.000,82	0.006,27	<0.000,3	0.017,8

Medium	Rain water	Rain water
	Coastal, Norway unfiltered ICP-MS	Inland, Norway unfiltered ICP-MS
Median	0.000,29	0.002,5
Min	0.000,16	0.000,1
Max	0.003,8	0.042

Concentrations in Air (ng/m^3) and Yearly Deposition (kg/km^2/yr)

Medium	Air	Air	Bulk deposition	Throughfall deposition	Bulk deposition	Bulk deposition
	World, remote	World, polluted	West Germany	West Germany	Kola, remote	Kola, polluted
Median	10 / 5.8	1500	-	-	0.562*	5.85*
Min	0.027	45	0.9	1.7	-	-
Max	21 (97)	13,000	25	35.3	-	-

*Estimated value

Concentrations in Plants (mg/kg)

Medium	Moss	Moss	Crustose lichen	Lichen	Dandelion	Spruce bark
	Norway conc. HNO$_3$ (1990)	Germany conc. HNO$_3$ + H$_2$O$_2$ (1995)	Germany conc. HNO$_3$	Northwest Territories oven-dried total (XRF)	Europe oven-dried conc. HNO$_3$ (AAS)	Canada ashed aqua regia
Median	9.3	7.715	66	4.2*	2.15#	349
Min	1.5	1.7	18	-	-	89
Max	79	144	1860	-	-	6959

*Mean   #Geometric mean

Concentrations in Human Fluids (mg/l)

Medium	Human blood	Human serum	Human urine
	Lombardy, Italy	Lombardy, Italy	Lombardy, Italy
Mean	0.157,7	0.000,3	0.017
Min	0.03	0.000,25	0.004
Max	0.39	0.000,54	0.039

LEAD   Pb

Environmental Geochemistry

Biological impacts	Considered non-essential. Toxic
Uses	Batteries, antiknock agent in leaded gasoline, pigments, stabiliser in plastic, ammunition, special alloys, cable sheathing, sheets, pipes, solder
Environmental pathways	Traffic, Cu, Zn, Pb smelting, steel works, battery factories, sewage sludge, coal combustion, waste incineration, geogenic dust
Environmental mobility	Oxidising conditions: low Acid conditions: low Reducing conditions: very low Neutral to alkaline conditions: low Comments: Pb binds strongly to organic material in soils, and does not readily migrate to ground water
Geochemical barriers	Sulphate, sulphide, carbonate, adsorption, pH
Natural association	Ag-Zn-Cd-Cu-Pb (lead deposits), Ag-Zn-Cd-Cu-Ba-Sr-V-Cr-Mn-Fe-Ga-In-Ta-Ge-Sn-As-Sb-Bi-Se-Hg-Te-Pb (sulphide deposits), Zn-Pb-Cd (Mississippi Valley type deposits)
Action levels	Drinking water, MAC: 0.02 mg/l (in running water) (Norway Shd), 0.015 mg/l (US EPA), 0.05 mg/l (Canada CWQG); recommended: 0.01 mg/l (WHO); MAC: 0.03 mg/l (Russia MoH). Ground water, background: 0.015 mg/l; remediate: 0.075 mg/l (Netherlands VROM). Soil, background: 85 mg/kg; remediate: 530 mg/kg (Netherlands VROM). Agricultural soil, maximum tolerable concentration: 100 mg/kg (Germany)
Remarks	Pb generally strongly immobilised in humic fraction of soil, leading to strong Pb enrichment in the upper few cm of soil. Plants can tolerate rather high Pb levels. Generally, low uptake by most vegetables. Pb uptake of plants from soil increases at pH<5. Soil micro-organisms are more sensitive to Pb poisoning than plants. $PbBr_2$ and $PbCl_2$ are emitted by cars using leaded fuel. Car exhaust is the main source of anthropogenic Pb in the environment, in those parts of the world where alkyl-Pb additives have not yet been prohibited. $Pb_3(AsO_4)_2$ is used in agriculture, and $PbCrO_4$ is an important yellow pigment. The US EPA has set a Maximum Contaminant Level Goal for Pb in drinking water of zero mg/l. The US EPA suggests that drinking water exceeding action levels can lead to interferences with red blood cell chemistry, delays in normal physical and mental development in babies and young children, slight deficits in the attention span, hearing and learning abilities of children, and slight increases in the blood pressure of some adults in the short term, and to stroke, kidney disease, and cancer in the long term
Suggested analytical method(s)	GF-AAS, ICP-MS

World Yearly Production (t/y)	2,629,000 (in 1995)
Price of Purest Form, in Small Quantities ($/kg)	613 (99.999%)
Market Price ($/kg)	0.88

Pd   PALLADIUM

Physico-Chemical Properties

Atomic number	46	Melting/boiling point (K)	1828.05 / 3236
Atomic mass	106.42(1)	Isotopes & isomers	6 stable + 28 unstable
Atomic radius (pm)	178	Acid/base of oxide	base
Main oxidation state(s)	+2 (+1,+3,+4)	State (at 300 K, 1 atm.)	solid
Ionic radius (pm)	73 (+1), 78-100 (+2), 90 (+3), 75.5 (+4)	Metallic character	pred. metal
		Element group(s)	heavy metal, noble metal, PGE
Electronegativity (Pauling)	2.2		
Density (g/cm^3)	12.02	Affinity	chalcophile, siderophile

Naturally occurring isotopes	Natural abundance (%)	Atomic mass	Half-life
^{102}Pd	1.02	101.905,61	stable
^{104}Pd	11.14	103.904,03	stable
^{105}Pd	22.33	104.905,08	stable
^{106}Pd	27.33	105.903,48	stable
^{108}Pd	26.46	107.903,895	stable
^{110}Pd	11.72	109.905,15	stable

Unstable isotopes	Longest half-life
^{92}Pd to ^{120}Pd	6,500,000 y

Concentrations in Crust/Rocks (mg/kg)

Bulk continental crust	0.000,4 / 0.015 / 0.001
Upper continental crust	- / 0.000,5
Ultramafic rock	0.002
Ocean ridge basalt	0.000,5
Gabbro, basalt	0.000,6
Granite, granodiorite	0.000,2
Sandstone	0.000,2
Greywacke	0.000,4
Shale, schist	0.000,5
Limestone	0.000,1
Coal	<0.1

Typical Minerals
arsenopalladinite (Pd$_3$As), michenerite (PdBi$_2$), froodite (PdBi$_2$), braggite ((Pt,Pd,Ni)S), cooperite ((Pt,Pd)S), merenskyite ((Pd,Pt)(Te,Bi)$_2$), stibiopalladinite (Pd$_3$Sb)

Possible Host Minerals
olivine, ilmenite, zircon, chromite, gadolinite

Mass (kg) in

Continental crust	9.44x10^{+12}	Oceans	-	Plants	1.84x10^{+5}

Concentrations in Media Sampled for the Kola Project (mg/kg)

Medium	Moss conc. HNO$_3$	Humus <2 mm	Humus <2 mm amm. acet.	Topsoil (0-5 cm) <2 mm	B horizon <2 mm aqua regia	B horizon <2 mm total (XRF)
Median	-	-	-	-	-	-
Min	-	-	-	-	-	-
Max	-	-	-	-	-	-

Medium	C horizon <2 mm aqua regia	C horizon <2 mm total (XRF)	C horizon <2 mm total (INAA)		Lake water unfiltered (mg/l)
Median	-	-	-		-
Min	-	-	-		-
Max	-	-	-		-

PALLADIUM — Pd

Concentrations in Soils and Sediments (mg/kg)

Medium	Soil	Agricultural soil - Ap horizon	Agricultural soil - Top (0-25 cm)	Agricultural Soil - Bottom (50-75 cm)	Topsoil (0-15 cm)	Urban soil (0-2 cm)
	World	Canada	Finland	Finland	England & Wales	Trondheim
	<2 mm	<2 mm	<2 mm	<2 mm	<2 mm	<2 mm
	total		aqua regia	aqua regia	aqua regia	aqua regia
Median	0.000,4*	-	-	-	-	-
Min	-	-	-	-	-	-
Max	-	-	-	-	-	-

*Estimated mean

Medium	Forest soil - Humus	Forest soil - B horizon	Forest soil - C horizon	Till (C horizon)	Till (C horizon)	Laterite (25 ±15 cm)
	Norway	Norway	Norway	Finland	Finland	Australia
	<2 mm	<2 mm	<2 mm	<0.063 mm	<0.063 mm	0.45-2 mm
	7N HNO$_3$	7N HNO$_3$	7N HNO$_3$	aqua regia	total	total
Median	-	-	-	-	0.000,2	<0.001
Min	-	-	-	-	-	<0.001
Max	-	-	-	-	-	0.004

Medium	Stream sediment	Stream sediment Southern Scotland	Stream sediment Harz, Germany	Overbank sediment	Overbank sediment	Organic stream sediment
	Austria			Norway	Norway	Finland
	<0.18 mm	<0.10 mm	<0.063 mm	<0.063 mm	<0.063 mm	<2 mm
		total	total	total (XRF)	7N HNO$_3$	Conc. HNO$_3$
Median	-	-	-	-	-	-
Min	-	-	-	-	-	-
Max	-	-	-	-	-	-

#2nd percentile ^98th percentile

Concentrations in Waters (mg/l)

Medium	Ocean water	Ocean water	Ocean water	Stream water	Stream water	Stream water
	World (1)	World (2)	North Pacific	World	Nova Scotia, Canada	Finland
					<0.45 μm	<0.45 μm
Median	-	-	0.000,000,06*	-	-	-
Min	-	-	-	-	-	-
Max	-	-	-	-	-	-

*Estimated mean #2nd percentile ^98th percentile

Medium	Stream water	Stream water	Stream water	Lake water	Ground water	
	Romania	Eastern India	Harz, Germany	Norway	Southern Norway	
	unfiltered	<0.2 μm	unfiltered	unfiltered	unfiltered	
	ICP-MS	ICP-MS		ICP-MS		
Median	-	<0.000,1	-	<0.000,03	-	
Min	-	<0.000,1	-	<0.000,03	-	
Max	-	<0.000,1	-	0.000,162	-	

Pd PALLADIUM

Concentrations in Precipitation (mg/l)

Medium	Rain water Kola, remote <0.45 µm	Rain water Kola, polluted <0.45 µm	Snow melt-water Kola, remote <0.45 µm	Snow melt-water Kola, polluted <0.45 µm	Snow filter residue Kola, remote >0.45 µm	Snow filter residue Kola, polluted >0.45 µm
Median	-	-	-	-	-	-
Min	-	-	-	-	-	-
Max	-	-	-	-	-	-

Medium	Rain water Coastal, Norway unfiltered ICP-MS	Rain water Inland, Norway unfiltered ICP-MS				
Median	-	-				
Min	-	-				
Max	-	-				

Concentrations in Air (ng/m^3) and Yearly Deposition (kg/km^2/yr)

Medium	Air World, remote	Air World, polluted	Bulk deposition West Germany	Throughfall deposition West Germany	Bulk deposition Kola, remote	Bulk deposition Kola, polluted
Median	-	-	-	-	-	-
Min	-	-	-	-	-	-
Max	-	-	-	-	-	-

*Estimated value

Concentrations in Plants (mg/kg)

Medium	Moss Norway conc. HNO$_3$ (1990)	Moss Germany conc. HNO$_3$ + H$_2$O$_2$ (1995)	Crustose lichen Germany conc. HNO$_3$	Lichen Northwest Territories oven-dried	Dandelion Europe oven-dried	Spruce bark Canada ashed
Median	-	-	-	-	-	-
Min	-	-	-	-	-	-
Max	-	-	-	-	-	-

*Mean   #Geometric mean

Concentrations in Human Fluids (mg/l)

Medium	Human blood Lombardy, Italy	Human serum Lombardy, Italy	Human urine Lombardy, Italy			
Mean	-	-	<0.000,15			
Min	-	-	-			
Max	-	-	-			

PALLADIUM — Pd

Environmental Geochemistry

Biological impacts	Considered non-essential. Toxic
Uses	Catalyst, dentistry, jewellery
Environmental pathways	Traffic (Pt-Pd catalysts)
Environmental mobility	Oxidising conditions: very low Acid conditions: very low Reducing conditions: very low Neutral to alkaline conditions: very low Comments: -
Geochemical barriers	Mechanical
Natural association	PGE-Ni-Cu-Co-As-Ag-Au-Te-Bi-Sn-Sb (massive Ni-Cu sulphide ores, e.g., Sudbury), PGE-Ag-Au-Cr-Fe-Cu-Ni-Co-S (sulphides, e.g., Merensky reef)
Action levels	-
Remarks	Very rich ore deposits contain about 8 ppm PGE. PGE occur often in Ni ores. PGE usually occur together in the following proportions: Pt 25%, Ru 23%, Os 21%, Pd 18%, Ir 7%, Rh 6%. Cosmic input via meteorites, cosmic dust, etc. significant. Some marine algae enrich PGE
Suggested analytical method(s)	ICP-MS

World Yearly Production (t/y)	300 all PGE (in 1995)
Price of Purest Form, in Small Quantities ($/kg)	286,750 (99.8%)
Market Price ($/kg)	4000

Po   POLONIUM

Physico-Chemical Properties

Atomic number	84	Melting/boiling point (K)	527 / 1235
Atomic mass	[209]	Isotopes & isomers	0 stable + 36 unstable
Atomic radius (pm)	153	Acid/base of oxide	amphoteric
Main oxidation state(s)	+4 (-2,0,+2,+3,+6)	State (at 300 K, 1 atm.)	solid
Ionic radius (pm)	108-122 (+4), 81 (+6)	Metallic character	pred. metal
Electronegativity (Pauling)	2	Element group(s)	heavy metal
Density (g/cm^3)	9.32 (alpha)	Affinity	-

Naturally occurring isotopes	Natural abundance (%)	Atomic mass	Half-life	Unstable isotopes	Longest half-life
-	-	-	-	^{192}Po to ^{218}Po	102 y

Concentrations in Crust/Rocks (mg/kg)

Bulk continental crust	- / 2x10^{-10} / -
Upper continental crust	- / -
Ultramafic rock	-
Ocean ridge basalt	-
Gabbro, basalt	-
Granite, granodiorite	-
Sandstone	-
Greywacke	-
Shale, schist	-
Limestone	-
Coal	1

Typical Minerals
-

Possible Host Minerals
uraninite/pitchblende (ca. 0.000,01 mg/kg)

Mass (kg) in

Continental crust	4.72x10^{+6}	Oceans	-	Plants	?

Concentrations in Media Sampled for the Kola Project (mg/kg)

Medium	Moss conc. HNO$_3$	Humus <2 mm amm. acet.	Humus <2 mm	Topsoil (0-5 cm) <2 mm	B horizon <2 mm aqua regia	B horizon <2 mm total (XRF)
Median	-	-	-	-	-	-
Min	-	-	-	-	-	-
Max	-	-	-	-	-	-

Medium	C horizon <2 mm aqua regia	C horizon <2 mm total (XRF)	C horizon <2 mm total (INAA)		Lake water unfiltered (mg/l)
Median	-	-	-		-
Min	-	-	-		-
Max	-	-	-		-

POLONIUM   Po

Concentrations in Soils and Sediments (mg/kg)

Medium	Soil	Agricultural soil - Ap horizon	Agricultural soil - Top (0-25 cm)	Agricultural Soil - Bottom (50-75 cm)	Topsoil (0-15 cm)	Urban soil (0-2 cm)
	World	Canada	Finland	Finland	England & Wales	Trondheim
	<2 mm total	<2 mm	<2 mm aqua regia	<2 mm aqua regia	<2 mm aqua regia	<2 mm aqua regia
Median	-	-	-	-	-	-
Min	-	-	-	-	-	-
Max	-	-	-	-	-	-

*Estimated mean

Medium	Forest soil - Humus	Forest soil - B horizon	Forest soil - C horizon	Till (C horizon)	Till (C horizon)	Laterite (25 ±15 cm)
	Norway	Norway	Norway	Finland	Finland	Australia
	<2 mm $7N\ HNO_3$	<2 mm $7N\ HNO_3$	<2 mm $7N\ HNO_3$	<0.063 mm aqua regia	<0.063 mm total	0.45-2 mm total
Median	-	-	-	-	-	-
Min	-	-	-	-	-	-
Max	-	-	-	-	-	-

Medium	Stream sediment	Stream sediment	Stream sediment	Overbank sediment	Overbank sediment	Organic stream sediment
	Austria	Southern Scotland	Harz, Germany	Norway	Norway	Finland
	<0.18 mm	<0.10 mm total	<0.063 mm total	<0.063 mm total (XRF)	<0.063 mm $7N\ HNO_3$	<2 mm Conc. HNO_3
Median	-	-	-	-	-	-
Min	-	-	-	-	-	-
Max	-	-	-	-	-	-

#2nd percentile   ^98th percentile

Concentrations in Waters (mg/l)

Medium	Ocean water	Ocean water	Ocean water	Stream water	Stream water	Stream water
	World (1)	World (2)	North Pacific	World	Nova Scotia, Canada	Finland
					<0.45 µm	<0.45 µm
Median	1.5×10^{-14}*	-	-	-	-	-
Min	-	-	-	-	-	-
Max	-	-	-	-	-	-

*Estimated mean   #2nd percentile   ^98th percentile

Medium	Stream water	Stream water	Stream water	Lake water	Ground water
	Romania	Eastern India	Harz, Germany	Norway	Southern Norway
	unfiltered ICP-MS	<0.2 µm ICP-MS	unfiltered	unfiltered ICP-MS	unfiltered
Median	-	-	-	-	-
Min	-	-	-	-	-
Max	-	-	-	-	-

Po   POLONIUM

Concentrations in Precipitation (mg/l)

Medium	Rain water Kola, remote <0.45 μm	Rain water Kola, polluted <0.45 μm	Snow melt- water Kola, remote <0.45 μm	Snow melt- water Kola, polluted <0.45 μm	Snow filter residue Kola, remote >0.45 μm	Snow filter residue Kola, polluted >0.45 μm
Median	-	-	-	-	-	-
Min	-	-	-	-	-	-
Max	-	-	-	-	-	-

Medium	Rain water Coastal, Norway unfiltered ICP-MS	Rain water Inland, Norway unfiltered ICP-MS				
Median	-	-				
Min	-	-				
Max	-	-				

Concentrations in Air (ng/m^3) and Yearly Deposition (kg/km^2/yr)

Medium	Air World, remote	Air World, polluted	Bulk deposition West Germany	Throughfall deposition West Germany	Bulk deposition Kola, remote	Bulk deposition Kola, polluted
Median	-	-	-	-	-	-
Min	-	-	-	-	-	-
Max	-	-	-	-	-	-

*Estimated value

Concentrations in Plants (mg/kg)

Medium	Moss Norway conc. HNO$_3$ (1990)	Moss Germany conc. HNO$_3$ + H$_2$O$_2$ (1995)	Crustose lichen Germany conc. HNO$_3$	Lichen Northwest Territories oven-dried	Dandelion Europe oven-dried	Spruce bark Canada ashed
Median	-	-	-	-	-	-
Min	-	-	-	-	-	-
Max	-	-	-	-	-	-

*Mean   #Geometric mean

Concentrations in Human Fluids (mg/l)

Medium	Human blood Lombardy, Italy	Human serum Lombardy, Italy	Human urine Lombardy, Italy			
Mean	-	-	-			
Min	-	-	-			
Max	-	-	-			

POLONIUM   Po

Environmental Geochemistry

Biological impacts	Considered non-essential. Very radiotoxic. Carcinogenic effects are discussed
Uses	^{210}Po in thermo-nuclear batteries
Environmental pathways	Poorly understood
Environmental mobility	Oxidising conditions: - Acid conditions: - Reducing conditions: - Neutral to alkaline conditions: - Comments: -
Geochemical barriers	-
Natural association	U-Po (pitchblende deposits)
Action levels	-
Remarks	Po enrichment has been noted in lungs of smokers. ^{209}Po costs 3200 $/microCi (closed marked)
Suggested analytical method(s)	-

World Yearly Production (t/y)	-
Price of Purest Form, in Small Quantities ($/kg)	-
Market Price ($/kg)	-

Pr PRASEODYMIUM

Physico-Chemical Properties

Atomic number	59	Melting/boiling point (K)	1204 / 3793
Atomic mass	140.907,65(2)	Isotopes & isomers	1 stable + 35 unstable
Atomic radius (pm)	267	Acid/base of oxide	base
Main oxidation state(s)	+3 (+4)	State (at 300 K, 1 atm.)	solid
Ionic radius (pm)	113-132 (+3), 99-110 (+4)	Metallic character	metal
Electronegativity (Pauling)	1.13	Element group(s)	heavy metal, REE
Density (g/cm^3)	6.773	Affinity	lithophile

Naturally occurring isotopes	Natural abundance (%)	Atomic mass	Half-life
^{141}Pr	100	140.907,65	stable

Unstable isotopes	Longest half-life
^{124}Pr to ^{154}Pr	13.57 d

Concentrations in Crust/Rocks (mg/kg)

Bulk continental crust	6.7 / 9.2 / 3.9
Upper continental crust	6.3 / 7.1
Ultramafic rock	-
Ocean ridge basalt	-
Gabbro, basalt	-
Granite, granodiorite	-
Sandstone	-
Greywacke	6.1
Shale, schist	-
Limestone	-
Coal	3

Typical Minerals
monazite ((Ce,La,Pr,Th)(PO$_4$,SiO$_4$)),
bastnaesite ((Ce,La,Pr)CO$_3$(F,OH)),
cerite ((Ce,La,Pr)$_9$(Mg,Fe)Si$_7$(O,OH,F)$_{28}$),
allanite ((Ca,Ce,La,Pr)$_2$FeAl$_2$OSi$_3$O$_{11}$(OH))

Possible Host Minerals
biotite, apatite, pyroxenes, feldspars, zircon

Mass (kg) in

Continental crust	1.58x10^{+17}	Oceans	7.93x10^{+8}	Plants	9.20x10^{+7}

Concentrations in Media Sampled for the Kola Project (mg/kg)

| Medium | Moss | Humus <2 mm | Humus <2 mm | Topsoil (0-5 cm) <2 mm | B horizon <2 mm | B horizon <2 mm |
	conc. HNO$_3$		amm. acet.		aqua regia	total (XRF)
Median	-	-	-	-	-	-
Min	-	-	-	-	-	-
Max	-	-	-	-	-	-

Medium	C horizon <2 mm aqua regia	C horizon <2 mm total (XRF)	C horizon <2 mm total (INAA)	Lake water unfiltered (mg/l)
Median	-	-	-	-
Min	-	-	-	-
Max	-	-	-	-

PRASEODYMIUM — Pr

Concentrations in Soils and Sediments (mg/kg)

Medium	Soil	Agricultural soil - Ap horizon	Agricultural soil - Top (0-25 cm)	Agricultural Soil - Bottom (50-75 cm)	Topsoil (0-15 cm)	Urban soil (0-2 cm)
	World	Canada	Finland	Finland	England & Wales	Trondheim
	<2 mm	<2 mm	<2 mm	<2 mm	<2 mm	<2 mm
	total	aqua regia	aqua regia	aqua regia	aqua regia	aqua regia
Median	-	-	-	-	-	-
Min	-	-	-	-	-	-
Max	-	-	-	-	-	-

*Estimated mean

Medium	Forest soil - Humus	Forest soil - B horizon	Forest soil - C horizon	Till (C horizon)	Till (C horizon)	Laterite (25 ±15 cm)
	Norway	Norway	Norway	Finland	Finland	Australia
	<2 mm	<2 mm	<2 mm	<0.063 mm	<0.063 mm	0.45-2 mm
	7N HNO$_3$	7N HNO$_3$	7N HNO$_3$	aqua regia	total	total
Median	-	-	-	-	-	-
Min	-	-	-	-	-	-
Max	-	-	-	-	-	-

Medium	Stream sediment	Stream sediment	Stream sediment	Overbank sediment	Overbank sediment	Organic stream sediment
	Austria	Southern Scotland	Harz, Germany	Norway	Norway	Finland
	<0.18 mm	<0.10 mm	<0.063 mm	<0.063 mm	<0.063 mm	<2 mm
	total	total	total	total (XRF)	7N HNO$_3$	Conc. HNO$_3$
Median	-	-	-	-	-	-
Min	-	-	-	-	-	-
Max	-	-	-	-	-	-

#2nd percentile   ^98th percentile

Concentrations in Waters (mg/l)

Medium	Ocean water	Ocean water	Ocean water	Stream water	Stream water	Stream water
	World (1)	World (2)	North Pacific	World	Nova Scotia, Canada	Finland
					<0.45 µm	<0.45 µm
					ICP-MS	
Median	0.000,000,64*	0.000,000,6*	0.000,000,7*	-	0.000,033	-
Min	-	-	-	-	<0.000,005	-
Max	-	-	-	-	0.000,2	-

*Estimated mean   #2nd percentile   ^98th percentile

Medium	Stream water	Stream water	Stream water	Lake water	Ground water	
	Romania	Eastern India	Harz, Germany	Norway	Southern Norway	
	unfiltered	<0.2 µm	unfiltered	unfiltered	unfiltered	
	ICP-MS	ICP-MS	ICP-MS	ICP-MS	ICP-MS	
Median	-	0.001,39	0.000,12	0.000,38	0.000,039,5	
Min	-	0.000,15	<0.000,002	<0.000,003	<0.000,001	
Max	-	0.017,2	0.001,9	0.000,703	0.014,3	

261

Pr PRASEODYMIUM

Concentrations in Precipitation (mg/l)

Medium	Rain water *Kola, remote* *<0.45 µm*	Rain water *Kola, polluted* *<0.45 µm*	Snow melt-water *Kola, remote* *<0.45 µm*	Snow melt-water *Kola, polluted* *<0.45 µm*	Snow filter residue *Kola, remote* *>0.45 µm*	Snow filter residue *Kola, polluted* *>0.45 µm*
Median	-	-	-	-	-	-
Min	-	-	-	-	-	-
Max	-	-	-	-	-	-

Medium	Rain water *Coastal,* *Norway* *unfiltered* *ICP-MS*	Rain water *Inland,* *Norway* *unfiltered* *ICP-MS*
Median	-	-
Min	-	-
Max	-	-

Concentrations in Air (ng/m^3) and Yearly Deposition (kg/km^2/yr)

Medium	Air *World,* *remote*	Air *World,* *polluted*	Bulk deposition *West* *Germany*	Throughfall deposition *West* *Germany*	Bulk deposition *Kola, remote*	Bulk deposition *Kola, polluted*
Median	-	-	-	-	-	-
Min	-	-	-	-	-	-
Max	-	-	-	-	-	-

*Estimated value

Concentrations in Plants (mg/kg)

Medium	Moss *Norway* *conc. HNO$_3$* *(1990)*	Moss *Germany* *conc. HNO$_3$ +* *H$_2$O$_2$ (1995)*	Crustose lichen *Germany* *conc. HNO$_3$*	Lichen *Northwest* *Territories* *oven-dried*	Dandelion *Europe* *oven-dried*	Spruce bark *Canada* *ashed*
Median	-	-	1.9	-	-	-
Min	-	-	0.5	-	-	-
Max	-	-	20	-	-	-

*Mean   #Geometric mean

Concentrations in Human Fluids (mg/l)

Medium	Human blood *Lombardy,* *Italy*	Human serum *Lombardy,* *Italy*	Human urine *Lombardy,* *Italy*
Mean	-	-	-
Min	-	-	-
Max	-	-	-

PRASEODYMIUM — Pr

Environmental Geochemistry

Biological impacts	Considered non-essential. Generally low toxicity, but data to assess the health relevance of REE are scarce
Uses	Magnets, lighters, glass additives, yellow pigment tile glazes
Environmental pathways	Geogenic dust, mining and processing of alkaline rocks. Generally, geogenic sources more important than anthropogenic ones
Environmental mobility	Oxidising conditions: very low — Acid conditions: very low Reducing conditions: very low — Neutral to alkaline conditions: very low Comments: -
Geochemical barriers	Mechanical
Natural association	REE-Li-Rb-Cs-Be-Nb-Ta-Zr-B-Th-U-F (pegmatites), REE-Th-P-Zr-Fe-Cu (monazite veins), REE-Th-Ba-Sr-P-F-N-C (carbonatites), REE-U-P-F (phosphorites), REE-Au-Ti-Sn-Sr-Th (placers)
Action levels	-
Remarks	Toxicity of REE decreases as atomic number increases. Inhaled REE probably cause pneumoconiosis. REE taken up orally (e.g., via drinking water) accumulate in the skeleton, teeth, lungs, liver and kidneys. Some trees tend to accumulate Pr (e.g., hickory (*Carya*)). In 1991, 10 t Pr-oxide were used in the glass industry, and 150 t in the ceramics industry
Suggested analytical method(s)	ICP-MS

World Yearly Production (t/y) — 54,000 REE-minerals (in 1991)
Price of Purest Form, in Small Quantities ($/kg) — 153,906 (99.99%)
Market Price ($/kg) — -

Pt   PLATINUM

Physico-Chemical Properties

Atomic number	78	Melting/boiling point (K)	2041.55 / 4098
Atomic mass	195.078(2)	Isotopes & isomers	5 stable + 36 unstable
Atomic radius (pm)	183	Acid/base of oxide	base
Main oxidation state(s)	+4 (+2,+3,+5)	State (at 300 K, 1 atm.)	solid
Ionic radius (pm)	74-94 (+2), 76.5 (+4), 71 (+5)	Metallic character	pred. metal
Electronegativity (Pauling)	2.28	Element group(s)	heavy metal, noble metal, PGE
Density (g/cm^3)	21.45	Affinity	siderophile

Naturally occurring isotopes	Natural abundance (%)	Atomic mass	Half-life	Unstable isotopes	Longest half-life
^{190}Pt	0.01	189.959,93	6.5x10^{+11} y	^{168}Pt to ^{202}Pt	6.5x10^{+11} y
^{192}Pt	0.79	191.961,04	stable		
^{194}Pt	32.9	193.962,66	stable		
^{195}Pt	33.8	194.964,77	stable		
^{196}Pt	25.3	195.964,93	stable		
^{198}Pt	7.2	197.967,88	stable		

Concentrations in Crust/Rocks (mg/kg)

Bulk continental crust	0.000,4 / 0.005 / -
Upper continental crust	- / -
Ultramafic rock	0.003
Ocean ridge basalt	0.000,3
Gabbro, basalt	0.000,2
Granite, granodiorite	0.000,05
Sandstone	0.000,1
Greywacke	0.000,4
Shale, schist	0.000,2
Limestone	0.000,05
Coal	-

Typical Minerals
sperrylite (PtAs$_2$), braggite ((Pt,Pd,Ni)S), cooperite (PtS), stumpflite (Pt(Sb,Bi)), nigglite (PtSn)

Possible Host Minerals
olivine, ilmenite, zircon, chromite, gadolinite

Mass (kg) in

Continental crust	9.44x10^{+12}	Oceans	-	Plants	92,000

Concentrations in Media Sampled for the Kola Project (mg/kg)

Medium	Moss conc. HNO$_3$	Humus <2 mm	Humus <2 mm amm. acet.	Topsoil (0-5 cm) <2 mm	B horizon <2 mm aqua regia	B horizon <2 mm total (XRF)
Median	-	-	-	-	-	-
Min	-	-	-	-	-	-
Max	-	-	-	-	-	-

Medium	C horizon <2 mm aqua regia	C horizon <2 mm total (XRF)	C horizon <2 mm total (INAA)		Lake water unfiltered (mg/l)
Median	-	-	-		-
Min	-	-	-		-
Max	-	-	-		-

PLATINUM Pt

Concentrations in Soils and Sediments (mg/kg)

Medium	Soil	Agricultural soil - Ap horizon	Agricultural soil - Top (0-25 cm)	Agricultural Soil - Bottom (50-75 cm)	Topsoil (0-15 cm)	Urban soil (0-2 cm)
	World	Canada	Finland	Finland	England & Wales	Trondheim
	<2 mm	<2 mm	<2 mm	<2 mm	<2 mm	<2 mm
	total	aqua regia	aqua regia	aqua regia	aqua regia	aqua regia
Median	-	-	-	-	-	-
Min	-	-	-	-	-	-
Max	-	-	-	-	-	-

*Estimated mean

Medium	Forest soil - Humus	Forest soil - B horizon	Forest soil - C horizon	Till (C horizon)	Till (C horizon)	Laterite (25 ±15 cm)
	Norway	Norway	Norway	Finland	Finland	Australia
	<2 mm	<2 mm	<2 mm	<0.063 mm	<0.063 mm	0.45-2 mm
	7N HNO$_3$	7N HNO$_3$	7N HNO$_3$	aqua regia	total	total
Median	-	-	-	-	-	<0.005
Min	-	-	-	-	-	<0.005
Max	-	-	-	-	-	0.007

Medium	Stream sediment	Stream sediment	Stream sediment	Overbank sediment	Overbank sediment	Organic stream sediment
	Austria	Southern Scotland	Harz, Germany	Norway	Norway	Finland
	<0.18 mm	<0.10 mm	<0.063 mm	<0.063 mm	<0.063 mm	<2 mm
		total	total	total (XRF)	7N HNO$_3$	Conc. HNO$_3$
Median	-	-	-	-	-	-
Min	-	-	-	-	-	-
Max	-	-	-	-	-	-

#2nd percentile ^98th percentile

Concentrations in Waters (mg/l)

Medium	Ocean water	Ocean water	Ocean water	Stream water	Stream water	Stream water
	World (1)	World (2)	North Pacific	World	Nova Scotia, Canada	Finland
					<0.45 µm	<0.45 µm
Median	-	-	0.000,000,05*	-	-	-
Min	-	-	-	-	-	-
Max	-	-	-	-	-	-

*Estimated mean #2nd percentile ^98th percentile

Medium	Stream water	Stream water	Stream water	Lake water	Ground water
	Romania	Eastern India	Harz, Germany	Norway	Southern Norway
	unfiltered	<0.2 µm	unfiltered	unfiltered	unfiltered
	ICP-MS	ICP-MS		ICP-MS	
Median	-	<0.000,01	-	<0.000,02	-
Min	-	<0.000,01	-	<0.000,02	-
Max	-	<0.000,01	-	0.000,021	-

Pt PLATINUM

Concentrations in Precipitation (mg/l)

Medium	Rain water Kola, remote <0.45 μm	Rain water Kola, polluted <0.45 μm	Snow melt-water Kola, remote <0.45 μm	Snow melt-water Kola, polluted <0.45 μm	Snow filter residue Kola, remote >0.45 μm	Snow filter residue Kola, polluted >0.45 μm
Median	-	-	-	-	-	-
Min	-	-	-	-	-	-
Max	-	-	-	-	-	-

Medium	Rain water Coastal, Norway unfiltered ICP-MS	Rain water Inland, Norway unfiltered ICP-MS				
Median	-	-				
Min	-	-				
Max	-	-				

Concentrations in Air (ng/m^3) and Yearly Deposition (kg/km^2/yr)

Medium	Air World, remote	Air World, polluted	Bulk deposition West Germany	Throughfall deposition West Germany	Bulk deposition Kola, remote	Bulk deposition Kola, polluted
Median	-	-	-	-	-	-
Min	-	-	-	-	-	-
Max	-	-	-	-	-	-

*Estimated value

Concentrations in Plants (mg/kg)

Medium	Moss Norway conc. HNO$_3$ (1990)	Moss Germany conc. HNO$_3$ + H$_2$O$_2$ (1995)	Crustose lichen Germany conc. HNO$_3$	Lichen Northwest Territories oven-dried	Dandelion Europe oven-dried	Spruce bark Canada ashed
Median	-	-	-	-	-	-
Min	-	-	-	-	-	-
Max	-	-	-	-	-	-

*Mean   #Geometric mean

Concentrations in Human Fluids (mg/l)

Medium	Human blood Lombardy, Italy	Human serum Lombardy, Italy	Human urine Lombardy, Italy			
Mean	-	-	<0.001			
Min	-	-	-			
Max	-	-	-			

PLATINUM — Pt

Environmental Geochemistry

Biological impacts	Considered non-essential. Considered toxic
Uses	Catalyst, laboratory equipment, jewellery
Environmental pathways	Traffic, chemical industry, waste water from nitric acid production, Cu-Ni smelters, cosmic sources
Environmental mobility	Oxidising conditions: very low Acid conditions: very low Reducing conditions: very low Neutral to alkaline conditions: very low Comments: -
Geochemical barriers	Mechanical
Natural association	Pt-Ni-Cu-Co-As-Ag-Au-Te-Bi-Sn-Sb (massive Ni-Cu sulphide ores, e.g., Sudbury), Pt-Ag-Au-Cr-Fe-Cu-Ni-Co-S (sulphides, e.g., Merensky reef), other platinum group elements (PGE deposits)
Action levels	Waste water, MAC: 3 mg/l (Germany)
Remarks	Very rich ore deposits contain about 8 ppm PGE. PGE occur often in Ni ores. PGE usually occur together in the following proportions: Pt 25%, Ru 23%, Os 21%, Pd 18%, Ir 7%, Rh 6%. Cosmic input via meteorites, cosmic dust, etc. significant. Some marine algae enrich PGE. Certain Pt compounds used in cancer therapy. Allergenic
Suggested analytical method(s)	ICP-MS

World Yearly Production (t/y)	300 all PGE (in 1995)
Price of Purest Form, in Small Quantities ($/kg)	209,500 (99.999%)
Market Price ($/kg)	11,600

Pu PLUTONIUM

Physico-Chemical Properties

Atomic number	94	Melting/boiling point (K)	913 / 3501
Atomic mass	[244]	Isotopes & isomers	0 stable + 18 unstable
Atomic radius (pm)	-	Acid/base of oxide	amphoteric
Main oxidation state(s)	+4 (+3,+5,+6,+7)	State (at 300 K, 1 atm.)	solid
Ionic radius (pm)	114 (+3), 100-110 (+4), 88 (+5), 85 (+6)	Metallic character	metal
		Element group(s)	actinide, heavy metal
Electronegativity (Pauling)	1.28	Affinity	-
Density (g/cm³)	19.84 (alpha)		

Naturally occurring isotopes	Natural abundance (%)	Atomic mass	Half-life	Unstable isotopes	Longest half-life
-	-	-	-	^{228}Pu to ^{246}Pu	$8.2 \times 10^{+7}$ y

Concentrations in Crust/Rocks (mg/kg)

Bulk continental crust	- / - / -
Upper continental crust	- / -
Ultramafic rock	-
Ocean ridge basalt	-
Gabbro, basalt	-
Granite, granodiorite	-
Sandstone	-
Greywacke	-
Shale, schist	-
Limestone	-
Coal	-

Typical Minerals
-

Possible Host Minerals
uraninite/pitchblende, carnotite

Mass (kg) in

Continental crust	-	Oceans	-	Plants	-

Concentrations in Media Sampled for the Kola Project (mg/kg)

Medium	Moss	Humus <2 mm	Humus <2 mm	Topsoil (0-5 cm) <2 mm	B horizon <2 mm	B horizon <2 mm
	conc. HNO₃		amm. acet.		aqua regia	total (XRF)
Median	-	-	-	-	-	-
Min	-	-	-	-	-	-
Max	-	-	-	-	-	-

Medium	C horizon <2 mm	C horizon <2 mm	C horizon <2 mm		Lake water unfiltered
	aqua regia	total (XRF)	total (INAA)		(mg/l)
Median	-	-	-		-
Min	-	-	-		-
Max	-	-	-		-

PLUTONIUM   Pu

Environmental Geochemistry

Biological impacts	Considered non-essential. Chemotoxic and radiotoxic
Uses	Nuclear industry, atomic bombs
Environmental pathways	Nuclear industry, atomic bomb tests
Environmental mobility	Oxidising conditions: - Acid conditions: - Reducing conditions: - Neutral to alkaline conditions: - Comments: -
Geochemical barriers	-
Natural association	U-Pu-Np
Action levels	-
Remarks	Very reactive element. Burns in contact with air or water. Micrograms of Pu are toxic for humans. Half-life in the human body is 200 yr
Suggested analytical method(s)	-

World Yearly Production (t/y)	several
Price of Purest Form, in Small Quantities ($/kg)	-
Market Price ($/kg)	-

Ra RADIUM

Physico-Chemical Properties

Atomic number	88
Atomic mass	[226]
Atomic radius (pm)	-
Main oxidation state(s)	+2
Ionic radius (pm)	162-184
Electronegativity (Pauling)	0.9
Density (g/cm^3)	5

Melting/boiling point (K)	973 / 1413
Isotopes & isomers	0 stable + 30 unstable
Acid/base of oxide	strong base
State (at 300 K, 1 atm.)	solid
Metallic character	metal
Element group(s)	alkaline earth, heavy metal
Affinity	lithophile

Naturally occurring isotopes	Natural abundance (%)	Atomic mass	Half-life
-	-	-	-

Unstable isotopes	Longest half-life
^{206}Ra to ^{234}Ra	1599 y

Concentrations in Crust/Rocks (mg/kg)

Bulk continental crust	- / 0.000,000,9 / -
Upper continental crust	- / -
Ultramafic rock	-
Ocean ridge basalt	-
Gabbro, basalt	-
Granite, granodiorite	-
Sandstone	-
Greywacke	-
Shale, schist	-
Limestone	-
Coal	-

Typical Minerals
radiobarite ((Ba,Ra)SO$_4$)

Possible Host Minerals
uraninite/pitchblende, carnotite

Mass (kg) in

Continental crust	2.12x10^{+10}	Oceans	92,500	Plants	?

Concentrations in Media Sampled for the Kola Project (mg/kg)

Medium	Moss	Humus <2 mm	Humus <2 mm	Topsoil (0-5 cm) <2 mm	B horizon <2 mm	B horizon <2 mm
	conc. HNO$_3$		amm. acet.		aqua regia	total (XRF)
Median	-	-	-	-	-	-
Min	-	-	-	-	-	-
Max	-	-	-	-	-	-

Medium	C horizon <2 mm	C horizon <2 mm	C horizon <2 mm	Lake water unfiltered
	aqua regia	total (XRF)	total (INAA)	(mg/l)
Median	-	-	-	-
Min	-	-	-	-
Max	-	-	-	-

RADIUM Ra

Concentrations in Soils and Sediments (mg/kg)

Medium	Soil World <2 mm total	Agricultural soil - Ap horizon Canada <2 mm	Agricultural soil - Top (0-25 cm) Finland <2 mm aqua regia	Agricultural Soil - Bottom (50-75 cm) Finland <2 mm aqua regia	Topsoil (0-15 cm) England & Wales <2 mm aqua regia	Urban soil (0-2 cm) Trondheim <2 mm aqua regia
Median	-	-	-	-	-	-
Min	-	-	-	-	-	-
Max	-	-	-	-	-	-

*Estimated mean

Medium	Forest soil - Humus Norway <2 mm $7N\ HNO_3$	Forest soil - B horizon Norway <2 mm $7N\ HNO_3$	Forest soil - C horizon Norway <2 mm $7N\ HNO_3$	Till (C horizon) Finland <0.063 mm aqua regia	Till (C horizon) Finland <0.063 mm total	Laterite (25 ±15 cm) Australia 0.45-2 mm total
Median	-	-	-	-	-	-
Min	-	-	-	-	-	-
Max	-	-	-	-	-	-

Medium	Stream sediment Austria <0.18 mm	Stream sediment Southern Scotland <0.10 mm total	Stream sediment Harz, Germany <0.063 mm total	Overbank sediment Norway <0.063 mm total (XRF)	Overbank sediment Norway <0.063 mm $7N\ HNO_3$	Organic stream sediment Finland <2 mm Conc. HNO_3
Median	-	-	-	-	-	-
Min	-	-	-	-	-	-
Max	-	-	-	-	-	-

#2nd percentile ^98th percentile

Concentrations in Waters (mg/l)

Medium	Ocean water World (1)	Ocean water World (2)	Ocean water North Pacific	Stream water World	Stream water Nova Scotia, Canada <0.45 µm	Stream water Finland <0.45 µm
Median	8.9×10^{-11}*	7×10^{-11}*	1.3×10^{-10}*	-	-	-
Min	-	-	-	-	-	-
Max	-	-	-	-	-	-

*Estimated mean #2nd percentile ^98th percentile

Medium	Stream water Romania unfiltered ICP-MS	Stream water Eastern India <0.2 µm ICP-MS	Stream water Harz, Germany unfiltered	Lake water Norway unfiltered ICP-MS	Ground water Southern Norway unfiltered
Median	-	-	-	-	-
Min	-	-	-	-	-
Max	-	-	-	-	-

Ra — RADIUM

Concentrations in Precipitation (mg/l)

Medium	Rain water *Kola, remote* *<0.45 μm*	Rain water *Kola, polluted* *<0.45 μm*	Snow melt- water *Kola, remote* *<0.45 μm*	Snow melt- water *Kola, polluted* *<0.45 μm*	Snow filter residue *Kola, remote* *>0.45 μm*	Snow filter residue *Kola, polluted* *>0.45 μm*
Median	-	-	-	-	-	-
Min	-	-	-	-	-	-
Max	-	-	-	-	-	-

Medium	Rain water *Coastal,* *Norway* *unfiltered* *ICP-MS*	Rain water *Inland,* *Norway* *unfiltered* *ICP-MS*
Median	-	-
Min	-	-
Max	-	-

Concentrations in Air (ng/m^3) and Yearly Deposition (kg/km^2/yr)

Medium	Air *World,* *remote*	Air *World,* *polluted*	Bulk deposition *West* *Germany*	Throughfall deposition *West* *Germany*	Bulk deposition *Kola, remote*	Bulk deposition *Kola, polluted*
Median	-	-	-	-	-	-
Min	-	-	-	-	-	-
Max	-	-	-	-	-	-

*Estimated value

Concentrations in Plants (mg/kg)

Medium	Moss *Norway* *conc. HNO$_3$* *(1990)*	Moss *Germany* *conc. HNO$_3$ +* *H$_2$O$_2$ (1995)*	Crustose lichen *Germany* *conc. HNO$_3$*	Lichen *Northwest* *Territories* *oven-dried*	Dandelion *Europe* *oven-dried*	Spruce bark *Canada* *ashed*
Median	-	-	-	-	-	-
Min	-	-	-	-	-	-
Max	-	-	-	-	-	-

*Mean   #Geometric mean

Concentrations in Human Fluids (mg/l)

Medium	Human blood *Lombardy,* *Italy*	Human serum *Lombardy,* *Italy*	Human urine *Lombardy,* *Italy*
Mean	-	-	-
Min	-	-	-
Max	-	-	-

RADIUM Ra

Environmental Geochemistry

Biological impacts	Considered non-essential. Highly radiotoxic and carcinogenic
Uses	Radiotherapy, nuclear research, fluorescent materials
Environmental pathways	Mostly geogenic
Environmental mobility	Oxidising conditions: high Acid conditions: high Reducing conditions: high Neutral to alkaline conditions: high Comments: -
Geochemical barriers	Co-crystallisation (Ba, Ca), co-precipitation (Mn, Fe), adsorption (Fe-Mn oxides, clay, organic matter)
Natural association	U-Th-Ra-Rn-Bi-Pb (U deposits), Ra-Ba (gossans), Ra-Mn-Fe (spring precipitates), Ra-organic matter (bogs)
Action levels	Drinking water, MAC: 1.2×10^{-10} mg ^{226}Ra/l (Russia MoH)
Remarks	Occurs together with U and Th minerals. Recovered (rather for environmental than economic reasons) as a by-product of U and Th processing. Glows even in daylight. Accumulates with organic matter. Similarity to Ba and Ca leads to incorporation of Ra in bones, causing radiotoxicity
Suggested analytical method(s)	Scintillation counter

World Yearly Production (t/y)   -
Price of Purest Form, in Small Quantities ($/kg)   -
Market Price ($/kg)   -

Rb   RUBIDIUM

Physico-Chemical Properties

Atomic number	37	Melting/boiling point (K)	312.46 / 961
Atomic mass	85.467,8(3)	Isotopes & isomers	1 stable + 34 unstable
Atomic radius (pm)	298	Acid/base of oxide	strong base
Main oxidation state(s)	+1 (+2,+3,+4)	State (at 300 K, 1 atm.)	solid
Ionic radius (pm)	166-197 (+1)	Metallic character	metal
Electronegativity (Pauling)	0.82	Element group(s)	alkali metal, light metal
Density (g/cm^3)	1.532	Affinity	lithophile

Naturally occurring isotopes	Natural abundance (%)	Atomic mass	Half-life
^{85}Rb	72.165	84.911,79	stable
^{87}Rb	27.835	86.909,19	4.88x10^{+10} y

Unstable isotopes	Longest half-life
^{74}Rb to ^{102}Rb	4.88x10^{+10} y

Concentrations in Crust/Rocks (mg/kg)

Bulk continental crust	78 / 90 / 32
Upper continental crust	110 / 112
Ultramafic rock	2
Ocean ridge basalt	8
Gabbro, basalt	30
Granite, granodiorite	120
Sandstone	40
Greywacke	72
Shale, schist	140
Limestone	4
Coal	15

Typical Minerals
-

Possible Host Minerals
replaces K in K-silicates (e.g., K-feldspar, biotite, lepidolite, zinnwaldite), in K-salts (carnallite), replaces Cs in pollucite (CsAlSi2O6.H2O)

Mass (kg) in

Continental crust 1.84x10^{+18}	Oceans 1.59x10^{+14}	Plants 9.20x10^{+10}

Concentrations in Media Sampled for the Kola Project (mg/kg)

Medium	Moss	Humus	Humus	Topsoil (0-5 cm)	B horizon	B horizon
		<2 mm	<2 mm	<2 mm	<2 mm	<2 mm
	conc. HNO$_3$	conc. HNO$_3$	amm. acet.	total (INAA)	aqua regia	total (XRF)
Median	11.5	5.77	-	29	-	-
Min	1.39	0.68	-	<15	-	-
Max	33.5	33.1	-	180	-	-

Medium	C horizon	C horizon	C horizon		Lake water
	<2 mm	<2 mm	<2 mm		unfiltered
	aqua regia	total (XRF)	total (INAA)		(mg/l)
Median	-	-	54		0.000,53
Min	-	-	<15		0.000,07
Max	-	-	270		0.002,93

RUBIDIUM   Rb

Concentrations in Soils and Sediments (mg/kg)

Medium	Soil	Agricultural soil - Ap horizon	Agricultural soil - Top (0-25 cm)	Agricultural Soil - Bottom (50-75 cm)	Topsoil (0-15 cm)	Urban soil (0-2 cm)
	World	Canada	Finland	Finland	England & Wales	Trondheim
	<2 mm	<2 mm	<2 mm	<2 mm	<2 mm	<2 mm
	total	total (INAA)	aqua regia	aqua regia	aqua regia	aqua regia
Median	65*	69	-	-	-	-
Min	-	2	-	-	-	-
Max	-	190	-	-	-	-

*Estimated mean

Medium	Forest soil - Humus	Forest soil - B horizon	Forest soil - C horizon	Till (C horizon)	Till (C horizon)	Laterite (25 ±15 cm)
	Norway	Norway	Norway	Finland	Finland	Australia
	<2 mm	<2 mm	<2 mm	<0.063 mm	<0.063 mm	0.45-2 mm
	7N HNO$_3$	7N HNO$_3$	7N HNO$_3$	aqua regia	total	total
Median	-	-	-	-	73	43
Min	-	-	-	-	-	2.4
Max	-	-	-	-	-	260

Medium	Stream sediment	Stream sediment	Stream sediment	Overbank sediment	Overbank sediment	Organic stream sediment
	Austria	Southern Scotland	Harz, Germany	Norway	Norway	Finland
	<0.18 mm	<0.10 mm	<0.063 mm	<0.063 mm	<0.063 mm	<2 mm
	total (XRF)	total (DCES)	total (XRF)	total (XRF)	7N HNO$_3$	Conc. HNO$_3$
Median	91	94	91	105	-	-
Min	-	4	16	8	-	-
Max	597	381	277	286	-	-

#2nd percentile ^98th percentile

Concentrations in Waters (mg/l)

Medium	Ocean water	Ocean water	Ocean water	Stream water	Stream water	Stream water
	World (1)	World (2)	North Pacific	World	Nova Scotia, Canada	Finland
					<0.45 μm	<0.45 μm
					ICP-MS	
Median	0.12*	0.12*	0.12*	0.001,1*	0.000,34	-
Min	-	-	-	-	<0.000,05	-
Max	-	-	-	-	0.001,9	-

*Estimated mean #2nd percentile ^98th percentile

Medium	Stream water	Stream water	Stream water	Lake water	Ground water
	Romania	Eastern India	Harz, Germany	Norway	Southern Norway
	unfiltered	<0.2 μm	unfiltered	unfiltered	unfiltered
	ICP-MS	ICP-MS	ICP-MS	ICP-MS	ICP-MS
Median	0.002,6	0.008,6	0.002,1	0.000,49	0.002,255
Min	0.000,4	0.002	<0.000,1	0.000,052	0.000,14
Max	0.117	0.047,3	0.019	0.071,1	0.017,1

Rb RUBIDIUM

Concentrations in Precipitation (mg/l)

Medium	Rain water	Rain water	Snow melt-water	Snow melt-water	Snow filter residue	Snow filter residue
	Kola, remote <0.45 µm ICP-MS	Kola, polluted <0.45 µm ICP-MS	Kola, remote <0.45 µm ICP-MS	Kola, polluted <0.45 µm ICP-MS	Kola, remote >0.45 µm	Kola, polluted >0.45 µm
Median	0.000,17	0.000,19	<0.000,05	0.000,195	-	-
Min	0.000,05	0.000,06	<0.000,05	0.000,18	-	-
Max	0.000,68	0.000,52	0.000,09	0.000,33	-	-

Medium	Rain water	Rain water				
	Coastal, Norway unfiltered ICP-MS	Inland, Norway unfiltered ICP-MS				
Median	0.000,08	0.000,15				
Min	<0.000,01	<0.000,01				
Max	0.003,6	0.005,4				

Concentrations in Air (ng/m^3) and Yearly Deposition (kg/km^2/yr)

Medium	Air	Air	Bulk deposition	Throughfall deposition	Bulk deposition	Bulk deposition
	World, remote	World, polluted	West Germany	West Germany	Kola, remote	Kola, polluted
Median	<1 / 0.38	2	-	-	-	-
Min	0.000,2	<1	-	-	-	-
Max	<1 (5.5)	6.6	-	-	-	-

*Estimated value

Concentrations in Plants (mg/kg)

Medium	Moss	Moss	Crustose lichen	Lichen	Dandelion	Spruce bark
	Norway conc. HNO$_3$ (1990)	Germany conc. HNO$_3$ + H$_2$O$_2$ (1995)	Germany conc. HNO$_3$	Northwest Territories oven-dried	Europe oven-dried total (INAA)	Canada ashed total (INAA)
Median	10	22.1	18	-	84.4#	63.5
Min	1.1	1.9	3.4	-	-	<5
Max	62	151	94	-	-	270

*Mean   #Geometric mean

Concentrations in Human Fluids (mg/l)

Medium	Human blood	Human serum	Human urine
	Lombardy, Italy	Lombardy, Italy	Lombardy, Italy
Mean	2.805	0.23	2.19
Min	0.9	0.078	0.24
Max	6.8	0.511	4.45

RUBIDIUM Rb

Environmental Geochemistry

Biological impacts	Considered non-essential. Considered non-toxic
Uses	Electronics, semi-conductors, glass
Environmental pathways	Poorly understood. Probably mostly windblown dust, weathering of Rb-bearing minerals. Generally, geogenic sources much more important than anthropogenic ones
Environmental mobility	Oxidising conditions: low Acid conditions: low Reducing conditions: very low Neutral to alkaline conditions: low Comments: -
Geochemical barriers	adsorption
Natural association	K-Ba-Rb (substitution in many minerals)
Action levels	-
Remarks	Invariably associated with K, which is essential for all organisms (see high Rb concentrations in plants, human blood and urine). K/Rb ratio in most common continental rocks is in the range of 160-300. K/Rb ratio decreases with weathering. K/Rb ratio in plants is about 1000
Suggested analytical method(s)	XRF, ICP-AES, ICP-MS (water)

World Yearly Production (t/y)	3
Price of Purest Form, in Small Quantities ($/kg)	117,250 (99.8%)
Market Price ($/kg)	-

Re — RHENIUM

Physico-Chemical Properties

Atomic number	75	Melting/boiling point (K)	3459 / 5869
Atomic mass	186.207(1)	Isotopes & isomers	1 stable + 40 unstable
Atomic radius (pm)	197	Acid/base of oxide	weak acid
Main oxidation state(s)	+7 (-1,+1,+2,+3,+4,+5,+6)	State (at 300 K, 1 atm.)	solid
Ionic radius (pm)	77 (+4), 72 (+5), 69 (+6), 52-67 (+7)	Metallic character	pred. non-metal
Electronegativity (Pauling)	1.9	Element group(s)	heavy non-metal, noble (non)metal
Density (g/cm^3)	21.02	Affinity	siderophile

Naturally occurring isotopes	Natural abundance (%)	Atomic mass	Half-life	Unstable isotopes	Longest half-life
^{185}Re	37.4	184.952,96	stable	^{160}Re to ^{192}Re	4.40x10^{+10} y
^{187}Re	62.6	186.955,75	4.40x10^{+10} y		

Concentrations in Crust/Rocks (mg/kg)

Bulk continental crust	0.000,4 / 0.000,7 / 0.000,4
Upper continental crust	- / 0.000,4
Ultramafic rock	<0.000,1
Ocean ridge basalt	<0.000,9
Gabbro, basalt	0.000,4
Granite, granodiorite	0.000,6
Sandstone	0.000,1
Greywacke	-
Shale, schist	0.000,5
Limestone	0.000,1
Coal	<0.1

Typical Minerals
dzhezkazganite (CuReS$_4$)

Possible Host Minerals
molybdenite, zircon, gadolinite

Mass (kg) in

Continental crust	9.44x10^{+12}	Oceans	5.29x10^{+9}	Plants	?

Concentrations in Media Sampled for the Kola Project (mg/kg)

Medium	Moss	Humus <2 mm	Humus <2 mm	Topsoil (0-5 cm) <2 mm	B horizon <2 mm	B horizon <2 mm
	conc. HNO$_3$		amm. acet.		aqua regia	total (XRF)
Median	-	-	-	-	-	-
Min	-	-	-	-	-	-
Max	-	-	-	-	-	-

Medium	C horizon <2 mm aqua regia	C horizon <2 mm total (XRF)	C horizon <2 mm total (INAA)	Lake water unfiltered (mg/l)
Median	-	-	-	-
Min	-	-	-	-
Max	-	-	-	-

RHENIUM — Re

Concentrations in Soils and Sediments (mg/kg)

Medium	Soil	Agricultural soil - Ap horizon	Agricultural soil - Top (0-25 cm)	Agricultural Soil - Bottom (50-75 cm)	Topsoil (0-15 cm)	Urban soil (0-2 cm)
	World	*Canada*	*Finland*	*Finland*	*England & Wales*	*Trondheim*
	<2 mm	<2 mm	<2 mm	<2 mm	<2 mm	<2 mm
	total		aqua regia	aqua regia	aqua regia	aqua regia
Median	-	-	-	-	-	-
Min	-	-	-	-	-	-
Max	-	-	-	-	-	-

*Estimated mean

Medium	Forest soil - Humus	Forest soil - B horizon	Forest soil - C horizon	Till (C horizon)	Till (C horizon)	Laterite (25 ±15 cm)
	Norway	*Norway*	*Norway*	*Finland*	*Finland*	*Australia*
	<2 mm	<2 mm	<2 mm	<0.063 mm	<0.063 mm	0.45-2 mm
	7N HNO$_3$	7N HNO$_3$	7N HNO$_3$	aqua regia	total	total
Median	-	-	-	-	-	-
Min	-	-	-	-	-	-
Max	-	-	-	-	-	-

Medium	Stream sediment	Stream sediment	Stream sediment	Overbank sediment	Overbank sediment	Organic stream sediment
	Austria	*Southern Scotland*	*Harz, Germany*	*Norway*	*Norway*	*Finland*
	<0.18 mm	<0.10 mm	<0.063 mm	<0.063 mm	<0.063 mm	<2 mm
		total	total	total (XRF)	7N HNO$_3$	Conc. HNO$_3$
Median	-	-	-	-	-	-
Min	-	-	-	-	-	-
Max	-	-	-	-	-	-

#2nd percentile   ^98th percentile

Concentrations in Waters (mg/l)

Medium	Ocean water	Ocean water	Ocean water	Stream water	Stream water	Stream water
	World (1)	*World (2)*	*North Pacific*	*World*	*Nova Scotia, Canada*	*Finland*
					<0.45 µm	<0.45 µm
Median	0.000,004*	0.000,004*	0.000,007,8*	-	-	-
Min	-	-	-	-	-	-
Max	-	-	-	-	-	-

*Estimated mean   #2nd percentile   ^98th percentile

Medium	Stream water	Stream water	Stream water	Lake water	Ground water	
	Romania	*Eastern India*	*Harz, Germany*	*Norway*	*Southern Norway*	
	unfiltered	<0.2 µm	unfiltered	unfiltered	unfiltered	
	ICP-MS	ICP-MS		ICP-MS		
Median	-	<0.000,01	-	<0.000,003	-	
Min	-	<0.000,01	-	<0.000,003	-	
Max	-	<0.000,01	-	0.000,272	-	

Re RHENIUM

Concentrations in Precipitation (mg/l)

Medium	Rain water Kola, remote <0.45 µm	Rain water Kola, polluted <0.45 µm	Snow melt-water Kola, remote <0.45 µm	Snow melt-water Kola, polluted <0.45 µm	Snow filter residue Kola, remote >0.45 µm	Snow filter residue Kola, polluted >0.45 µm
Median	-	-	-	-	-	-
Min	-	-	-	-	-	-
Max	-	-	-	-	-	-

Medium	Rain water Coastal, Norway unfiltered ICP-MS	Rain water Inland, Norway unfiltered ICP-MS				
Median	-	-				
Min	-	-				
Max	-	-				

Concentrations in Air (ng/m^3) and Yearly Deposition (kg/km^2/yr)

Medium	Air World, remote	Air World, polluted	Bulk deposition West Germany	Throughfall deposition West Germany	Bulk deposition Kola, remote	Bulk deposition Kola, polluted
Median	-	-	-	-	-	-
Min	-	-	-	-	-	-
Max	-	-	-	-	-	-

*Estimated value

Concentrations in Plants (mg/kg)

Medium	Moss Norway conc. HNO$_3$ (1990)	Moss Germany conc. HNO$_3$ + H$_2$O$_2$ (1995)	Crustose lichen Germany conc. HNO$_3$	Lichen Northwest Territories oven-dried	Dandelion Europe oven-dried	Spruce bark Canada ashed
Median	-	-	-	-	-	-
Min	-	-	-	-	-	-
Max	-	-	-	-	-	-

*Mean   #Geometric mean

Concentrations in Human Fluids (mg/l)

Medium	Human blood Lombardy, Italy	Human serum Lombardy, Italy	Human urine Lombardy, Italy			
Mean	-	-	-			
Min	-	-	-			
Max	-	-	-			

RHENIUM   Re

Environmental Geochemistry

Biological impacts	Considered non-essential
Uses	Lamps, thermostats, bulbs for photographic flashes, catalysts
Environmental pathways	Poorly understood
Environmental mobility	Oxidising conditions: high Acid conditions: high Reducing conditions: very low Neutral to alkaline conditions: very high Comments: -
Geochemical barriers	Reduction
Natural association	Cu-Mo-Re-Fe-Au-Ag-Zn (some porphyry Cu deposits)
Action levels	-
Remarks	Re is a by-product of Mo production
Suggested analytical method(s)	ICP-MS, INAA

World Yearly Production (t/y)	28.2 (in 1995)
Price of Purest Form, in Small Quantities ($/kg)	52,500 (99.999%)
Market Price ($/kg)	-

Rh RHODIUM

Physico-Chemical Properties

Atomic number	45	Melting/boiling point (K)	2237 / 3968
Atomic mass	102.905,50(2)	Isotopes & isomers	1 stable + 45 unstable
Atomic radius (pm)	183	Acid/base of oxide	amphoteric
Main oxidation state(s)	+3 (+2,+4,+5,+6)	State (at 300 K, 1 atm.)	solid
Ionic radius (pm)	80.5 (+3), 74 (+4), 69 (+5)	Metallic character	pred. metal
Electronegativity (Pauling)	2.28	Element group(s)	heavy metal, noble metal, PGE
Density (g/cm^3)	12.41	Affinity	chalcophile, siderophile

Naturally occurring isotopes	Natural abundance (%)	Atomic mass	Half-life	Unstable isotopes	Longest half-life
^{103}Rh	100	102.905,5	stable	^{90}Rh to ^{117}Rh	3.3 y

Concentrations in Crust/Rocks (mg/kg)

Bulk continental crust	0.000,06 / 0.001 / -	Typical Minerals	hollingworthite ((Rh,Pd,Pt)AsS), irarsite ((Ir,Ru,Rh,Pt)AsS)
Upper continental crust	- / -		
Ultramafic rock	-	Possible Host Minerals	olivine, ilmenite, zircon, chromite, gadolinite
Ocean ridge basalt	-		
Gabbro, basalt	-		
Granite, granodiorite	-		
Sandstone	-		
Greywacke	0.000,06		
Shale, schist	-		
Limestone	-		
Coal	<0.1		

Mass (kg) in

Continental crust	1.42x10^{+12}	Oceans	-	Plants	18,400

Concentrations in Media Sampled for the Kola Project (mg/kg)

Medium	Moss	Humus <2 mm	Humus <2 mm	Topsoil (0-5 cm) <2 mm	B horizon <2 mm	B horizon <2 mm
	conc. HNO$_3$		amm. acet.		aqua regia	total (XRF)
Median	-	-	-	-	-	-
Min	-	-	-	-	-	-
Max	-	-	-	-	-	-

Medium	C horizon <2 mm aqua regia	C horizon <2 mm total (XRF)	C horizon <2 mm total (INAA)	Lake water unfiltered (mg/l)
Median	-	-	-	-
Min	-	-	-	-
Max	-	-	-	-

RHODIUM Rh

Concentrations in Soils and Sediments (mg/kg)

Medium	Soil	Agricultural soil -	Agricultural soil - Top	Agricultural Soil - Bottom	Topsoil (0-15 cm)	Urban soil (0-2 cm)
	World	Ap horizon	(0-25 cm)	(50-75 cm)	England &	Trondheim
		Canada	Finland	Finland	Wales	
	<2 mm	<2 mm	<2 mm	<2 mm	<2 mm	<2 mm
	total		aqua regia	aqua regia	aqua regia	aqua regia
Median	-	-	-	-	-	-
Min	-	-	-	-	-	-
Max	-	-	-	-	-	-

*Estimated mean

Medium	Forest soil - Humus	Forest soil - B horizon	Forest soil - C horizon	Till (C horizon)	Till (C horizon)	Laterite (25 ±15 cm)
	Norway	Norway	Norway	Finland	Finland	Australia
	<2 mm	<2 mm	<2 mm	<0.063 mm	<0.063 mm	0.45-2 mm
	7N HNO$_3$	7N HNO$_3$	7N HNO$_3$	aqua regia	total	total
Median	-	-	-	-	-	-
Min	-	-	-	-	-	-
Max	-	-	-	-	-	-

Medium	Stream sediment	Stream sediment	Stream sediment	Overbank sediment	Overbank sediment	Organic stream sediment
		Southern Scotland	Harz, Germany	Norway	Norway	Finland
	Austria					
	<0.18 mm	<0.10 mm	<0.063 mm	<0.063 mm	<0.063 mm	<2 mm
		total	total	total (XRF)	7N HNO$_3$	Conc. HNO$_3$
Median	-	-	-	-	-	-
Min	-	-	-	-	-	-
Max	-	-	-	-	-	-

#2nd percentile   ^98th percentile

Concentrations in Waters (mg/l)

Medium	Ocean water	Ocean water	Ocean water	Stream water	Stream water	Stream water
	World (1)	World (2)	North Pacific	World	Nova Scotia, Canada	Finland
					<0.45 μm	<0.45 μm
Median	-	-	0.000,000,08*	-	-	-
Min	-	-	-	-	-	-
Max	-	-	-	-	-	-

*Estimated mean   #2nd percentile   ^98th percentile

Medium	Stream water	Stream water	Stream water	Lake water	Ground water
	Romania	Eastern India	Harz, Germany	Norway	Southern Norway
	unfiltered	<0.2 μm	unfiltered	unfiltered	unfiltered
	ICP-MS	ICP-MS		ICP-MS	
Median	-	<0.000,1	-	-	-
Min	-	<0.000,1	-	-	-
Max	-	<0.000,1	-	-	-

Rh RHODIUM

Concentrations in Precipitation (mg/l)

Medium	Rain water Kola, remote <0.45 µm	Rain water Kola, polluted <0.45 µm	Snow melt- water Kola, remote <0.45 µm	Snow melt- water Kola, polluted <0.45 µm	Snow filter residue Kola, remote >0.45 µm	Snow filter residue Kola, polluted >0.45 µm
Median	-	-	-	-	-	-
Min	-	-	-	-	-	-
Max	-	-	-	-	-	-

Medium	Rain water Coastal, Norway unfiltered ICP-MS	Rain water Inland, Norway unfiltered ICP-MS				
Median	-	-				
Min	-	-				
Max	-	-				

Concentrations in Air (ng/m^3) and Yearly Deposition (kg/km^2/yr)

Medium	Air World, remote	Air World, polluted	Bulk deposition West Germany	Throughfall deposition West Germany	Bulk deposition Kola, remote	Bulk deposition Kola, polluted
Median	-	-	-	-	-	-
Min	-	-	-	-	-	-
Max	-	-	-	-	-	-

*Estimated value

Concentrations in Plants (mg/kg)

Medium	Moss Norway conc. HNO$_3$ (1990)	Moss Germany conc. HNO$_3$ + H$_2$O$_2$ (1995)	Crustose lichen Germany conc. HNO$_3$	Lichen Northwest Territories oven-dried	Dandelion Europe oven-dried	Spruce bark Canada ashed
Median	-	-	-	-	-	-
Min	-	-	-	-	-	-
Max	-	-	-	-	-	-

*Mean #Geometric mean

Concentrations in Human Fluids (mg/l)

Medium	Human blood Lombardy, Italy	Human serum Lombardy, Italy	Human urine Lombardy, Italy			
Mean	-	-	-			
Min	-	-	-			
Max	-	-	-			

RHODIUM   Rh

Environmental Geochemistry

Biological impacts	Considered non-essential. Toxic
Uses	Catalysts, jewellery, decoration, thermostats
Environmental pathways	Catalyst wear (e.g., traffic)
Environmental mobility	Oxidising conditions: very low Acid conditions: very low Reducing conditions: very low Neutral to alkaline conditions: very low Comments: -
Geochemical barriers	Mechanical
Natural association	PGE-Ni-Cu-Co-As-Ag-Au-Te-Bi-Sn-Sb (massive Ni-Cu sulphide ores, e.g., Sudbury), PGE-Ag-Au-Cr-Fe-Cu-Ni-Co-S (sulphides, e.g., Merensky reef)
Action levels	-
Remarks	Very rich ore deposits contain about 8 ppm PGE. PGE occur often in Ni ores. PGE usually occur together in the following proportions: Pt 25%, Ru 23%, Os 21%, Pd 18%, Ir 7%, Rh 6%. Cosmic input via meteorites, cosmic dust, etc. significant. Some marine algae enrich PGE
Suggested analytical method(s)	ICP-MS

World Yearly Production (t/y)	300 all PGE (in 1995)
Price of Purest Form, in Small Quantities ($/kg)	120,313 (99.99%)
Market Price ($/kg)	7556

Rn   RADON

Physico-Chemical Properties

Atomic number	86	Melting/boiling point (K)	202 / 211.45
Atomic mass	[222]	Isotopes & isomers	0 stable + 34 unstable
Atomic radius (pm)	134	Acid/base of oxide	-
Main oxidation state(s)	0	State (at 300 K, 1 atm.)	gas
Ionic radius (pm)	-	Metallic character	-
Electronegativity (Pauling)	-	Element group(s)	noble gas
Density (g/cm^3)	0.009,73	Affinity	atmophile

Naturally occurring isotopes	Natural abundance (%)	Atomic mass	Half-life
-	-	-	-

Unstable isotopes	Longest half-life
^{198}Rn to ^{228}Rn	3.823,5 d

Concentrations in Crust/Rocks (mg/kg)

Bulk continental crust	- / 4x10^{-13} / -
Upper continental crust	- / -
Ultramafic rock	-
Ocean ridge basalt	-
Gabbro, basalt	-
Granite, granodiorite	-
Sandstone	-
Greywacke	-
Shale, schist	-
Limestone	-
Coal	-

Typical Minerals
-

Possible Host Minerals
-

Mass (kg) in

Continental crust	9.44x10^{+3}	Oceans	0.793	Plants	-

Concentrations in Media Sampled for the Kola Project (mg/kg)

| Medium | Moss | Humus | Humus | Topsoil (0-5 cm) | B horizon | B horizon |
| | | <2 mm | <2 mm | <2 mm | <2 mm | <2 mm |
	conc. HNO$_3$		amm. acet.		aqua regia	total (XRF)
Median	-	-	-	-	-	-
Min	-	-	-	-	-	-
Max	-	-	-	-	-	-

| Medium | C horizon | C horizon | C horizon | | | Lake water |
| | <2 mm | <2 mm | <2 mm | | | unfiltered |
	aqua regia	total (XRF)	total (INAA)			(mg/l)
Median	-	-	-			-
Min	-	-	-			-
Max	-	-	-			-

RADON — Rn

Concentrations in Soils and Sediments (mg/kg)

Medium	Soil	Agricultural soil -	Agricultural soil - Top	Agricultural Soil - Bottom	Topsoil (0-15 cm)	Urban soil (0-2 cm)
	World	Ap horizon Canada	(0-25 cm) Finland	(50-75 cm) Finland	England & Wales	Trondheim
	<2 mm total	<2 mm	<2 mm aqua regia	<2 mm aqua regia	<2 mm aqua regia	<2 mm aqua regia
Median	-	-	-	-	-	-
Min	-	-	-	-	-	-
Max	-	-	-	-	-	-

*Estimated mean

Medium	Forest soil - Humus	Forest soil - B horizon	Forest soil - C horizon	Till (C horizon)	Till (C horizon)	Laterite (25 ±15 cm)
	Norway	Norway	Norway	Finland	Finland	Australia
	<2 mm $7N\ HNO_3$	<2 mm $7N\ HNO_3$	<2 mm $7N\ HNO_3$	<0.063 mm aqua regia	<0.063 mm total	0.45-2 mm total
Median	-	-	-	-	-	-
Min	-	-	-	-	-	-
Max	-	-	-	-	-	-

Medium	Stream sediment	Stream sediment	Stream sediment	Overbank sediment	Overbank sediment	Organic stream sediment
	Austria	Southern Scotland	Harz, Germany	Norway	Norway	Finland
	<0.18 mm	<0.10 mm total	<0.063 mm total	<0.063 mm total (XRF)	<0.063 mm $7N\ HNO_3$	<2 mm Conc. HNO_3
Median	-	-	-	-	-	-
Min	-	-	-	-	-	-
Max	-	-	-	-	-	-

#2nd percentile   ^98th percentile

Concentrations in Waters (mg/l)

Medium	Ocean water	Ocean water	Ocean water	Stream water	Stream water	Stream water
	World (1)	World (2)	North Pacific	World	Nova Scotia, Canada	Finland
					<0.45 µm	<0.45 µm
Median	6×10^{-16}*	6×10^{-16}*	-	-	-	-
Min	-	-	-	-	-	-
Max	-	-	-	-	-	-

*Estimated mean   #2nd percentile   ^98th percentile

Medium	Stream water	Stream water	Stream water	Lake water	Ground water
	Romania	Eastern India	Harz, Germany	Norway	Southern Norway
	unfiltered ICP-MS	<0.2 µm ICP-MS	unfiltered	unfiltered ICP-MS	unfiltered Scintill. counter
Median	-	-	-	-	3.54×10^{-11} (200 Bq/l)
Min	-	-	-	-	1.77×10^{-12} (10 Bq/l)
Max	-	-	-	-	1.21×10^{-9} (6840 Bq/l)

Rn   RADON

Concentrations in Precipitation (mg/l)

Medium	Rain water *Kola, remote* *<0.45 μm*	Rain water *Kola, polluted* *<0.45 μm*	Snow melt- water *Kola, remote* *<0.45 μm*	Snow melt- water *Kola, polluted* *<0.45 μm*	Snow filter residue *Kola, remote* *>0.45 μm*	Snow filter residue *Kola, polluted* *>0.45 μm*
Median	-	-	-	-	-	-
Min	-	-	-	-	-	-
Max	-	-	-	-	-	-

Medium	Rain water *Coastal,* *Norway* *unfiltered* *ICP-MS*	Rain water *Inland,* *Norway* *unfiltered* *ICP-MS*				
Median	-	-				
Min	-	-				
Max	-	-				

Concentrations in Air (ng/m^3) and Yearly Deposition (kg/km^2/yr)

Medium	Air *World,* *remote*	Air *World,* *polluted*	Bulk deposition *West* *Germany*	Throughfall deposition *West* *Germany*	Bulk deposition *Kola, remote*	Bulk deposition *Kola, polluted*
Median	-	-	-	-	-	-
Min	-	-	-	-	-	-
Max	-	-	-	-	-	-

*Estimated value

Concentrations in Plants (mg/kg)

Medium	Moss *Norway* *conc. HNO$_3$* *(1990)*	Moss *Germany* *conc. HNO$_3$ +* *H$_2$O$_2$ (1995)*	Crustose lichen *Germany* *conc. HNO$_3$*	Lichen *Northwest* *Territories* *oven-dried*	Dandelion *Europe* *oven-dried*	Spruce bark *Canada* *ashed*
Median	-	-	-	-	-	-
Min	-	-	-	-	-	-
Max	-	-	-	-	-	-

*Mean   #Geometric mean

Concentrations in Human Fluids (mg/l)

Medium	Human blood *Lombardy,* *Italy*	Human serum *Lombardy,* *Italy*	Human urine *Lombardy,* *Italy*			
Mean	-	-	-			
Min	-	-	-			
Max	-	-	-			

RADON Rn

Environmental Geochemistry

Biological impacts	Considered non-essential. Carcinogenic when inhaled
Uses	Has been used in mineral waters, and therapeutic spas
Environmental pathways	Natural trace component of atmosphere (short-lived radioisotope), uraniferous granites, alum shales, fault zones
Environmental mobility	Oxidising conditions: very high Acid conditions: very high Reducing conditions: very high Neutral to alkaline conditions: very high Comments: Inert gas, short-lived radioisotope
Geochemical barriers	-
Natural association	U-Th-Ra-Rn-Bi-Pb (U deposits), Ra-Rn-Ba (gossans), Ra-Rn-Mn-Fe (spring precipitates), Ra-Rn-organic matter (bogs)
Action levels	Drinking water, MAC: 500 Bq/l or 8.8×10^{-11} mg ^{222}Rn/l (Norway Shd)
Remarks	Rn is a daughter product of Ra, and is released by minerals like uraninite/pitchblende and carnotite. Short-lived radioisotope
Suggested analytical method(s)	Scintillation counter

World Yearly Production (t/y)   -
Price of Purest Form, in Small Quantities ($/kg)   -
Market Price ($/kg)   -

Ru — RUTHENIUM

Physico-Chemical Properties

Atomic number	44
Atomic mass	101.07(2)
Atomic radius (pm)	189
Main oxidation state(s)	+3,+4 (0,+1,+2,+5,+6,+7,+8)
Ionic radius (pm)	82 (+3), 76 (+4), 70.5 (+5), 52 (+7), 50 (+8)
Electronegativity (Pauling)	2.2
Density (g/cm^3)	12.41

Melting/boiling point (K)	2607 / 4423
Isotopes & isomers	7 stable + 22 unstable
Acid/base of oxide	weak acid
State (at 300 K, 1 atm.)	solid
Metallic character	pred. metal
Element group(s)	heavy metal, noble metal, PGE
Affinity	siderophile

Naturally occurring isotopes	Natural abundance (%)	Atomic mass	Half-life
^{96}Ru	5.52	95.907,6	stable
^{98}Ru	1.88	97.905,29	stable
^{99}Ru	12.7	98.905,94	stable
^{100}Ru	12.6	99.904,22	stable
^{101}Ru	17	100.905,58	stable
^{102}Ru	31.6	101.904,35	stable
^{104}Ru	18.7	103.905,43	stable

Unstable isotopes	Longest half-life
^{88}Ru to ^{115}Ru	1.02 y

Concentrations in Crust/Rocks (mg/kg)

Bulk continental crust	0.000,1 / 0.001 / -
Upper continental crust	- / -
Ultramafic rock	-
Ocean ridge basalt	-
Gabbro, basalt	-
Granite, granodiorite	-
Sandstone	-
Greywacke	0.000,1
Shale, schist	-
Limestone	-
Coal	<0.1

Typical Minerals
laurite ((Ru,Os)S$_2$), irarsite ((Ir,Ru,Rh,Pt)AsS)

Possible Host Minerals
olivine, ilmenite, zircon, chromite, gadolinite

Mass (kg) in

Continental crust	2.36x10^{+12}	Oceans	-	Plants	18,400

Concentrations in Media Sampled for the Kola Project (mg/kg)

| Medium | Moss | Humus | Humus | Topsoil (0-5 cm) | B horizon | B horizon |
| | | <2 mm | <2 mm | <2 mm | <2 mm | <2 mm |
	conc. HNO$_3$		amm. acet.		aqua regia	total (XRF)
Median	-	-	-	-	-	-
Min	-	-	-	-	-	-
Max	-	-	-	-	-	-

| Medium | C horizon | C horizon | C horizon | | Lake water |
| | <2 mm | <2 mm | <2 mm | | unfiltered |
	aqua regia	total (XRF)	total (INAA)		(mg/l)
Median	-	-	-		-
Min	-	-	-		-
Max	-	-	-		-

RUTHENIUM — Ru

Concentrations in Soils and Sediments (mg/kg)

Medium	Soil	Agricultural soil - Ap horizon	Agricultural soil - Top (0-25 cm)	Agricultural Soil - Bottom (50-75 cm)	Topsoil (0-15 cm)	Urban soil (0-2 cm)
	World	Canada	Finland	Finland	England & Wales	Trondheim
	<2 mm total	<2 mm	<2 mm aqua regia	<2 mm aqua regia	<2 mm aqua regia	<2 mm aqua regia
Median	-	-	-	-	-	-
Min	-	-	-	-	-	-
Max	-	-	-	-	-	-

*Estimated mean

Medium	Forest soil - Humus	Forest soil - B horizon	Forest soil - C horizon	Till (C horizon)	Till (C horizon)	Laterite (25 ±15 cm)
	Norway	Norway	Norway	Finland	Finland	Australia
	<2 mm	<2 mm	<2 mm	<0.063 mm	<0.063 mm	0.45-2 mm
	7N HNO$_3$	7N HNO$_3$	7N HNO$_3$	aqua regia	total	total
Median	-	-	-	-	-	-
Min	-	-	-	-	-	-
Max	-	-	-	-	-	-

Medium	Stream sediment	Stream sediment	Stream sediment	Overbank sediment	Overbank sediment	Organic stream sediment
	Austria	Southern Scotland	Harz, Germany	Norway	Norway	Finland
	<0.18 mm	<0.10 mm	<0.063 mm	<0.063 mm	<0.063 mm	<2 mm
		total	total	total (XRF)	7N HNO$_3$	Conc. HNO$_3$
Median	-	-	-	-	-	-
Min	-	-	-	-	-	-
Max	-	-	-	-	-	-

#2nd percentile ^98th percentile

Concentrations in Waters (mg/l)

Medium	Ocean water	Ocean water	Ocean water	Stream water	Stream water	Stream water
	World (1)	World (2)	North Pacific	World	Nova Scotia, Canada	Finland
					<0.45 µm	<0.45 µm
Median	0.000,000,7*	-	<5x10^{-9}*	-	-	-
Min	-	-	-	-	-	-
Max	-	-	-	-	-	-

*Estimated mean   #2nd percentile   ^98th percentile

Medium	Stream water	Stream water	Stream water	Lake water	Ground water
	Romania	Eastern India	Harz, Germany	Norway	Southern Norway
	unfiltered	<0.2 µm	unfiltered	unfiltered	unfiltered
	ICP-MS	ICP-MS		ICP-MS	
Median	-	<0.000,1	-	<0.000,02	-
Min	-	<0.000,1	-	<0.000,02	-
Max	-	<0.000,1	-	0.000,346	-

Ru   RUTHENIUM

Concentrations in Precipitation (mg/l)

Medium	Rain water Kola, remote <0.45 µm	Rain water Kola, polluted <0.45 µm	Snow melt-water Kola, remote <0.45 µm	Snow melt-water Kola, polluted <0.45 µm	Snow filter residue Kola, remote >0.45 µm	Snow filter residue Kola, polluted >0.45 µm
Median	-	-	-	-	-	-
Min	-	-	-	-	-	-
Max	-	-	-	-	-	-

Medium	Rain water Coastal, Norway unfiltered ICP-MS	Rain water Inland, Norway unfiltered ICP-MS				
Median	-	-				
Min	-	-				
Max	-	-				

Concentrations in Air (ng/m^3) and Yearly Deposition (kg/km^2/yr)

Medium	Air World, remote	Air World, polluted	Bulk deposition West Germany	Throughfall deposition West Germany	Bulk deposition Kola, remote	Bulk deposition Kola, polluted
Median	-	-	-	-	-	-
Min	-	-	-	-	-	-
Max	-	-	-	-	-	-

*Estimated value

Concentrations in Plants (mg/kg)

Medium	Moss Norway conc. HNO$_3$ (1990)	Moss Germany conc. HNO$_3$ + H$_2$O$_2$ (1995)	Crustose lichen Germany conc. HNO$_3$	Lichen Northwest Territories oven-dried	Dandelion Europe oven-dried	Spruce bark Canada ashed
Median	-	-	-	-	-	-
Min	-	-	-	-	-	-
Max	-	-	-	-	-	-

*Mean #Geometric mean

Concentrations in Human Fluids (mg/l)

Medium	Human blood Lombardy, Italy	Human serum Lombardy, Italy	Human urine Lombardy, Italy			
Mean	-	-	-			
Min	-	-	-			
Max	-	-	-			

RUTHENIUM   Ru

Environmental Geochemistry

Biological impacts	-
Uses	Catalyst, alloys (little used)
Environmental pathways	-
Environmental mobility	Oxidising conditions: very low Acid conditions: very low Reducing conditions: very low Neutral to alkaline conditions: very low Comments: -
Geochemical barriers	-
Natural association	PGE-Ni-Cu-Co-As-Ag-Au-Te-Bi-Sn-Sb (massive Ni-Cu sulphide ores, e.g., Sudbury), PGE-Ag-Au-Cr-Fe-Cu-Ni-Co-S (sulphides, e.g., Merensky reef)
Action levels	-
Remarks	Very rich ore deposits contain about 8 ppm PGE. PGE occur often in Ni ores. PGE usually occur together in the following proportions: Pt 25%, Ru 23%, Os 21%, Pd 18%, Ir 7%, Rh 6%. Cosmic input via meteorites, cosmic dust, etc. significant. Some marine algae enrich PGE
Suggested analytical method(s)	ICP-MS, INAA

World Yearly Production (t/y)	300 all PGE (in 1995)
Price of Purest Form, in Small Quantities ($/kg)	240,000 (99.999%)
Market Price ($/kg)	-

S   SULPHUR

Physico-Chemical Properties

Atomic number	16	Melting/boiling point (K)	388.36 / 717.75
Atomic mass	32.066(6)	Isotopes & isomers	4 stable + 13 unstable
Atomic radius (pm)	109	Acid/base of oxide	strong acid
Main oxidation state(s)	+6 (-2,0,+2,+4)	State (at 300 K, 1 atm.)	solid
Ionic radius (pm)	170 (-2), 51 (+4), 26-43 (+6)	Metallic character	non-metal
Electronegativity (Pauling)	2.58	Element group(s)	light non-metal
Density (g/cm^3)	2.07 (rhombic), 1.96 (monoclinic)	Affinity	biophile, chalcophile

Naturally occurring isotopes	Natural abundance (%)	Atomic mass	Half-life
^{32}S	95.02	31.972,07	stable
^{33}S	0.75	32.971,46	stable
^{34}S	4.21	33.967,87	stable
^{36}S	0.02	35.967,08	stable

Unstable isotopes	Longest half-life
^{27}S to ^{44}S	87.2 d

Concentrations in Crust/Rocks (mg/kg)

Bulk continental crust	697 / 350 / -
Upper continental crust	953 / 300
Ultramafic rock	600
Ocean ridge basalt	800
Gabbro, basalt	900
Granite, granodiorite	100
Sandstone	200
Greywacke	-
Shale, schist	1100
Limestone	500
Coal	20,000

Typical Minerals
native sulphur (S), pyrite (FeS$_2$), galena (PbS), sphalerite (ZnS), other sulphides, gypsum (CaSO$_4 \cdot$2H$_2$O), anhydrite (CaSO$_4$), other sulphates

Possible Host Minerals
biotite, hornblende

Mass (kg) in

Continental crust	1.64x10^{+19}	Oceans	1.20x10^{+18}	Plants	5.52x10^{+13}

Concentrations in Media Sampled for the Kola Project (mg/kg)

Medium	Moss	Humus	Humus	Topsoil (0-5 cm)	B horizon	B horizon
	<2 mm	<2 mm	<2 mm	<2 mm	<2 mm	<2 mm
	conc. HNO$_3$	conc. HNO$_3$	amm. acet.		aqua regia	total (XRF)
Median	863	1530	115	-	162	-
Min	543	400	39.8	-	11	-
Max	2090	3830	592	-	777	-

Medium	C horizon	C horizon	C horizon			Lake water
	<2 mm	<2 mm	<2 mm			unfiltered
	aqua regia	total (XRF)	total (INAA)			(mg/l)
Median	30	-	-			1
Min	<5	-	-			0.35
Max	531	-	-			6.18

SULPHUR S

Concentrations in Soils and Sediments (mg/kg)

Medium	Soil	Agricultural soil -	Agricultural soil - Top	Agricultural Soil - Bottom	Topsoil	Urban soil
	World	Ap horizon	(0-25 cm)	(50-75 cm)	(0-15 cm)	(0-2 cm)
		Canada	Finland	Finland	England & Wales	Trondheim
	<2 mm	<2 mm	<2 mm	<2 mm	<2 mm	<2 mm
	total		aqua regia	aqua regia	aqua regia	aqua regia
Median	800*	-	354	78.6	-	476
Min	-	-	<50	<50	-	37
Max	-	-	3520	10,800	-	4850

*Estimated mean

Medium	Forest soil - Humus	Forest soil - B horizon	Forest soil - C horizon	Till (C horizon)	Till (C horizon)	Laterite (25 ±15 cm)
	Norway	Norway	Norway	Finland	Finland	Australia
	<2 mm	<2 mm	<2 mm	<0.063 mm	<0.063 mm	0.45-2 mm
	7N HNO$_3$	7N HNO$_3$	7N HNO$_3$	aqua regia	total	total
Median	-	-	-	-	140	<100
Min	-	-	-	-	-	<100
Max	-	-	-	-	-	179,600

Medium	Stream sediment	Stream sediment	Stream sediment	Overbank sediment	Overbank sediment	Organic stream sediment
		Southern	Harz,			
	Austria	Scotland	Germany	Norway	Norway	Finland
	<0.18 mm	<0.10 mm	<0.063 mm	<0.063 mm	<0.063 mm	<2 mm
		total	total	total (XRF)	7N HNO$_3$	Conc. HNO$_3$
Median	-	-	-	400	-	1260
Min	-	-	-	100	-	261#
Max	-	-	-	7100	-	6044^

#2nd percentile ^98th percentile

Concentrations in Waters (mg/l)

Medium	Ocean water	Ocean water	Ocean water	Stream water	Stream water	Stream water
	World (1)	World (2)	North Pacific	World	Nova Scotia, Canada	Finland
					<0.45 μm	<0.45 μm
Median	905*	905*	898*	4*	-	-
Min	-	-	-	-	-	-
Max	-	-	-	-	-	-

*Estimated mean   #2nd percentile   ^98th percentile

Medium	Stream water	Stream water	Stream water	Lake water	Ground water
	Romania	Eastern India	Harz, Germany	Norway	Southern Norway
	unfiltered	<0.2 μm	unfiltered	unfiltered	unfiltered
	ICP-MS	ICP-MS	IC	ICP-MS	
Median	-	-	5	-	-
Min	-	-	0.73	-	-
Max	-	-	48	-	-

S SULPHUR

Concentrations in Precipitation (mg/l)

Medium	Rain water	Rain water	Snow melt-water	Snow melt-water	Snow filter residue	Snow filter residue
	Kola, remote <0.45 μm ICP-AES	Kola, polluted <0.45 μm ICP-AES	Kola, remote <0.45 μm ICP-AES	Kola, polluted <0.45 μm ICP-AES	Kola, remote >0.45 μm ICP-AES	Kola, polluted >0.45 μm ICP-AES
Median	0.29	1.63	0.256	0.825,5	<0.005	0.181,5
Min	0.09	0.87	0.234	0.534	<0.005	0.089
Max	0.56	3.92	0.327	1.65	<0.005	0.568

Medium	Rain water	Rain water				
	Coastal, Norway unfiltered ICP-MS	Inland, Norway unfiltered ICP-MS				
Median	-	-				
Min	-	-				
Max	-	-				

Concentrations in Air (ng/m^3) and Yearly Deposition (kg/km^2/yr)

Medium	Air	Air	Bulk deposition	Throughfall deposition	Bulk deposition	Bulk deposition
	World, remote	World, polluted	West Germany	West Germany	Kola, remote	Kola, polluted
Median	- / 290	5000	-	-	1.89*	654*
Min	27	2000	-	-	-	-
Max	4500	13,000	-	-	-	-

*Estimated value

Concentrations in Plants (mg/kg)

Medium	Moss	Moss	Crustose lichen	Lichen	Dandelion	Spruce bark
	Norway conc. HNO$_3$ (1990)	Germany conc. HNO$_3$ + H$_2$O$_2$ (1995)	Germany conc. HNO$_3$	Northwest Territories oven-dried total (XRF)	Europe oven-dried	Canada ashed
Median	-	-	-	221*	-	-
Min	-	-	-	-	-	-
Max	-	-	-	-	-	-

*Mean   #Geometric mean

Concentrations in Human Fluids (mg/l)

Medium	Human blood	Human serum	Human urine			
	Lombardy, Italy	Lombardy, Italy	Lombardy, Italy			
Mean	-	-	-			
Min	-	-	-			
Max	-	-	-			

SULPHUR S

Environmental Geochemistry

Biological impacts	Essential for all organisms. Pure S non-toxic, but many compounds are toxic (SO_2, H_2S)
Uses	Chemical industry, rubber industry, matches, fungicide, pharmacy, gunpowder
Environmental pathways	SO_2 emissions from power plants, cement factories, metal smelters, roasting plants
Environmental mobility	Oxidising conditions: very high Acid conditions: very high Reducing conditions: very low Neutral to alkaline conditions: very high Comments: Mobility can vary greatly with compound (e.g., sulphate versus sulphide)
Geochemical barriers	Reduction (sulphide formation with many metals), sulphate precipitation with some ions (e.g., Ba)
Natural association	S occurs ubiquitously with many metals, e.g., as sulphides (FeS_2, FeAsS, ZnS, PbS, etc.)
Action levels	Drinking water, MAC: 100 mg SO_4/l (=33 mg S/l) (Norway Shd), 500 mg SO_4/l (=167 mg S/l) (Canada CWQG), 0.05 mg H_2S/l (Canada CWQG)
Remarks	S deficiency is rare in organisms, but has been reported in soils. High sulphate in soil is usually tolerated by plants (SO_4 bound in gypsum). H_2S formation in saturated soil can lead to plant damage. Natural emission of SO_2 estimated at 20 million tons per year. Anthropogenic emission of SO_2 estimated at 150 million tons per year. Isotopic fractionation important in nature. Concentrations of 0.05 mg H_2S/l in drinking water may give rise to consumer complaints due to odour and taste. Concentrations of 250 mg SO_4^{2-}/l in drinking water may give rise to consumer complaints due to taste and corrosion
Suggested analytical method(s)	ICP-AES, S elemental analyser

World Yearly Production (t/y)	56,000,000 (in 1995)
Price of Purest Form, in Small Quantities ($/kg)	4925 (99.999%)
Market Price ($/kg)	0.04 (solid), 0.07 (liquid)

Sb ANTIMONY

Physico-Chemical Properties

Atomic number	51	Melting/boiling point (K)	903.78 / 1860
Atomic mass	121.760(1)	Isotopes & isomers	2 stable + 42 unstable
Atomic radius (pm)	153	Acid/base of oxide	weak acid
Main oxidation state(s)	+3 (-3,+4,+5)	State (at 300 K, 1 atm.)	solid
Ionic radius (pm)	90-94 (+3), 74 (+5)	Metallic character	pred. non-metal
Electronegativity (Pauling)	2.05	Element group(s)	heavy non-metal
Density (g/cm^3)	6.691	Affinity	chalcophile

Naturally occurring isotopes	Natural abundance (%)	Atomic mass	Half-life
^{121}Sb	57.21	120.903,82	stable
^{123}Sb	42.79	122.904,22	stable

Unstable isotopes	Longest half-life
^{106}Sb to ^{136}Sb	2.758 y

Concentrations in Crust/Rocks (mg/kg)

Bulk continental crust	0.3 / 0.2 / 0.2
Upper continental crust	0.31 / 0.2
Ultramafic rock	0.1
Ocean ridge basalt	0.1
Gabbro, basalt	0.2
Granite, granodiorite	0.3
Sandstone	0.05
Greywacke	-
Shale, schist	1
Limestone	0.15
Coal	2

Typical Minerals
stibnite (Sb$_2$S$_3$), kermesite (2Sb$_2$S$_3$.Sb$_2$O$_3$), valentinite (Sb$_2$O$_3$), cervantite (Sb$_2$O$_4$)

Possible Host Minerals
ilmenite, Mg-olivine, galena, sphalerite, pyrite

Mass (kg) in

Continental crust 7.08x10^{+15}	Oceans 3.17x10^{+11}	Plants 1.84x10^{+8}

Concentrations in Media Sampled for the Kola Project (mg/kg)

Medium	Moss	Humus	Humus	Topsoil (0-5 cm)	B horizon	B horizon
		<2 mm	<2 mm	<2 mm	<2 mm	<2 mm
	conc. HNO$_3$	conc. HNO$_3$	amm. acet.	total (INAA)	aqua regia	total (XRF)
Median	0.041	0.183	<0.5	<0.1	<0.01	-
Min	<0.02	0.016	<0.5	<0.1	<0.01	-
Max	0.623	0.962	0.9	4.4	1.33	-

Medium	C horizon	C horizon	C horizon			Lake water
	<2 mm	<2 mm	<2 mm			unfiltered
	aqua regia	total (XRF)	total (INAA)			(mg/l)
Median	<0.01	-	<0.1			<0.000,02
Min	<0.01	-	<0.1			<0.000,02
Max	1.42	-	3			0.000,15

ANTIMONY   Sb

Concentrations in Soils and Sediments (mg/kg)

Medium	Soil	Agricultural soil - Ap horizon	Agricultural soil - Top (0-25 cm)	Agricultural Soil - Bottom (50-75 cm)	Topsoil (0-15 cm)	Urban soil (0-2 cm)
	World	Canada	Finland	Finland	England & Wales	Trondheim
	<2 mm total	<2 mm total (INAA)	<2 mm aqua regia	<2 mm aqua regia	<2 mm aqua regia	<2 mm aqua regia
Median	0.5*	0.6	0.058	0.048	-	<7
Min	-	<0.2	<0.04	<0.04	-	<7
Max	-	6.8	0.908	0.192	-	18

*Estimated mean

Medium	Forest soil - Humus	Forest soil - B horizon	Forest soil - C horizon	Till (C horizon)	Till (C horizon)	Laterite (25 ±15 cm)
	Norway	Norway	Norway	Finland	Finland	Australia
	<2 mm	<2 mm	<2 mm	<0.063 mm	<0.063 mm	0.45-2 mm
	7N HNO$_3$	7N HNO$_3$	7N HNO$_3$	aqua regia	total	total
Median	-	-	-	-	0.3	<0.5
Min	-	-	-	-	-	<0.5
Max	-	-	-	-	-	5.1

Medium	Stream sediment	Stream sediment	Stream sediment	Overbank sediment	Overbank sediment	Organic stream sediment
	Austria	Southern Scotland	Harz, Germany	Norway	Norway	Finland
	<0.18 mm	<0.10 mm	<0.063 mm	<0.063 mm	<0.063 mm	<2 mm
	"total" (ICP-AES)	total (AAS/XRF)	total (ICP-MS)	total (XRF)	7N HNO$_3$	Conc. HNO$_3$
Median	2	<2	2.7	-	-	0.052
Min	-	<2	0.8	-	-	<0.02#
Max	180	66	10	-	-	0.179^

#2nd percentile   ^98th percentile

Concentrations in Waters (mg/l)

Medium	Ocean water	Ocean water	Ocean water	Stream water	Stream water	Stream water
	World (1)	World (2)	North Pacific	World	Nova Scotia, Canada	Finland
					<0.45 µm	<0.45 µm
					ICP-MS-Hydride	ICP-MS
Median	0.000,24*	0.000,24*	0.000,2*	<0.000,1*	0.000,012	0.000,028
Min	-	-	-	-	<0.000,004	<0.000,02#
Max	-	-	-	-	0.000,105	0.000,093^

*Estimated mean   #2nd percentile   ^98th percentile

Medium	Stream water	Stream water	Stream water	Lake water	Ground water
	Romania	Eastern India	Harz, Germany	Norway	Southern Norway
	unfiltered	<0.2 µm	unfiltered	unfiltered	unfiltered
	ICP-MS	ICP-MS	ICP-MS	ICP-MS	ICP-MS
Median	0.000,14	0.000,1	0.000,13	0.000,025	0.000,032
Min	0.000,02	<0.000,1	<0.000,01	<0.000,01	0.000,002
Max	0.000,89	0.001	0.001,8	0.000,358	0.000,81

Sb ANTIMONY

Concentrations in Precipitation (mg/l)

Medium	Rain water	Rain water	Snow melt-water	Snow melt-water	Snow filter residue	Snow filter residue
	Kola, remote <0.45 μm ICP-MS	Kola, polluted <0.45 μm ICP-MS	Kola, remote <0.45 μm ICP-MS	Kola, polluted <0.45 μm ICP-MS	Kola, remote >0.45 μm ICP-AES	Kola, polluted >0.45 μm ICP-AES
Median	<0.000,025	0.000,32	<0.000,03	0.000,055	<0.000,5	0.000,65
Min	<0.000,025	0.000,12	<0.000,03	<0.000,03	<0.000,5	<0.000,5
Max	0.000,36	0.001,78	0.000,05	0.000,07	0.000,9	0.004

Medium	Rain water	Rain water
	Coastal, Norway unfiltered ICP-MS	Inland, Norway unfiltered ICP-MS
Median	0.000,02	0.000,05
Min	<0.000,01	<0.000,01
Max	0.000,37	0.001

Concentrations in Air (ng/m^3) and Yearly Deposition (kg/km^2/yr)

Medium	Air	Air	Bulk deposition	Throughfall deposition	Bulk deposition	Bulk deposition
	World, remote	World, polluted	West Germany	West Germany	Kola, remote	Kola, polluted
Median	0.3 / 0.03	10	-	-	0.318*	0.687*
Min	0.000,45	0.08	-	-	-	-
Max	0.93	55	-	-	-	-

*Estimated value

Concentrations in Plants (mg/kg)

Medium	Moss	Moss	Crustose lichen	Lichen	Dandelion	Spruce bark
	Norway conc. HNO$_3$ (1990)	Germany conc. HNO$_3$ + H$_2$O$_2$ (1995)	Germany conc. HNO$_3$	Northwest Territories oven-dried total (INAA)	Europe oven-dried total (INAA)	Canada ashed total (INAA)
Median	0.09	0.171	1	0.08*	0.08#	3.6
Min	<0.01	0.04	0.3	-	-	0.6
Max	0.64	3.1	7.3	-	-	88

*Mean #Geometric mean

Concentrations in Human Fluids (mg/l)

Medium	Human blood	Human serum	Human urine			
	Lombardy, Italy	Lombardy, Italy	Lombardy, Italy			
Mean	0.002,16	0.000,5	0.000,79			
Min	0.000,03	0.000,01	0.000,1			
Max	0.005	0.003,1	0.003,6			

ANTIMONY   Sb

Environmental Geochemistry

Biological impacts	Considered non-essential. Toxic. At high concentrations, Sb is more toxic than As or Pb. Some compounds are carcinogenic
Uses	Alloys, batteries, paint, ceramics, semi-conductors, ammunitions, flame retardant, rubber manufacturing, pharmaceutical industry, bactericide
Environmental pathways	Cu-Pb smelters, coal combustion, car exhaust, sewage sludge
Environmental mobility	Oxidising conditions: low Acid conditions: low
Reducing conditions: very low   Neutral to alkaline conditions: low
Comments: -
Geochemical barriers	Sulphide, adsorption (Fe-Mn oxides)
Natural association	As-Bi-Pb-Ag-Cu, particularly enriched in some Pb-Zn-Ag deposits
Action levels	Drinking water, MAC: 0.01 mg/l (Norway Shd), 0.006 mg/l (US EPA); recommended: 0.005 mg/l (WHO). Agricultural soil, maximum tolerable concentration: 5 mg/kg (Germany)
Remarks	Sb^{3+} compounds generally more toxic than Sb^{5+} compounds. Spruce needles, leaves of fruit trees, grass, moss, lichen and fungi can accumulate Sb. Growth rates can increase in the presence of Sb at low concentration. The US EPA suggests that drinking water exceeding action levels can lead to nausea, vomiting, and diarrhea in the short term; reliable data for assessing long term effects are not available, but Sb is potentially carcinogenic
Suggested analytical method(s)	ICP-MS, hydride AAS, INAA

World Yearly Production (t/y)	119,000 (in 1995)
Price of Purest Form, in Small Quantities ($/kg)	6050 (99.999,9%)
Market Price ($/kg)	2.2

Sc SCANDIUM

Physico-Chemical Properties

Atomic number	21	Melting/boiling point (K)	1814 / 3109
Atomic mass	44.955,910(8)	Isotopes & isomers	1 stable + 18 unstable
Atomic radius (pm)	209	Acid/base of oxide	weak base
Main oxidation state(s)	+3	State (at 300 K, 1 atm.)	solid
Ionic radius (pm)	88.5-101	Metallic character	pred. metal
Electronegativity (Pauling)	1.36	Element group(s)	light metal
Density (g/cm^3)	2.989	Affinity	lithophile

Naturally occurring isotopes	Natural abundance (%)	Atomic mass	Half-life	Unstable isotopes	Longest half-life
^{45}Sc	100	44.955,91	stable	^{40}Sc to ^{55}Sc	83.81 d

Concentrations in Crust/Rocks (mg/kg)

Bulk continental crust	16 / 22 / 30
Upper continental crust	7 / 11
Ultramafic rock	10
Ocean ridge basalt	38
Gabbro, basalt	35
Granite, granodiorite	5
Sandstone	3
Greywacke	16
Shale, schist	15
Limestone	1
Coal	5

Typical Minerals
thorveitite (Sc$_2$Si$_2$O$_7$), kolbeckite (ScPO$_4$.2H$_2$O), bazzite (Be$_3$(Sc,Fe)$_2$Al$_6$O$_{18}$)

Possible Host Minerals
pyroxenes, amphiboles, biotite, garnets, xenotine, zircon, monazite

Mass (kg) in

Continental crust	3.78x10^{+17}	Oceans	7.93x10^{+8}	Plants	3.68x10^{+7}

Concentrations in Media Sampled for the Kola Project (mg/kg)

Medium	Moss conc. HNO$_3$	Humus <2 mm conc. HNO$_3$	Humus <2 mm amm. acet.	Topsoil (0-5 cm) <2 mm total (INAA)	B horizon <2 mm aqua regia	B horizon <2 mm total (XRF)
Median	<0.1	0.5	-	5.4	2.6	-
Min	<0.1	<0.1	-	0.3	<0.1	-
Max	1.1	4.1	-	27	11	-

Medium	C horizon <2 mm aqua regia	C horizon <2 mm total (XRF)	C horizon <2 mm total (INAA)		Lake water unfiltered (mg/l)
Median	2.3	-	13		-
Min	<0.1	-	1.7		-
Max	15.4	-	36		-

SCANDIUM   Sc

Concentrations in Soils and Sediments (mg/kg)

Medium	Soil	Agricultural soil - Ap horizon	Agricultural soil - Top (0-25 cm)	Agricultural Soil - Bottom (50-75 cm)	Topsoil (0-15 cm)	Urban soil (0-2 cm)
	World	Canada	Finland	Finland	England & Wales	Trondheim
	<2 mm	<2 mm	<2 mm	<2 mm	<2 mm	<2 mm
	total	total (INAA)	aqua regia	aqua regia	aqua regia	aqua regia
Median	12*	7.8	-	-	-	3.3
Min	-	0.6	-	-	-	0.1
Max	-	24.1	-	-	-	9.2

*Estimated mean

Medium	Forest soil - Humus	Forest soil - B horizon	Forest soil - C horizon	Till (C horizon)	Till (C horizon)	Laterite (25 ±15 cm)
	Norway	Norway	Norway	Finland	Finland	Australia
	<2 mm	<2 mm	<2 mm	<0.063 mm	<0.063 mm	0.45-2 mm
	7N HNO$_3$	7N HNO$_3$	7N HNO$_3$	aqua regia	total	total
Median	0.5	3.3	4.7	2.7	15.7	4.5
Min	<0.1	0.5	0.8	-	-	<0.8
Max	10.8	17.1	38.8	-	-	50

Medium	Stream sediment	Stream sediment	Stream sediment	Overbank sediment	Overbank sediment	Organic stream sediment
	Austria	Southern Scotland	Harz, Germany	Norway	Norway	Finland
	<0.18 mm	<0.10 mm	<0.063 mm	<0.063 mm	<0.063 mm	<2 mm
	"total" (ICP-AES)	total	total (XRF)	total (XRF)	7N HNO$_3$ ICP-AES	Conc. HNO$_3$
Median	14	-	10	-	1.6	4.6
Min	-	-	<2	-	0.5	1.4#
Max	77	-	28	-	12	12.8^

#2nd percentile   ^98th percentile

Concentrations in Waters (mg/l)

Medium	Ocean water	Ocean water	Ocean water	Stream water	Stream water	Stream water
	World (1)	World (2)	North Pacific	World	Nova Scotia, Canada	Finland
					<0.45 µm	<0.45 µm
Median	0.000,000,6*	0.000,000,6*	0.000,000,7*	0.000,004*	-	-
Min	-	-	-	-	-	-
Max	-	-	-	-	-	-

*Estimated mean   #2nd percentile   ^98th percentile

Medium	Stream water	Stream water	Stream water	Lake water	Ground water
	Romania	Eastern India	Harz, Germany	Norway	Southern Norway
	unfiltered	<0.2 µm	unfiltered	unfiltered	unfiltered
	ICP-MS	ICP-MS	ICP-MS	ICP-MS	ICP-MS
Median	0.002	0.000,5	0.003,8	-	0.001,96
Min	0.000,4	0.000,3	0.000,1	-	0.000,54
Max	0.057	0.002,5	0.025	-	0.009,61

Sc SCANDIUM

Concentrations in Precipitation (mg/l)

Medium	Rain water	Rain water	Snow melt-water	Snow melt-water	Snow filter residue	Snow filter residue
	Kola, remote <0.45 µm	Kola, polluted <0.45 µm	Kola, remote <0.45 µm	Kola, polluted <0.45 µm	Kola, remote >0.45 µm ICP-AES	Kola, polluted >0.45 µm ICP-AES
Median	-	-	-	-	<0.000,01	0.000,07
Min	-	-	-	-	<0.000,01	0.000,04
Max	-	-	-	-	0.000,01	0.000,52

Medium	Rain water	Rain water				
	Coastal, Norway unfiltered ICP-MS	Inland, Norway unfiltered ICP-MS				
Median	-	-				
Min	-	-				
Max	-	-				

Concentrations in Air (ng/m^3) and Yearly Deposition (kg/km^2/yr)

Medium	Air	Air	Bulk deposition	Throughfall deposition	Bulk deposition	Bulk deposition
	World, remote	World, polluted	West Germany	West Germany	Kola, remote	Kola, polluted
Median	0.03 / 0.019	0.7	-	-	-	-
Min	0.000,04	0.03	-	-	-	-
Max	0.044 (0.13)	3	-	-	-	-

*Estimated value

Concentrations in Plants (mg/kg)

Medium	Moss	Moss	Crustose lichen	Lichen	Dandelion	Spruce bark
	Norway conc. HNO$_3$ (1985)	Germany conc. HNO$_3$ + H$_2$O$_2$ (1995)	Germany conc. HNO$_3$	Northwest Territories oven-dried total (INAA)	Europe oven-dried total (INAA)	Canada ashed total (INAA)
Median	0.17	0.195	-	0.18*	0.053#	5.3
Min	0.03	<0.001	-	-	-	0.6
Max	4.2	2.3	-	-	-	15

*Mean #Geometric mean

Concentrations in Human Fluids (mg/l)

Medium	Human blood	Human serum	Human urine			
	Lombardy, Italy	Lombardy, Italy	Lombardy, Italy			
Mean	0.000,061	0.000,043	0.000,038			
Min	0.000,002	0.000,003	0.000,000,3			
Max	0.000,18	0.000,1	0.000,16			

SCANDIUM — Sc

Environmental Geochemistry

Biological impacts	Considered non-essential. Little is known about the toxicity of Sc
Uses	Cathode-ray tubes, lasers, fluorescent materials
Environmental pathways	U production, windblown geogenic dust. Generally, natural sources more important than anthropogenic ones
Environmental mobility	Oxidising conditions: -　　　Acid conditions: - Reducing conditions: -　　　Neutral to alkaline conditions: - Comments: -
Geochemical barriers	-
Natural association	Sc-Y-REE, Fe-Mg-Mn-V-Ti-Sc-S (many silicates, sulphides)
Action levels	-
Remarks	Thorveitite is the main Sc mineral of which only very few occurrences are known in the world (e.g., Norway, 35% Sc_2O_3, and Madagascar, 20%). Chemically very similar to Al, and to a lesser extent to Ca, Ba, Sr, Ra. Sc is in the same group as Y, La and Ac. Sc is as common as Pb and Co in the crust, but it is generally thought to be much rarer due to its widely dispersed occurrence, and surprisingly little is known about this element's behaviour and importance in the environment
Suggested analytical method(s)	ICP-AES, INAA, ICP-MS (water)

World Yearly Production (t/y)　　-
Price of Purest Form, in Small Quantities ($/kg)　　536,750 (99.99%)
Market Price ($/kg)　　-

Se SELENIUM

Physico-Chemical Properties

Atomic number	34	Melting/boiling point (K)	494 / 958
Atomic mass	78.96(3)	Isotopes & isomers	6 stable + 24 unstable
Atomic radius (pm)	122	Acid/base of oxide	strong acid
Main oxidation state(s)	-2,+4 (0,+6)	State (at 300 K, 1 atm.)	solid
Ionic radius (pm)	184 (-2), 64 (+4), 42-56 (+6)	Metallic character	pred. non-metal
		Element group(s)	heavy non-metal
Electronegativity (Pauling)	2.55	Affinity	chalcophile
Density (g/cm^3)	4.79 (grey), 4.5 (red)		

Naturally occurring isotopes	Natural abundance (%)	Atomic mass	Half-life
^{74}Se	0.89	73.922,48	stable
^{76}Se	9.36	75.919,21	stable
^{77}Se	7.63	76.919,92	stable
^{78}Se	23.78	77.917,31	stable
^{80}Se	49.61	79.916,52	stable
^{82}Se	8.73	81.916,7	stable

Unstable isotopes	Longest half-life
^{65}Se to ^{91}Se	65,000 y

Concentrations in Crust/Rocks (mg/kg)

Bulk continental crust	0.12 / 0.05 / 50 (?)
Upper continental crust	0.083 / 50 (?)
Ultramafic rock	0.07
Ocean ridge basalt	0.16
Gabbro, basalt	0.12
Granite, granodiorite	0.025
Sandstone	0.01
Greywacke	-
Shale, schist	0.3
Limestone	0.025
Coal	3

(?) unit conversion error

Typical Minerals
crookesite ((Cu,Tl,Ag)$_2$Se), clausthalite (PbSe), berzelianite (Cu$_2$Se), tiemannite (HgSe), native Se

Possible Host Minerals
pyrite, chalcopyrite, pyrrhotite, sphalerite

Mass (kg) in

Continental crust	2.83x10^{+15}	Oceans	2.64x10^{+11}	Plants	3.68x10^{+7}

Concentrations in Media Sampled for the Kola Project (mg/kg)

Medium	Moss	Humus	Humus	Topsoil (0-5 cm)	B horizon	B horizon
		<2 mm	<2 mm	<2 mm	<2 mm	<2 mm
	conc. HNO$_3$	conc. HNO$_3$	amm. acet.	total (INAA)	aqua regia	total (XRF)
Median	<0.8	<0.8	-	<3	0.166	-
Min	<0.8	<0.8	-	<3	0.015	-
Max	1.23	7.38	-	6	2.26	-

Medium	C horizon	C horizon	C horizon	Lake water
	<2 mm	<2 mm	<2 mm	unfiltered
	aqua regia	total (XRF)	total (INAA)	(mg/l)
Median	0.046	-	<3	<0.000,5
Min	<0.01	-	<3	<0.000,5
Max	1.155	-	12	0.000,9

SELENIUM Se

Concentrations in Soils and Sediments (mg/kg)

Medium	Soil	Agricultural soil -	Agricultural soil - Top	Agricultural Soil - Bottom	Topsoil (0-15 cm)	Urban soil (0-2 cm)
	World	Ap horizon Canada	(0-25 cm) Finland	(50-75 cm) Finland	England & Wales	Trondheim
	<2 mm total	<2 mm total (AAS)	<2 mm aqua regia	<2 mm aqua regia	<2 mm aqua regia	<2 mm aqua regia
Median	0.3*	0.5	0.146	<0.1	-	0.21
Min	-	<0.2	<0.1	<0.1	-	0.04
Max	-	4.7	8.3	5.88	-	3.69

*Estimated mean

Medium	Forest soil - Humus	Forest soil - B horizon	Forest soil - C horizon	Till (C horizon)	Till (C horizon)	Laterite (25 ±15 cm)
	Norway	Norway	Norway	Finland	Finland	Australia
	<2 mm	<2 mm	<2 mm	<0.063 mm	<0.063 mm	0.45-2 mm
	7N HNO$_3$	7N HNO$_3$	7N HNO$_3$	aqua regia	total	total
Median	-	-	-	-	-	0.9
Min	-	-	-	-	-	<0.5
Max	-	-	-	-	-	6.4

Medium	Stream sediment	Stream sediment	Stream sediment	Overbank sediment	Overbank sediment	Organic stream sediment
	Austria	Southern Scotland	Harz, Germany	Norway	Norway	Finland
	<0.18 mm	<0.10 mm	<0.063 mm	<0.063 mm	<0.063 mm	<2 mm
		total	total (ICP-MS)	total (XRF)	7N HNO$_3$ AAS	Conc. HNO$_3$
Median	-	-	2.3	-	0.37	<1
Min	-	-	0.1	-	<0.05	<1#
Max	-	-	8.6	-	3	3.63^

#2nd percentile   ^98th percentile

Concentrations in Waters (mg/l)

Medium	Ocean water	Ocean water	Ocean water	Stream water	Stream water	Stream water
	World (1)	World (2)	North Pacific	World	Nova Scotia, Canada	Finland
					<0.45 µm ICP-MS	<0.45 µm ICP-MS
Median	0.000,2*	0.000,2*	0.000,155*	<0.000,2*	<0.001	0.000,067
Min	-	-	-	-	<0.001	0.000,035#
Max	-	-	-	-	0.005	0.000,153^

*Estimated mean   #2nd percentile   ^98th percentile

Medium	Stream water	Stream water	Stream water	Lake water	Ground water	
	Romania	Eastern India	Harz, Germany	Norway	Southern Norway	
	unfiltered ICP-MS	<0.2 µm ICP-MS	unfiltered ICP-MS	unfiltered ICP-MS	unfiltered ICP-MS	
Median	0.000,46	<0.000,1	0.000,6	-	0.000,295	
Min	0.000,01	<0.000,1	<0.000,1	-	<0.000,01	
Max	0.023	0.003,1	0.005,9	-	0.004,82	

Se   SELENIUM

Concentrations in Precipitation (mg/l)

Medium	Rain water	Rain water	Snow melt-water	Snow melt-water	Snow filter residue	Snow filter residue
	Kola, remote <0.45 μm ICP-MS	Kola, polluted <0.45 μm ICP-MS	Kola, remote <0.45 μm ICP-MS	Kola, polluted <0.45 μm ICP-MS	Kola, remote >0.45 μm	Kola, polluted >0.45 μm
Median	<0.000,5	0.000,93	<0.000,5	<0.000,5	-	-
Min	<0.000,5	<0.000,5	<0.000,5	<0.000,5	-	-
Max	<0.000,5	0.006,6	<0.000,5	0.000,54	-	-

Medium	Rain water	Rain water				
	Coastal, Norway unfiltered ICP-MS	Inland, Norway unfiltered ICP-MS				
Median	-	-				
Min	-	-				
Max	-	-				

Concentrations in Air (ng/m^3) and Yearly Deposition (kg/km^2/yr)

Medium	Air	Air	Bulk deposition	Throughfall deposition	Bulk deposition	Bulk deposition
	World, remote	World, polluted	West Germany	West Germany	Kola, remote	Kola, polluted
Median	- / 0.2	4	-	-	-	-
Min	0.006,3	0.06	-	-	-	-
Max	1.4	30	-	-	-	-

*Estimated value

Concentrations in Plants (mg/kg)

Medium	Moss	Moss	Crustose lichen	Lichen	Dandelion	Spruce bark
	Norway conc. HNO$_3$ (1985)	Germany conc. HNO$_3$ + H$_2$O$_2$ (1995)	Germany conc. HNO$_3$	Northwest Territories oven-dried	Europe oven-dried total (INAA)	Canada ashed total (INAA)
Median	0.43	0.237	-	-	0.095#	<2
Min	0.04	<0.001	-	-	-	<2
Max	2.18	1	-	-	-	16

*Mean   #Geometric mean

Concentrations in Human Fluids (mg/l)

Medium	Human blood	Human serum	Human urine			
	Lombardy, Italy	Lombardy, Italy	Lombardy, Italy			
Mean	0.107,5	0.081	0.022,1			
Min	0.04	0.033	0.002,1			
Max	0.18	0.121	0.068			

SELENIUM Se

Environmental Geochemistry

Biological impacts	Essential for many organisms. Toxic. There exists a narrow range of optimal Se intake: deficiency and toxicity are thus both important. Thought to be anti-carcinogenic
Uses	Glass, galvanising, semi-conductors, agriculture, pigments, vulcanising agent
Environmental pathways	Coal combustion, Cu-Pb smelters, waste water, some P fertilisers can contain Se
Environmental mobility	Oxidising conditions: high Acid conditions: high Reducing conditions: very low Neutral to alkaline conditions: very high Comments: -
Geochemical barriers	Reduction, adsorption
Natural association	Se-Hg-As-Sb-Ag-Cu-Zn-Cd-Pb (polymetallic sulphide ores), Cu-Ni-Se-Ag-Co (Cu-pyrite ores, e.g., Sudbury), U-V-Se-Cu-Mo (sandstone U deposits), Au-Ag-Se (Au-Ag selenide deposits)
Action levels	Drinking water, MAC: 0.01 mg/l (Norway Shd), 0.05 mg/l (US EPA), 0.01 mg/l (Canada CWQG); recommended: 0.01 mg/l (WHO); MAC: 0.001 mg/l (Russia MoH). Agricultural soil, maximum tolerable concentration: 10 mg/kg (Germany)
Remarks	Se can replace S; it is mostly recovered from Cu-sulphides (Se/S ratio in magmatic sulphides is 0.1, while it is only 0.000,1 in fossil fuel). Toxicity of As, Cd, Hg, Tl can be reduced by Se intake. Speciation of Se is very important to assess risks. Se-rich soils develop primarily in arid climates. Some plants enrich Se. Some plants can tolerate extremely Se-rich soils. Horses, cows, pigs and hens are quite sensitive to Se poisoning. Keshan and Kaschin-Beck diseases (China) caused by Se deficiency. The US EPA suggests that drinking water exceeding action levels can lead to hair and fingernail changes, damage to the peripheral nervous system, fatigue and irritability in the short term, and to hair and fingernail loss, damage to kidney and liver tissues, and to the nervous and circulatory systems in the long term
Suggested analytical method(s)	Hydride-AAS, ICP-MS

World Yearly Production (t/y)	2310 (in 1995)
Price of Purest Form, in Small Quantities ($/kg)	3548 (99.999%)
Market Price ($/kg)	6

Si SILICON

Physico-Chemical Properties

Atomic number	14	Melting/boiling point (K)	1687 / 3538
Atomic mass	28.085,5(3)	Isotopes & isomers	3 stable + 14 unstable
Atomic radius (pm)	146	Acid/base of oxide	amphoteric
Main oxidation state(s)	+4	State (at 300 K, 1 atm.)	solid
Ionic radius (pm)	40-54	Metallic character	non-metal
Electronegativity (Pauling)	1.9	Element group(s)	light non-metal
Density (g/cm^3)	2.33	Affinity	lithophile

Naturally occurring isotopes	Natural abundance (%)	Atomic mass	Half-life
^{28}Si	92.23	27.976,93	stable
^{29}Si	4.67	28.976,495	stable
^{30}Si	3.1	29.973,77	stable

Unstable isotopes	Longest half-life
^{22}Si to ^{39}Si	160 y

Concentrations in Crust/Rocks (mg/kg)

Bulk continental crust	288,000 / 282,000 / 268,000
Upper continental crust	303,480 / 308,000
Ultramafic rock	201,000
Ocean ridge basalt	232,000
Gabbro, basalt	227,000
Granite, granodiorite	337,000
Sandstone	403,000
Greywacke	322,973
Shale, schist	288,000
Limestone	31,000
Coal	34,000

Typical Minerals
all silicates, e.g., quartz (SiO_2), olivine (($Mg_1Fe)_2SiO_4$), muscovite ($kAl_2(Si_3AlO_{10})(OH_1Fl_2)$), orthoclase $k(Si,Al)_4O_8$
some non silicates, e.g., carborundum (SiC)

Possible Host Minerals
-

Mass (kg) in

Continental crust 6.80x10^{+21}	Oceans 2.64x10^{+15}	Plants 1.84x10^{+12}

Concentrations in Media Sampled for the Kola Project (mg/kg)

Medium	Moss	Humus <2 mm conc. HNO_3	Humus <2 mm amm. acet.	Topsoil (0-5 cm) <2 mm	B horizon <2 mm aqua regia	B horizon <2 mm total (XRF)
	conc. HNO_3					
Median	197	530	7	-	160	293,403
Min	24.9	290	1	-	70	123,140
Max	983	940	78	-	464	381,340

Medium	C horizon <2 mm aqua regia	C horizon <2 mm total (XRF)	C horizon <2 mm total (INAA)	Lake water unfiltered (mg/l)
Median	140	317,400	-	1.45
Min	50	170,500	-	0.05
Max	590	402,700	-	4.5

SILICON — Si

Concentrations in Soils and Sediments (mg/kg)

Medium	Soil World <2 mm total	Agricultural soil - Ap horizon Canada <2 mm	Agricultural soil - Top (0-25 cm) Finland <2 mm aqua regia	Agricultural Soil - Bottom (50-75 cm) Finland <2 mm aqua regia	Topsoil (0-15 cm) England & Wales <2 mm aqua regia	Urban soil (0-2 cm) Trondheim <2 mm aqua regia
Median	280,000*	-	-	-	-	2000
Min	-	-	-	-	-	1200
Max	-	-	-	-	-	11,100

*Estimated mean

Medium	Forest soil - Humus Norway <2 mm 7N HNO$_3$	Forest soil - B horizon Norway <2 mm 7N HNO$_3$	Forest soil - C horizon Norway <2 mm 7N HNO$_3$	Till (C horizon) Finland <0.063 mm aqua regia	Till (C horizon) Finland <0.063 mm total	Laterite (25 ±15 cm) Australia 0.45-2 mm total
Median	50	90	90	1500	302,000	382,777
Min	20	10	10	-	-	76,513
Max	250	750	510	-	-	443,796

Medium	Stream sediment Austria <0.18 mm	Stream sediment Southern Scotland <0.10 mm total	Stream sediment Harz, Germany <0.063 mm total (XRF)	Overbank sediment Norway <0.063 mm total (XRF)	Overbank sediment Norway <0.063 mm 7N HNO$_3$	Organic stream sediment Finland <2 mm Conc. HNO$_3$
Median	-	-	298,388	286,500	-	-
Min	-	-	50,713	107,600	-	-
Max	-	-	387,895	385,000	-	-

#2nd percentile   ^98th percentile

Concentrations in Waters (mg/l)

Medium	Ocean water World (1)	Ocean water World (2)	Ocean water North Pacific	Stream water World	Stream water Nova Scotia, Canada <0.45 µm ICP-AES	Stream water Finland <0.45 µm ICP-AES
Median	2.2*	2*	2.8*	6*	1.322	3.412
Min	-	-	-	-	<0.05	-
Max	-	-	-	-	4.644	-

*Estimated mean   #2nd percentile   ^98th percentile

Medium	Stream water Romania unfiltered ICP-MS	Stream water Eastern India <0.2 µm ICP-MS	Stream water Harz, Germany unfiltered	Lake water Norway unfiltered ICP-MS	Ground water Southern Norway unfiltered	
Median	-	-	-	-	-	
Min	-	-	-	-	-	
Max	-	-	-	-	-	

Si   SILICON

Concentrations in Precipitation (mg/l)

Medium	Rain water	Rain water	Snow melt-water	Snow melt-water	Snow filter residue	Snow filter residue
	Kola, remote <0.45 μm ICP-AES	Kola, polluted <0.45 μm ICP-AES	Kola, remote <0.45 μm	Kola, polluted <0.45 μm	Kola, remote >0.45 μm ICP-AES	Kola, polluted >0.45 μm ICP-AES
Median	<0.1	<0.1	-	-	<0.01	0.02
Min	<0.1	<0.1	-	-	<0.01	0.02
Max	<0.1	0.2	-	-	0.02	0.04

Medium	Rain water	Rain water				
	Coastal, Norway unfiltered ICP-MS	Inland, Norway unfiltered ICP-MS				
Median	-	-				
Min	-	-				
Max	-	-				

Concentrations in Air (ng/m^3) and Yearly Deposition (kg/km^2/yr)

Medium	Air	Air	Bulk deposition	Throughfall deposition	Bulk deposition	Bulk deposition
	World, remote	World, polluted	West Germany	West Germany	Kola, remote	Kola, polluted
Median	- / 350	-	-	-	-	-
Min	21	1000	-	-	-	-
Max	100 (3900)	63,000	-	-	-	-

*Estimated value

Concentrations in Plants (mg/kg)

Medium	Moss	Moss	Crustose lichen	Lichen	Dandelion	Spruce bark
	Norway conc. HNO$_3$ (1990)	Germany conc. HNO$_3$ + H$_2$O$_2$ (1995)	Germany conc. HNO$_3$	Northwest Territories oven-dried	Europe oven-dried	Canada ashed
Median	-	-	-	-	-	-
Min	-	-	-	-	-	-
Max	-	-	-	-	-	-

*Mean   #Geometric mean

Concentrations in Human Fluids (mg/l)

Medium	Human blood	Human serum	Human urine			
	Lombardy, Italy	Lombardy, Italy	Lombardy, Italy			
Mean	-	-	7.5			
Min	-	-	2			
Max	-	-	14.5			

SILICON — Si

Environmental Geochemistry

Biological impacts	Essential for many organisms. Si deficiency leads to growth disturbance. Some Si compounds are toxic
Uses	Semi-conductors, glass, china, cement
Environmental pathways	Windblown geogenic dust. Generally, natural sources more important than anthropogenic ones. Cement factories, Si and ferro-silicon production
Environmental mobility	Oxidising conditions: low Acid conditions: low Reducing conditions: low Neutral to alkaline conditions: low Comments: -
Geochemical barriers	-
Natural association	Si-Al (most rock forming minerals)
Action levels	Drinking water, MAC: 10 mg/l (Russia MoH)
Remarks	Some organisms use Si to build their skeletons. Certain plants (e.g., some grasses) accumulate Si
Suggested analytical method(s)	XRF, ICP-AES

World Yearly Production (t/y)	3,100,000 (in 1995)
Price of Purest Form, in Small Quantities ($/kg)	2073 (99.999,5%)
Market Price ($/kg)	0.02 (silica sand for glass production, ex-producer)

Sm   SAMARIUM

Physico-Chemical Properties

Atomic number	62
Atomic mass	150.36(3)
Atomic radius (pm)	259
Main oxidation state(s)	+3 (+2)
Ionic radius (pm)	136-142 (+2), 110-138 (+3)
Electronegativity (Pauling)	1.17
Density (g/cm^3)	7.52 (alpha)

Melting/boiling point (K)	1347 / 2067
Isotopes & isomers	4 stable + 28 unstable
Acid/base of oxide	base
State (at 300 K, 1 atm.)	solid
Metallic character	metal
Element group(s)	heavy metal, REE
Affinity	lithophile

Naturally occurring isotopes	Natural abundance (%)	Atomic mass	Half-life
^{144}Sm	3.1	143.911,996	stable
^{147}Sm	15	146.914,89	1.06x10^{+11} y
^{148}Sm	11.3	147.914,82	7x10^{+15} y
^{149}Sm	13.8	148.917,18	1x10^{+16} y
^{150}Sm	7.4	149.917,27	stable
^{152}Sm	26.7	151.919,73	stable
^{154}Sm	22.7	153.922,21	stable

Unstable isotopes	Longest half-life
^{131}Sm to ^{160}Sm	1x10^{+16} y

Concentrations in Crust/Rocks (mg/kg)

Bulk continental crust	5.3 / 7.05 / 3.5
Upper continental crust	4.7 / 4.5
Ultramafic rock	0.6
Ocean ridge basalt	3.3
Gabbro, basalt	3.5
Granite, granodiorite	8
Sandstone	3.2
Greywacke	4.6
Shale, schist	7.2
Limestone	2
Coal	2

Typical Minerals
monazite ((Ce,La,Nd,Th,Sm)(PO$_4$,SiO$_4$)),
bastnaesite ((Ce,La,Sm)CO$_3$(F,OH)),
cerite ((Ce,La,Sm)$_9$(Mg,Fe)Si$_7$(O,OH,F)$_{28}$),
allanite ((Ca,Ce,La,Sm)$_2$FeAl$_2$OSi$_3$O$_{11}$(OH))

Possible Host Minerals
biotite, apatite, pyroxenes, feldspars, zircon

Mass (kg) in

Continental crust	1.25x10^{+17}	Oceans	6.61x10^{+7}	Plants	7.36x10^{+7}

Concentrations in Media Sampled for the Kola Project (mg/kg)

Medium	Moss	Humus <2 mm	Humus <2 mm	Topsoil (0-5 cm) <2 mm	B horizon <2 mm	B horizon <2 mm
	conc. HNO$_3$		amm. acet.	total (INAA)	aqua regia	total (XRF)
Median	-	-	-	1.3	-	-
Min	-	-	-	<0.1	-	-
Max	-	-	-	16	-	-

Medium	C horizon <2 mm aqua regia	C horizon <2 mm total (XRF)	C horizon <2 mm total (INAA)		Lake water unfiltered (mg/l)
Median	-	-	3.4		-
Min	-	-	0.9		-
Max	-	-	37		-

SAMARIUM   Sm

Concentrations in Soils and Sediments (mg/kg)

Medium	Soil *World* *<2 mm* *total*	Agricultural soil - Ap horizon *Canada* *<2 mm* *total (INAA)*	Agricultural soil - Top (0-25 cm) *Finland* *<2 mm* *aqua regia*	Agricultural Soil - Bottom (50-75 cm) *Finland* *<2 mm* *aqua regia*	Topsoil (0-15 cm) *England & Wales* *<2 mm* *aqua regia*	Urban soil (0-2 cm) *Trondheim* *<2 mm* *aqua regia*
Median	6.1*	3.7	-	-	-	-
Min	-	0.3	-	-	-	-
Max	-	6.6	-	-	-	-

*Estimated mean

Medium	Forest soil - Humus *Norway* *<2 mm* *7N HNO$_3$*	Forest soil - B horizon *Norway* *<2 mm* *7N HNO$_3$*	Forest soil - C horizon *Norway* *<2 mm* *7N HNO$_3$*	Till (C horizon) *Finland* *<0.063 mm* *aqua regia*	Till (C horizon) *Finland* *<0.063 mm* *total*	Laterite (25 ±15 cm) *Australia* *0.45-2 mm* *total*
Median	-	-	-	-	6.5	-
Min	-	-	-	-	-	-
Max	-	-	-	-	-	-

Medium	Stream sediment *Austria* *<0.18 mm*	Stream sediment *Southern Scotland* *<0.10 mm* *total*	Stream sediment *Harz, Germany* *<0.063 mm* *total*	Overbank sediment *Norway* *<0.063 mm* *total (XRF)*	Overbank sediment *Norway* *<0.063 mm* *7N HNO$_3$*	Organic stream sediment *Finland* *<2 mm* *Conc. HNO$_3$*
Median	-	-	-	-	-	-
Min	-	-	-	-	-	-
Max	-	-	-	-	-	-

#2nd percentile ^98th percentile

Concentrations in Waters (mg/l)

Medium	Ocean water *World (1)*	Ocean water *World (2)*	Ocean water *North Pacific*	Stream water *World*	Stream water *Nova Scotia, Canada* *<0.45 μm* *ICP-MS*	Stream water *Finland* *<0.45 μm*
Median	0.000,000,45*	0.000,000,05*	0.000,000,57*	0.000,01*	0.000,033	-
Min	-	-	-	-	<0.000,005	-
Max	-	-	-	-	0.000,196	-

*Estimated mean   #2nd percentile   ^98th percentile

Medium	Stream water *Romania* *unfiltered* *ICP-MS*	Stream water *Eastern India* *<0.2 μm* *ICP-MS*	Stream water *Harz, Germany* *unfiltered* *ICP-MS*	Lake water *Norway* *unfiltered* *ICP-MS*	Ground water *Southern Norway* *unfiltered* *ICP-MS*
Median	-	0.000,84	0.000,11	0.000,03	0.000,028
Min	-	0.000,09	<0.000,002	<0.000,015	<0.000,001
Max	-	0.013,22	0.000,99	0.000,535	0.005,79

Sm SAMARIUM

Concentrations in Precipitation (mg/l)

Medium	Rain water Kola, remote <0.45 μm	Rain water Kola, polluted <0.45 μm	Snow melt- water Kola, remote <0.45 μm	Snow melt- water Kola, polluted <0.45 μm	Snow filter residue Kola, remote >0.45 μm	Snow filter residue Kola, polluted >0.45 μm
Median	-	-	-	-	-	-
Min	-	-	-	-	-	-
Max	-	-	-	-	-	-

Medium	Rain water Coastal, Norway unfiltered ICP-MS	Rain water Inland, Norway unfiltered ICP-MS				
Median	-	-				
Min	-	-				
Max	-	-				

Concentrations in Air (ng/m^3) and Yearly Deposition (kg/km^2/yr)

Medium	Air World, remote	Air World, polluted	Bulk deposition West Germany	Throughfall deposition West Germany	Bulk deposition Kola, remote	Bulk deposition Kola, polluted
Median	0.01 / 0.007	0.2	-	-	-	-
Min	0.000,09	0.02	-	-	-	-
Max	0.01 (0.07)	1	-	-	-	-

*Estimated value

Concentrations in Plants (mg/kg)

Medium	Moss Norway conc. HNO$_3$ (1990)	Moss Germany conc. HNO$_3$ + H$_2$O$_2$ (1995)	Crustose lichen Germany conc. HNO$_3$	Lichen Northwest Territories oven-dried	Dandelion Europe oven-dried total (INAA)	Spruce bark Canada ashed total (INAA)
Median	-	-	-	-	0.023#	2.4
Min	-	-	-	-	-	0.2
Max	-	-	-	-	-	12

*Mean   #Geometric mean

Concentrations in Human Fluids (mg/l)

Medium	Human blood Lombardy, Italy	Human serum Lombardy, Italy	Human urine Lombardy, Italy			
Mean	0.000,26	-	0.000,055			
Min	0.000,03	-	0.000,001,5			
Max	0.000,5	-	0.000,25			

SAMARIUM   Sm

Environmental Geochemistry

Biological impacts	Considered non-essential. Generally low toxicity, but data to assess the health relevance of REE are scarce
Uses	Lighters, magnets, condensers, nuclear reactor control rods, microwave controllers
Environmental pathways	Geogenic dust, mining and processing of alkaline rocks. Generally, geogenic sources more important than anthropogenic ones
Environmental mobility	Oxidising conditions: very low Acid conditions: very low Reducing conditions: very low Neutral to alkaline conditions: very low Comments: -
Geochemical barriers	Mechanical
Natural association	REE-Li-Rb-Cs-Be-Nb-Ta-Zr-B-Th-U-F (pegmatites), REE-Th-P-Zr-Fe-Cu (monazite veins), REE-Th-Ba-Sr-P-F-N-C (carbonatites), REE-U-P-F (phosphorites), REE-Au-Ti-Sn-Sr-Th (placers)
Action levels	-
Remarks	Toxicity of REE decreases as atomic number increases. Inhaled REE probably cause pneumoconiosis. REE taken up orally (e.g., via drinking water) accumulate in the skeleton, teeth, lungs, liver and kidneys. Some trees tend to accumulate Sm (e.g., hickory (*Carya*)). In 1991, 5 t Sm-oxide were used in the ceramics industry
Suggested analytical method(s)	ICP-MS, INAA

World Yearly Production (t/y)	54,000 REE-minerals (in 1991)
Price of Purest Form, in Small Quantities ($/kg)	27,369 (99.99%)
Market Price ($/kg)	-

Sn   TIN

Physico-Chemical Properties

Atomic number	50	Melting/boiling point (K)	505.08 / 2875
Atomic mass	118.710(7)	Isotopes & isomers	10 stable + 36 unstable
Atomic radius (pm)	172	Acid/base of oxide	amphoteric
Main oxidation state(s)	+4 (+2)	State (at 300 K, 1 atm.)	solid
Ionic radius (pm)	69-95 (+4)	Metallic character	pred. metal
Electronegativity (Pauling)	1.96	Element group(s)	heavy metal
Density (g/cm^3)	7.31 (white), 5.76 (grey)	Affinity	siderophile

Naturally occurring isotopes	Natural abundance (%)	Atomic mass	Half-life
^{110}Sn	0.97	111.904,82	stable
^{114}Sn	0.65	113.902,78	stable
^{115}Sn	0.34	114.903,35	stable
^{116}Sn	14.54	115.901,75	stable
^{117}Sn	7.68	116.902,96	stable
^{118}Sn	24.22	117.901,61	stable
^{119}Sn	8.59	118.903,31	stable
^{120}Sn	32.59	119.902,199	stable
^{122}Sn	4.63	121.903,44	stable
^{124}Sn	5.79	123.905,28	stable

Unstable isotopes	Longest half-life
^{100}Sn to ^{134}Sn	100,000 y

Concentrations in Crust/Rocks (mg/kg)

Bulk continental crust	2.3 / 2.3 / 2.5
Upper continental crust	2.5 / 5.5
Ultramafic rock	0.3
Ocean ridge basalt	0.9
Gabbro, basalt	0.9
Granite, granodiorite	3.6
Sandstone	0.6
Greywacke	-
Shale, schist	5
Limestone	0.3
Coal	8

Typical Minerals
cassiterite (SnO_2), stannite (Cu_2FeSnS_4)

Possible Host Minerals
biotite, muscovite, amphiboles, sphene, rutile, tourmaline, magnetite

Mass (kg) in

Continental crust	$5.43 \times 10^{+16}$	Oceans	$1.32 \times 10^{+10}$	Plants	$3.68 \times 10^{+8}$

Concentrations in Media Sampled for the Kola Project (mg/kg)

Medium	Moss conc. HNO_3	Humus <2 mm	Humus <2 mm amm. acet.	Topsoil (0-5 cm) <2 mm total (INAA)	B horizon <2 mm aqua regia	B horizon <2 mm total (XRF)
Median	-	-	-	<100	-	-
Min	-	-	-	<100	-	-
Max	-	-	-	800	-	-

Medium	C horizon <2 mm aqua regia	C horizon <2 mm total (XRF)	C horizon <2 mm total (INAA)			Lake water unfiltered (mg/l)
Median	-	-	<100			-
Min	-	-	<100			-
Max	-	-	100			-

TIN   Sn

Concentrations in Soils and Sediments (mg/kg)

Medium	Soil	Agricultural soil - Ap horizon	Agricultural soil - Top (0-25 cm)	Agricultural Soil - Bottom (50-75 cm)	Topsoil (0-15 cm)	Urban soil (0-2 cm)
	World	Canada	Finland	Finland	England & Wales	Trondheim
	<2 mm	<2 mm	<2 mm	<2 mm	<2 mm	<2 mm
	total	aqua regia	aqua regia	aqua regia	aqua regia	aqua regia
Median	4*	-	-	-	-	-
Min	-	-	-	-	-	-
Max	-	-	-	-	-	-

*Estimated mean

Medium	Forest soil - Humus	Forest soil - B horizon	Forest soil - C horizon	Till (C horizon)	Till (C horizon)	Laterite (25 ±15 cm)
	Norway	Norway	Norway	Finland	Finland	Australia
	<2 mm	<2 mm	<2 mm	<0.063 mm	<0.063 mm	0.45-2 mm
	7N HNO_3	7N HNO_3	7N HNO_3	aqua regia	total	total
Median	-	-	-	-	-	1.5
Min	-	-	-	-	-	<0.3
Max	-	-	-	-	-	6.5

Medium	Stream sediment	Stream sediment	Stream sediment	Overbank sediment	Overbank sediment	Organic stream sediment
	Austria	Southern Scotland	Harz, Germany	Norway	Norway	Finland
	<0.18 mm	<0.10 mm	<0.063 mm	<0.063 mm	<0.063 mm	<2 mm
	"total" (ICP-AES)	total (DCES)	total (XRF)	total (XRF)	7N HNO_3	Conc. HNO_3
Median	3.3	<6	6	-	-	-
Min	-	<6	<3	-	-	-
Max	285	677	57	-	-	-

#2nd percentile   ^98th percentile

Concentrations in Waters (mg/l)

Medium	Ocean water	Ocean water	Ocean water	Stream water	Stream water	Stream water
	World (1)	World (2)	North Pacific	World	Nova Scotia, Canada	Finland
					<0.45 μm	<0.45 μm
Median	0.000,004*	0.000,01*	0.000,000,5*	<0.000,01*	-	-
Min	-	-	-	-	-	-
Max	-	-	-	-	-	-

*Estimated mean   #2nd percentile   ^98th percentile

Medium	Stream water	Stream water	Stream water	Lake water	Ground water
	Romania	Eastern India	Harz, Germany	Norway	Southern Norway
	unfiltered	<0.2 μm	unfiltered	unfiltered	unfiltered
	ICP-MS	ICP-MS	ICP-MS	ICP-MS	ICP-MS
Median	0.000,005	0.001,2	0.000,03	0.000,061	<0.000,005
Min	0.000,005	0.000,55	<0.000,01	<0.000,04	<0.000,005
Max	0.000,17	0.004,12	0.001,3	0.003,52	0.002,33

Sn — TIN

Concentrations in Precipitation (mg/l)

Medium	Rain water	Rain water	Snow melt-water	Snow melt-water	Snow filter residue	Snow filter residue
	Kola, remote <0.45 µm	Kola, polluted <0.45 µm	Kola, remote <0.45 µm	Kola, polluted <0.45 µm	Kola, remote >0.45 µm	Kola, polluted >0.45 µm
Median	-	-	-	-	-	-
Min	-	-	-	-	-	-
Max	-	-	-	-	-	-

Medium	Rain water	Rain water				
	Coastal, Norway unfiltered ICP-MS	Inland, Norway unfiltered ICP-MS				
Median	-	-				
Min	-	-				
Max	-	-				

Concentrations in Air (ng/m^3) and Yearly Deposition (kg/km^2/yr)

Medium	Air	Air	Bulk deposition	Throughfall deposition	Bulk deposition	Bulk deposition
	World, remote	World, polluted	West Germany	West Germany	Kola, remote	Kola, polluted
Median	-	-	-	-	-	-
Min	-	1.5	-	-	-	-
Max	-	800	-	-	-	-

*Estimated value

Concentrations in Plants (mg/kg)

Medium	Moss	Moss	Crustose lichen	Lichen	Dandelion	Spruce bark
	Norway conc. HNO$_3$ (1990)	Germany conc. HNO$_3$ + H$_2$O$_2$ (1995)	Germany conc. HNO$_3$	Northwest Territories oven-dried	Europe oven-dried	Canada ashed aqua regia
Median	-	0.244	2.6	-	-	<2
Min	-	0.05	1.1	-	-	<2
Max	-	15	120	-	-	52

*Mean #Geometric mean

Concentrations in Human Fluids (mg/l)

Medium	Human blood	Human serum	Human urine			
	Lombardy, Italy	Lombardy, Italy	Lombardy, Italy			
Mean	-	-	-			
Min	-	-	-			
Max	-	-	-			

TIN   Sn

Environmental Geochemistry

Biological impacts	Possibly essential for some organisms (e.g., humans). Many compounds are toxic for lower organisms. Metallic Sn practically non-toxic. Inorganic Sn does not appear to be teratogenic nor carcinogenic
Uses	Sn-plated steel, alloys (Sn-Cu, Sn-Bi-Cu), fungicide, insecticide, bactericide, dental amalgams (Ag-Sn-Hg), stabiliser in PVC, paint (esp. for ships), solder (Pb-Sb-Ag-Zn-In alloy)
Environmental pathways	Coal and wood combustion, waste incineration, sewage sludge
Environmental mobility	Oxidising conditions: very low Acid conditions: very low Reducing conditions: very low Neutral to alkaline conditions: very low Comments: -
Geochemical barriers	Mechanical (cassiterite), pH (Sn^{2+})
Natural association	Sn-W-Nb-Ta-Be-B-Li-Rb-Cs-REE (pegmatites), Sn-W-B-F-Be (veins and greisens), Sn-B-F-(As) (cassiterite pipes)
Action levels	Drinking water: health-based recommended value not deemed necessary (WHO). Agricultural soil, maximum tolerable concentration: 50 mg/kg (Germany)
Remarks	Accumulates in aquatic sediments. Application of organotin rapidly increasing
Suggested analytical method(s)	ICP-MS, hydride AAS

World Yearly Production (t/y)	189,000 (in 1995)
Price of Purest Form, in Small Quantities ($/kg)	9048 (99.999,9+%)
Market Price ($/kg)	6

Sr — STRONTIUM

Physico-Chemical Properties

Atomic number	38	Melting/boiling point (K)	1050 / 1655
Atomic mass	87.62(1)	Isotopes & isomers	4 stable + 26 unstable
Atomic radius (pm)	245	Acid/base of oxide	strong base
Main oxidation state(s)	+2	State (at 300 K, 1 atm.)	solid
Ionic radius (pm)	132-158	Metallic character	metal
Electronegativity (Pauling)	0.95	Element group(s)	alkaline earth, light metal
Density (g/cm^3)	2.54	Affinity	lithophile

Naturally occurring isotopes	Natural abundance (%)	Atomic mass	Half-life
^{84}Sr	0.56	83.913,43	stable
^{86}Sr	9.86	85.909,27	stable
^{87}Sr	7	86.908,88	stable
^{88}Sr	82.58	87.905,62	stable

Unstable isotopes	Longest half-life
^{76}Sr to ^{102}Sr	29.1 y

Concentrations in Crust/Rocks (mg/kg)

Bulk continental crust	333 / 370 / 260
Upper continental crust	316 / 350
Ultramafic rock	10
Ocean ridge basalt	180
Gabbro, basalt	400
Granite, granodiorite	220
Sandstone	100
Greywacke	201
Shale, schist	250
Limestone	500
Coal	150

Typical Minerals
strontianite (SrCO$_3$), celestite (SrSO$_4$)

Possible Host Minerals
feldspars, micas, gypsum, calcite, dolomite

Mass (kg) in

Continental crust	7.86x10^{+18}	Oceans	1.06x10^{+16}	Plants	9.20x10^{+10}

Concentrations in Media Sampled for the Kola Project (mg/kg)

Medium	Moss	Humus <2 mm conc. HNO$_3$	Humus <2 mm amm. acet.	Topsoil (0-5 cm) <2 mm total (INAA)	B horizon <2 mm aqua regia	B horizon <2 mm total (XRF)
	conc. HNO$_3$					
Median	9.4	28.8	15.4	<500	6.3	-
Min	2.47	6.1	2.25	<500	1.6	-
Max	435	1430	355	1200	879	-

Medium	C horizon <2 mm aqua regia	C horizon <2 mm total (XRF)	C horizon <2 mm total (INAA)			Lake water unfiltered (mg/l)
Median	7.7	-	<500			0.015,5
Min	1.6	-	<500			0.004,5
Max	1040	-	1700			0.104

STRONTIUM — Sr

Concentrations in Soils and Sediments (mg/kg)

Medium	Soil	Agricultural soil - Ap horizon	Agricultural soil - Top (0-25 cm)	Agricultural Soil - Bottom (50-75 cm)	Topsoil (0-15 cm)	Urban soil (0-2 cm)
	World	Canada	Finland	Finland	England & Wales	Trondheim
	<2 mm total	<2 mm	<2 mm aqua regia	<2 mm aqua regia	<2 mm aqua regia	<2 mm aqua regia
Median	240*	-	22.9	12.7	27	27
Min	-	-	<2	<2	3	5.8
Max	-	-	153	84.7	1445	255

*Estimated mean

Medium	Forest soil - Humus	Forest soil - B horizon	Forest soil - C horizon	Till (C horizon)	Till (C horizon)	Laterite (25 ±15 cm)
	Norway	Norway	Norway	Finland	Finland	Australia
	<2 mm 7N HNO₃	<2 mm 7N HNO₃	<2 mm 7N HNO₃	<0.063 mm aqua regia	<0.063 mm total	0.45-2 mm total
Median	23.7	12.9	24.2	7.1	240	27
Min	3.4	1.6	2.7	-	-	1.4
Max	102.6	174	446.6	-	-	260

Medium	Stream sediment	Stream sediment	Stream sediment	Overbank sediment	Overbank sediment	Organic stream sediment
	Austria	Southern Scotland	Harz, Germany	Norway	Norway	Finland
	<0.18 mm "total" (ICP-AES)	<0.10 mm total (DCES)	<0.063 mm total (XRF)	<0.063 mm total (XRF)	<0.063 mm 7N HNO₃ ICP-AES	<2 mm Conc. HNO₃
Median	187	125	76	250	26	35.7
Min	-	4	28	55	4	16.4#
Max	8695	490	262	956	311	69.2^

#2nd percentile   ^98th percentile

Concentrations in Waters (mg/l)

Medium	Ocean water	Ocean water	Ocean water	Stream water	Stream water	Stream water
	World (1)	World (2)	North Pacific	World	Nova Scotia, Canada	Finland
					<0.45 μm ICP-MS	<0.45 μm ICP-MS
Median	7.9*	8*	7.8*	0.07*	0.011,7	0.022,4
Min	-	-	-	-	<0.000,5	0.008,7#
Max	-	-	-	-	1.943,7	0.133^

*Estimated mean   #2nd percentile   ^98th percentile

Medium	Stream water	Stream water	Stream water	Lake water	Ground water
	Romania	Eastern India	Harz, Germany	Norway	Southern Norway
	unfiltered ICP-MS	<0.2 μm ICP-MS	unfiltered ICP-MS	unfiltered ICP-MS	unfiltered ICP-MS
Median	0.385	0.076,5	0.03	0.005,91	0.179,35
Min	0.014	0.037,7	<0.000,1	0.000,32	0.000,9
Max	4.1	0.309,6	1.53	3.861	1.870,9

Sr   STRONTIUM

Concentrations in Precipitation (mg/l)

Medium	Rain water	Rain water	Snow melt-water	Snow melt-water	Snow filter residue	Snow filter residue
	Kola, remote <0.45 μm ICP-MS	*Kola, polluted <0.45 μm ICP-MS*	*Kola, remote <0.45 μm ICP-MS*	*Kola, polluted <0.45 μm ICP-MS*	*Kola, remote >0.45 μm ICP-AES*	*Kola, polluted >0.45 μm ICP-AES*
Median	0.000,2	0.001,37	0.000,21	0.003,81	0.000,03	0.002,29
Min	0.000,1	0.000,43	0.000,16	0.001,98	0.000,02	0.001,24
Max	0.000,44	0.004,44	0.000,8	0.005,08	0.000,07	0.007,5

Medium	Rain water	Rain water
	Coastal, Norway unfiltered ICP-MS	*Inland, Norway unfiltered ICP-MS*
Median	0.000,64	0.000,44
Min	<0.000,01	<0.000,01
Max	0.012	0.037

Concentrations in Air (ng/m^3) and Yearly Deposition (kg/km^2/yr)

Medium	Air	Air	Bulk deposition	Throughfall deposition	Bulk deposition	Bulk deposition
	World, remote	*World, polluted*	*West Germany*	*West Germany*	*Kola, remote*	*Kola, polluted*
Median	0.81	-	-	-	0.234*	7.726*
Min	0.031	2	-	9	-	-
Max	4	50	-	14	-	-

*Estimated value

Concentrations in Plants (mg/kg)

Medium	Moss	Moss	Crustose lichen	Lichen	Dandelion	Spruce bark
	Norway conc. HNO$_3$ (1990)	*Germany conc. HNO$_3$ + H$_2$O$_2$ (1995)*	*Germany conc. HNO$_3$*	*Northwest Territories oven-dried*	*Europe oven-dried*	*Canada ashed total (INAA)*
Median	13	10.4	71	-	-	1100
Min	2.7	1.5	16	-	-	<300
Max	73	190	370	-	-	9600

*Mean   #Geometric mean

Concentrations in Human Fluids (mg/l)

Medium	Human blood	Human serum	Human urine
	Lombardy, Italy	*Lombardy, Italy*	*Lombardy, Italy*
Mean	-	-	-
Min	-	-	-
Max	-	-	-

STRONTIUM — Sr

Environmental Geochemistry

Biological impacts	Considered non-essential for most organisms. ^{90}Sr is highly radiotoxic
Uses	Alloys, colour television tubes (Sr-carbonate), pyrotechnic materials, ferrite magnets, zinc refining
Environmental pathways	Weathering, geogenic dust, sea spray. Radiogenic Sr from nuclear tests and accidents
Environmental mobility	Oxidising conditions: high Acid conditions: high Reducing conditions: high Neutral to alkaline conditions: high Comments: -
Geochemical barriers	-
Natural association	Sr-Ca (replacement in Ca minerals)
Action levels	Drinking water, MAC: 7 mg/l (Russia MoH), 4×10^{-10} mg ^{90}Sr/l (Russia MoH)
Remarks	Concerns exist about accumulation of radiogenic Sr (^{89}Sr and ^{90}Sr) from nuclear bomb tests and accidents, because of its uptake (instead of Ca) by organisms. Natural isotopes ^{86}Sr and ^{87}Sr used in hydrological and geochemical research
Suggested analytical method(s)	XRF, ICP-AES, ICP-MS (water)

World Yearly Production (t/y)	170,000 (in 1995)
Price of Purest Form, in Small Quantities ($/kg)	16,725 (99.95%)
Market Price ($/kg)	-

Ta   TANTALUM

Physico-Chemical Properties

Atomic number	73	Melting/boiling point (K)	3290 / 5731
Atomic mass	180.947,9(1)	Isotopes & isomers	1 stable + 34 unstable
Atomic radius (pm)	209	Acid/base of oxide	weak acid
Main oxidation state(s)	+5 (+2?,+3,+4?)	State (at 300 K, 1 atm.)	solid
Ionic radius (pm)	86 (+3), 82 (+4), 78-88 (+5)	Metallic character	pred. non-metal
		Element group(s)	heavy non-metal
Electronegativity (Pauling)	1.5	Affinity	lithophile, siderophile
Density (g/cm^3)	16.654		

Naturally occurring isotopes	Natural abundance (%)	Atomic mass	Half-life
^{180m}Ta	0.012	179.947,47	>1.2x10^{+15} y
^{181}Ta	99.988	180.947,996	stable

Unstable isotopes	Longest half-life
^{156}Ta to ^{186}Ta	>1.2x10^{+15} y

Concentrations in Crust/Rocks (mg/kg)

Bulk continental crust	1.1 / 2 / 1
Upper continental crust	1.5 / 2.2
Ultramafic rock	0.07
Ocean ridge basalt	0.3
Gabbro, basalt	1
Granite, granodiorite	2
Sandstone	1.5
Greywacke	-
Shale, schist	1.7
Limestone	0.01
Coal	0.2

Typical Minerals
columbite-tantalite ((Fe,Mn)(Nb-Ta)$_2$O$_6$)),
microlite ((Na,Ca)$_2$Ta$_2$O$_6$(O,OH,F))

Possible Host Minerals
biotite, zircon, rutile, ilmenite, sphene

Mass (kg) in

Continental crust	2.60x10^{+16}	Oceans	2.64x10^{+9}	Plants	1.84x10^{+6}

Concentrations in Media Sampled for the Kola Project (mg/kg)

| Medium | Moss | Humus <2 mm | Humus <2 mm | Topsoil (0-5 cm) <2 mm | B horizon <2 mm | B horizon <2 mm |
	conc. HNO$_3$		amm. acet.	total (INAA)	aqua regia	total (XRF)
Median	-	-	-	<0.5	-	-
Min	-	-	-	<0.5	-	-
Max	-	-	-	27	-	-

Medium	C horizon <2 mm aqua regia	C horizon <2 mm total (XRF)	C horizon <2 mm total (INAA)		Lake water unfiltered (mg/l)
Median	-	-	<0.5		-
Min	-	-	<0.5		-
Max	-	-	41		-

TANTALUM   Ta

Concentrations in Soils and Sediments (mg/kg)

Medium	Soil	Agricultural soil - Ap horizon	Agricultural soil - Top (0-25 cm)	Agricultural Soil - Bottom (50-75 cm)	Topsoil (0-15 cm)	Urban soil (0-2 cm)
	World	Canada	Finland	Finland	England & Wales	Trondheim
	<2 mm	<2 mm	<2 mm	<2 mm	<2 mm	<2 mm
	total	total (INAA)	aqua regia	aqua regia	aqua regia	aqua regia
Median	1.1*	0.7	-	-	-	-
Min	-	<0.4	-	-	-	-
Max	-	1.3	-	-	-	-

*Estimated mean

Medium	Forest soil - Humus	Forest soil - B horizon	Forest soil - C horizon	Till (C horizon)	Till (C horizon)	Laterite (25 ±15 cm)
	Norway	Norway	Norway	Finland	Finland	Australia
	<2 mm	<2 mm	<2 mm	<0.063 mm	<0.063 mm	0.45-2 mm
	7N HNO$_3$	7N HNO$_3$	7N HNO$_3$	aqua regia	total	total
Median	-	-	-	-	1.1	-
Min	-	-	-	-	-	-
Max	-	-	-	-	-	-

Medium	Stream sediment	Stream sediment	Stream sediment	Overbank sediment	Overbank sediment	Organic stream sediment
	Austria	Southern Scotland	Harz, Germany	Norway	Norway	Finland
	<0.18 mm	<0.10 mm	<0.063 mm	<0.063 mm	<0.063 mm	<2 mm
		total	total (ICP-MS)	total (XRF)	7N HNO$_3$	Conc. HNO$_3$
Median	-	-	2.7	-	-	-
Min	-	-	0.5	-	-	-
Max	-	-	9.8	-	-	-

#2nd percentile ^98th percentile

Concentrations in Waters (mg/l)

Medium	Ocean water	Ocean water	Ocean water	Stream water	Stream water	Stream water
	World (1)	World (2)	North Pacific	World	Nova Scotia, Canada	Finland
					<0.45 μm	<0.45 μm
Median	0.000,002*	0.000,002*	<0.000,002,5*	-	-	-
Min	-	-	-	-	-	-
Max	-	-	-	-	-	-

*Estimated mean   #2nd percentile   ^98th percentile

Medium	Stream water	Stream water	Stream water	Lake water	Ground water
	Romania	Eastern India	Harz, Germany	Norway	Southern Norway
	unfiltered	<0.2 μm	unfiltered	unfiltered	unfiltered
	ICP-MS	ICP-MS	ICP-MS	ICP-MS	ICP-MS
Median	0.000,005	<0.000,01	<0.000,005	<0.000,01	<0.000,001
Min	0.000,002	<0.000,01	<0.000,005	<0.000,01	<0.000,001
Max	0.000,13	0.000,02	0.000,1	0.000,036	0.000,089

Ta   TANTALUM

Concentrations in Precipitation (mg/l)

Medium	Rain water	Rain water	Snow melt-water	Snow melt-water	Snow filter residue	Snow filter residue
	Kola, remote <0.45 µm	Kola, polluted <0.45 µm	Kola, remote <0.45 µm	Kola, polluted <0.45 µm	Kola, remote >0.45 µm	Kola, polluted >0.45 µm
Median	-	-	-	-	-	-
Min	-	-	-	-	-	-
Max	-	-	-	-	-	-

Medium	Rain water	Rain water				
	Coastal, Norway unfiltered ICP-MS	Inland, Norway unfiltered ICP-MS				
Median	-	-				
Min	-	-				
Max	-	-				

Concentrations in Air (ng/m^3) and Yearly Deposition (kg/km^2/yr)

Medium	Air	Air	Bulk deposition	Throughfall deposition	Bulk deposition	Bulk deposition
	World, remote	World, polluted	West Germany	West Germany	Kola, remote	Kola, polluted
Median	- / 0.000,66	-	-	-	-	-
Min	0.000,06	-	-	-	-	-
Max	0.015	0.011	-	-	-	-

*Estimated value

Concentrations in Plants (mg/kg)

Medium	Moss	Moss	Crustose lichen	Lichen	Dandelion	Spruce bark
	Norway conc. HNO$_3$ (1990)	Germany conc. HNO$_3$ + H$_2$O$_2$ (1995)	Germany conc. HNO$_3$	Northwest Territories oven-dried	Europe oven-dried	Canada ashed total (INAA)
Median	-	0.011	-	-	-	<0.5
Min	-	<0.001	-	-	-	<0.5
Max	-	0.19	-	-	-	2.7

*Mean   #Geometric mean

Concentrations in Human Fluids (mg/l)

Medium	Human blood	Human serum	Human urine			
	Lombardy, Italy	Lombardy, Italy	Lombardy, Italy			
Mean	0.000,23	<0.000,1	0.000,16			
Min	0.000,04	<0.000,1	0.000,02			
Max	0.000,7	<0.000,1	0.000,9			

TANTALUM   Ta

Environmental Geochemistry

Biological impacts	Considered non-essential for animals. Does not seem to have a biological function. Very low toxicity
Uses	Aircraft and missile parts, steel alloys, lenses, electronics, chemical reactors, alloys (with Co, W, Mo), medicine (implants)
Environmental pathways	Geogenic dust is the main source of Ta in the environment
Environmental mobility	Oxidising conditions: very low Acid conditions: very low Reducing conditions: very low Neutral to alkaline conditions: very low Comments: -
Geochemical barriers	Mechanical
Natural association	Nb-Ta-Sn-W-Li-Be-Ti-Rb-Cs-U-Th-B-Zr-Hf-P-F-REE (granites and syenitic pegmatites), Nb-Ta-Na-K-Ba-Sr-Ti-Zr-U-Th-Cu-Zn-P-S-F-REE (carbonatites), Nb-Ta-Ti-Ga-Be-Al (bauxite developed on alkaline rocks)
Action levels	-
Remarks	Ta and Nb always occur together in nature. Reported to show little geographical variation (this is probably a misconception based on analytical problems). Resistant to chemical attack by acids and bases up to 150°C (exception: HF dissolves Ta). Some organisms tend to accumulate Ta, up to 400 mg/kg or more (e.g., ascidian (sea squirt))
Suggested analytical method(s)	ICP-MS, INAA

World Yearly Production (t/y)	18,300 Nb-Ta (in 1995)
Price of Purest Form, in Small Quantities ($/kg)	7938 (99.9%)
Market Price ($/kg)	70 (Ta-oxide, 60%)

Tb TERBIUM

Physico-Chemical Properties

Atomic number	65
Atomic mass	158.925,34(2)
Atomic radius (pm)	251
Main oxidation state(s)	+3 (+4)
Ionic radius (pm)	106-123.5 (+3), 90-102 (+4)
Electronegativity (Pauling)	1.2
Density (g/cm^3)	8.23
Melting/boiling point (K)	1629 / 3503
Isotopes & isomers	1 stable + 40 unstable
Acid/base of oxide	weak base
State (at 300 K, 1 atm.)	solid
Metallic character	metal
Element group(s)	heavy metal, REE
Affinity	lithophile

Naturally occurring isotopes	Natural abundance (%)	Atomic mass	Half-life
^{159}Tb	100	158.925,34	stable

Unstable isotopes	Longest half-life
^{140}Tb to ^{165}Tb	180 y

Concentrations in Crust/Rocks (mg/kg)

Bulk continental crust	0.65 / 1.2 / 0.6
Upper continental crust	0.5 / 0.64
Ultramafic rock	-
Ocean ridge basalt	-
Gabbro, basalt	-
Granite, granodiorite	-
Sandstone	-
Greywacke	0.63
Shale, schist	-
Limestone	-
Coal	0.2

Typical Minerals
monazite ((Ce,La,Nd,Th,Tb)(PO$_4$,SiO$_4$)),
xenotime ((Y,Tb)PO$_4$),
bastnaesite ((Ce,La,Tb)CO$_3$(F,OH)),
cerite ((Ce,La,Tb)$_9$(Mg,Fe)Si$_7$(O,OH,F)$_{28}$),
allanite ((Ca,Ce,La,Tb)$_2$FeAl$_2$OSi$_3$O$_{11}$(OH))

Possible Host Minerals
biotite, apatite, pyroxenes, feldspars, zircon

Mass (kg) in

Continental crust 1.53x10^{+16} Oceans 1.32x10^{+8} Plants 1.47x10^{+7}

Concentrations in Media Sampled for the Kola Project (mg/kg)

Medium	Moss conc. HNO$_3$	Humus <2 mm <2 mm	Humus <2 mm amm. acet.	Topsoil (0-5 cm) <2 mm total (INAA)	B horizon <2 mm aqua regia	B horizon <2 mm total (XRF)
Median	-	-	-	<0.5	-	-
Min	-	-	-	<0.5	-	-
Max	-	-	-	3.2	-	-

Medium	C horizon <2 mm aqua regia	C horizon <2 mm total (XRF)	C horizon <2 mm total (INAA)			Lake water unfiltered (mg/l)
Median	-	-	<0.5			-
Min	-	-	<0.5			-
Max	-	-	6.4			-

TERBIUM   Tb

Concentrations in Soils and Sediments (mg/kg)

Medium	Soil	Agricultural soil - Ap horizon	Agricultural soil - Top (0-25 cm)	Agricultural Soil - Bottom (50-75 cm)	Topsoil (0-15 cm)	Urban soil (0-2 cm)
	World	*Canada*	*Finland*	*Finland*	*England & Wales*	*Trondheim*
	<2 mm total	<2 mm	<2 mm aqua regia	<2 mm aqua regia	<2 mm aqua regia	<2 mm aqua regia
Median	-	-	-	-	-	-
Min	-	-	-	-	-	-
Max	-	-	-	-	-	-

*Estimated mean

Medium	Forest soil - Humus	Forest soil - B horizon	Forest soil - C horizon	Till (C horizon)	Till (C horizon)	Laterite (25 ±15 cm)
	Norway	*Norway*	*Norway*	*Finland*	*Finland*	*Australia*
	<2 mm 7N HNO$_3$	<2 mm 7N HNO$_3$	<2 mm 7N HNO$_3$	<0.063 mm aqua regia	<0.063 mm total	0.45-2 mm total
Median	-	-	-	-	-	-
Min	-	-	-	-	-	-
Max	-	-	-	-	-	-

Medium	Stream sediment	Stream sediment	Stream sediment	Overbank sediment	Overbank sediment	Organic stream sediment
	Austria	*Southern Scotland*	*Harz, Germany*	*Norway*	*Norway*	*Finland*
	<0.18 mm	<0.10 mm total	<0.063 mm total	<0.063 mm total (XRF)	<0.063 mm 7N HNO$_3$	<2 mm Conc. HNO$_3$
Median	-	-	-	-	-	-
Min	-	-	-	-	-	-
Max	-	-	-	-	-	-

#2nd percentile ^98th percentile

Concentrations in Waters (mg/l)

Medium	Ocean water	Ocean water	Ocean water	Stream water	Stream water	Stream water
	World (1)	*World (2)*	*North Pacific*	*World*	*Nova Scotia, Canada*	*Finland*
					<0.45 μm ICP-MS	<0.45 μm
Median	0.000,000,14*	0.000,000,1*	0.000,000,17*	-	<0.000,005	-
Min	-	-	-	-	<0.000,005	-
Max	-	-	-	-	0.000,027	-

*Estimated mean   #2nd percentile   ^98th percentile

Medium	Stream water	Stream water	Stream water	Lake water	Ground water	
	Romania	*Eastern India*	*Harz, Germany*	*Norway*	*Southern Norway*	
	unfiltered ICP-MS	<0.2 μm ICP-MS	unfiltered ICP-MS	unfiltered ICP-MS	unfiltered ICP-MS	
Median	-	0.000,16	0.000,02	0.000,003	0.000,004	
Min	-	0.000,02	<0.000,002	<0.000,002	<0.000,001	
Max	-	0.002,21	0.000,17	0.000,089	0.000,4	

Tb TERBIUM

Concentrations in Precipitation (mg/l)

Medium	Rain water Kola, remote <0.45 μm	Rain water Kola, polluted <0.45 μm	Snow melt- water Kola, remote <0.45 μm	Snow melt- water Kola, polluted <0.45 μm	Snow filter residue Kola, remote >0.45 μm	Snow filter residue Kola, polluted >0.45 μm
Median	-	-	-	-	-	-
Min	-	-	-	-	-	-
Max	-	-	-	-	-	-

Medium	Rain water Coastal, Norway unfiltered ICP-MS	Rain water Inland, Norway unfiltered ICP-MS				
Median	-	-				
Min	-	-				
Max	-	-				

Concentrations in Air (ng/m^3) and Yearly Deposition (kg/km^2/yr)

Medium	Air World, remote	Air World, polluted	Bulk deposition West Germany	Throughfall deposition West Germany	Bulk deposition Kola, remote	Bulk deposition Kola, polluted
Median	-	-	-	-	-	-
Min	-	-	-	-	-	-
Max	0.02	0.01	-	-	-	-

*Estimated value

Concentrations in Plants (mg/kg)

Medium	Moss Norway conc. HNO$_3$ (1990)	Moss Germany conc. HNO$_3$ + H$_2$O$_2$ (1995)	Crustose lichen Germany conc. HNO$_3$	Lichen Northwest Territories oven-dried	Dandelion Europe oven-dried	Spruce bark Canada ashed
Median	-	-	-	-	-	-
Min	-	-	-	-	-	-
Max	-	-	-	-	-	-

*Mean #Geometric mean

Concentrations in Human Fluids (mg/l)

Medium	Human blood Lombardy, Italy	Human serum Lombardy, Italy	Human urine Lombardy, Italy			
Mean	-	-	-			
Min	-	-	-			
Max	-	-	-			

TERBIUM   Tb

Environmental Geochemistry

Biological impacts	Considered non-essential. Data to assess the toxicity of REE are scarce
Uses	Fluorescent and magneto-optical materials
Environmental pathways	Geogenic dust, mining and processing of alkaline rocks. Generally, geogenic sources more important than anthropogenic ones
Environmental mobility	Oxidising conditions: very low Acid conditions: very low Reducing conditions: very low Neutral to alkaline conditions: very low Comments: -
Geochemical barriers	Mechanical
Natural association	REE-Li-Rb-Cs-Be-Nb-Ta-Zr-B-Th-U-F (pegmatites), REE-Th-P-Zr-Fe-Cu (monazite veins), REE-Th-Ba-Sr-P-F-N-C (carbonatites), REE-U-P-F (phosphorites), REE-Au-Ti-Sn-Sr-Th (placers)
Action levels	-
Remarks	Toxicity of REE decreases as atomic number increases. Inhaled REE probably cause pneumoconiosis. REE taken up orally (e.g., via drinking water) accumulate in the skeleton, teeth, lungs, liver and kidneys. Some trees tend to accumulate Tb (e.g., hickory (*Carya*))
Suggested analytical method(s)	ICP-MS

World Yearly Production (t/y)	54,000 REE-minerals (in 1991)
Price of Purest Form, in Small Quantities ($/kg)	145,531 (99.99%)
Market Price ($/kg)	-

Te TELLURIUM

Physico-Chemical Properties

Atomic number	52	Melting/boiling point (K)	722.66 / 1261
Atomic mass	127.60(3)	Isotopes & isomers	6 stable + 36 unstable
Atomic radius (pm)	142	Acid/base of oxide	weak acid
Main oxidation state(s)	+4 (-2,0,+2,6)	State (at 300 K, 1 atm.)	solid
Ionic radius (pm)	207 (-2), 66-111 (+4), 57-70 (+6)	Metallic character	pred. non-metal
		Element group(s)	heavy non-metal
Electronegativity (Pauling)	2.1	Affinity	chalcophile
Density (g/cm^3)	6.24		

Naturally occurring isotopes	Natural abundance (%)	Atomic mass	Half-life
^{120}Te	0.096	119.904,03	stable
^{122}Te	2.603	121.903,06	stable
^{123}Te	0.908	122.904,27	1.3x10^{+13} y
^{124}Te	4.816	123.902,82	stable
^{125}Te	7.139	124.904,42	stable
^{126}Te	18.952	125.903,31	stable
^{128}Te	31.687	127.904,46	stable
^{130}Te	33.799	129.906,22	2.5x10^{+21} y

Unstable isotopes	Longest half-life
^{106}Te to ^{138}Te	2.5x10^{+21} y

Concentrations in Crust/Rocks (mg/kg)

Bulk continental crust	(0.005) / 0.001 / -
Upper continental crust	- / 0.005
Ultramafic rock	0.001
Ocean ridge basalt	0.003
Gabbro, basalt	0.006
Granite, granodiorite	0.005
Sandstone	0.002
Greywacke	-
Shale, schist	0.009
Limestone	0.002
Coal	<0.1

Typical Minerals
tellurite (TeO$_2$), sylvanite ((Ag,Au)Te$_4$), hessite (Ag$_2$Te), nagyagite (Pb$_5$Au(Te,Sb)$_4$S$_{5-8}$), calaverite (AuTe$_2$)

Possible Host Minerals
pyrite, molybdenite, chalcopyrite, gold, silver, pentlandite

Mass (kg) in

Continental crust	(1.05x10^{+14})	Oceans	1.32x10^{+8}	Plants	9.20x10^{+7}

Concentrations in Media Sampled for the Kola Project (mg/kg)

Medium	Moss conc. HNO$_3$	Humus <2 mm	Humus <2 mm amm. acet.	Topsoil (0-5 cm) <2 mm	B horizon <2 mm aqua regia	B horizon <2 mm total (XRF)
Median	-	-	-	-	0.009	-
Min	-	-	-	-	<0.003	-
Max	-	-	-	-	0.221	-

Medium	C horizon <2 mm aqua regia	C horizon <2 mm total (XRF)	C horizon <2 mm total (INAA)			Lake water unfiltered (mg/l)
Median	0.008	-	-			-
Min	<0.003	-	-			-
Max	0.271	-	-			-

TELLURIUM   Te

Concentrations in Soils and Sediments (mg/kg)

Medium	Soil	Agricultural soil - Ap horizon	Agricultural soil - Top (0-25 cm)	Agricultural Soil - Bottom (50-75 cm)	Topsoil (0-15 cm)	Urban soil (0-2 cm)
	World	Canada	Finland	Finland	England & Wales	Trondheim
	<2 mm	<2 mm	<2 mm	<2 mm	<2 mm	<2 mm
	total		aqua regia	aqua regia	aqua regia	aqua regia
Median	0.006*	-	-	-	-	-
Min	-	-	-	-	-	-
Max	-	-	-	-	-	-

*Estimated mean

Medium	Forest soil - Humus	Forest soil - B horizon	Forest soil - C horizon	Till (C horizon)	Till (C horizon)	Laterite (25 ±15 cm)
	Norway	Norway	Norway	Finland	Finland	Australia
	<2 mm	<2 mm	<2 mm	<0.063 mm	<0.063 mm	0.45-2 mm
	7N HNO$_3$	7N HNO$_3$	7N HNO$_3$	aqua regia	total	total
Median	-	-	-	-	-	-
Min	-	-	-	-	-	-
Max	-	-	-	-	-	-

Medium	Stream sediment	Stream sediment	Stream sediment	Overbank sediment	Overbank sediment	Organic stream sediment
	Austria	Southern Scotland	Harz, Germany	Norway	Norway	Finland
	<0.18 mm	<0.10 mm	<0.063 mm	<0.063 mm	<0.063 mm	<2 mm
	total	total	total	total (XRF)	7N HNO$_3$	Conc. HNO$_3$
Median	-	-	-	-	-	-
Min	-	-	-	-	-	-
Max	-	-	-	-	-	-

#2nd percentile   ^98th percentile

Concentrations in Waters (mg/l)

Medium	Ocean water	Ocean water	Ocean water	Stream water	Stream water	Stream water
	World (1)	World (2)	North Pacific	World	Nova Scotia, Canada	Finland
					<0.45 µm	<0.45 µm
					ICP-MS-Hydride	
Median	-	0.000,000,1*	0.000,000,07*	<1x10^{-10}*	<0.000,004	-
Min	-	-	-	-	<0.000,004	-
Max	-	-	-	-	0.000,006	-

*Estimated mean   #2nd percentile   ^98th percentile

Medium	Stream water	Stream water	Stream water	Lake water	Ground water	
	Romania	Eastern India	Harz, Germany	Norway	Southern Norway	
	unfiltered	<0.2 µm	unfiltered	unfiltered	unfiltered	
	ICP-MS	ICP-MS	ICP-MS	ICP-MS	ICP-MS	
Median	0.000,005	<0.000,1	<0.000,01	-	0.000,009	
Min	0.000,003	<0.000,1	<0.000,01	-	<0.000,001	
Max	0.000,07	0.000,2	0.000,08	-	0.000,21	

Te TELLURIUM

Concentrations in Precipitation (mg/l)

Medium	Rain water	Rain water	Snow melt-water	Snow melt-water	Snow filter residue	Snow filter residue
	Kola, remote <0.45 μm	Kola, polluted <0.45 μm	Kola, remote <0.45 μm	Kola, polluted <0.45 μm	Kola, remote >0.45 μm	Kola, polluted >0.45 μm
Median	-	-	-	-	-	-
Min	-	-	-	-	-	-
Max	-	-	-	-	-	-

Medium	Rain water	Rain water				
	Coastal, Norway unfiltered ICP-MS	Inland, Norway unfiltered ICP-MS				
Median	-	-				
Min	-	-				
Max	-	-				

Concentrations in Air (ng/m^3) and Yearly Deposition (kg/km^2/yr)

Medium	Air	Air	Bulk deposition	Throughfall deposition	Bulk deposition	Bulk deposition
	World, remote	World, polluted	West Germany	West Germany	Kola, remote	Kola, polluted
Median	-	-	-	-	-	-
Min	-	-	-	-	-	-
Max	-	-	-	-	-	-

*Estimated value

Concentrations in Plants (mg/kg)

Medium	Moss	Moss	Crustose lichen	Lichen	Dandelion	Spruce bark
	Norway conc. HNO$_3$ (1990)	Germany conc. HNO$_3$ + H$_2$O$_2$ (1995)	Germany conc. HNO$_3$	Northwest Territories oven-dried	Europe oven-dried	Canada ashed
Median	<0.03	0.004	-	-	-	-
Min	<0.03	<0.001	-	-	-	-
Max	0.27	0.094	-	-	-	-

*Mean #Geometric mean

Concentrations in Human Fluids (mg/l)

Medium	Human blood	Human serum	Human urine			
	Lombardy, Italy	Lombardy, Italy	Lombardy, Italy			
Mean	-	-	<0.001			
Min	-	-	-			
Max	-	-	-			

TELLURIUM — Te

Environmental Geochemistry

Biological impacts	Essentiality for plants and animals not documented. Data on toxicity are sparse. Compounds generally more toxic than metal
Uses	Steel alloys, alloys with Cu and Pb, photography, pharmaceutical industry, catalyst, batteries, solar cells, semi-conductors
Environmental pathways	Coal combustion, Cu-Ni smelters
Environmental mobility	Oxidising conditions: very low Acid conditions: very low Reducing conditions: very low Neutral to alkaline conditions: very low Comments: -
Geochemical barriers	Mechanical
Natural association	Ni-Cu-Co-Te-etc. (pyrrhotite-pentlandite mafic sulphide ores with chalcopyrite, e.g., Sudbury), Au-Te (sulphide bearing veins), Au-Ag-Te-Hg (high-temperature veins), Cu-Mo-Te-S (some porphyry Cu deposits), Pb-Te (some polymetallic sulphide deposits)
Action levels	Drinking water, MAC: 0.01 mg/l (Russia MoH)
Remarks	Te may interact with Se. Accumulates in bones. H_2Te and TeF_6 are very toxic compounds. Te forms extensive haloes in rocks around ore deposits. Some plants tend to accumulate Te. Vegetable, fruit, corn can contain up to 2-3 mg/kg Te
Suggested analytical method(s)	ICP-MS, hydride-AAS

World Yearly Production (t/y)	284 (in 1995)
Price of Purest Form, in Small Quantities ($/kg)	4201 (99.999,9%)
Market Price ($/kg)	24

Th THORIUM

Physico-Chemical Properties

Atomic number	90	Melting/boiling point (K)	2023 / 5061
Atomic mass	232.038,1(1)	Isotopes & isomers	0 stable + 27 unstable
Atomic radius (pm)	-	Acid/base of oxide	weak base
Main oxidation state(s)	+4 (+2?,+3?)	State (at 300 K, 1 atm.)	solid
Ionic radius (pm)	108-135 (+4)	Metallic character	metal
Electronegativity (Pauling)	1.3	Element group(s)	actinide, heavy metal
Density (g/cm^3)	11.72	Affinity	lithophile

Naturally occurring isotopes	Natural abundance (%)	Atomic mass	Half-life	Unstable isotopes	Longest half-life
^{232}Th	100	232.038,05	1.4x10^{+10} y	^{212}Th to ^{237}Th	1.4x10^{+10} y

Concentrations in Crust/Rocks (mg/kg)

Bulk continental crust	8.5 / 9.6 / 3.5
Upper continental crust	10.3 / 10.7
Ultramafic rock	0.05
Ocean ridge basalt	0.3
Gabbro, basalt	2.2
Granite, granodiorite	15
Sandstone	5
Greywacke	9
Shale, schist	12
Limestone	2
Coal	2

Typical Minerals
thorite (ThSiO$_4$),
monazite ((Ce,La,Nd,Th)(PO$_4$,SiO$_4$)),
thorianite (ThO$_2$)

Possible Host Minerals
zircon, sphene, epidote, allanite, xenotine, uraninite

Mass (kg) in

Continental crust	2.01x10^{+17}	Oceans	1.32x10^{+10}	Plants	9.20x10^{+6}

Concentrations in Media Sampled for the Kola Project (mg/kg)

Medium	Moss <2 mm conc. HNO$_3$	Humus <2 mm conc. HNO$_3$	Humus <2 mm amm. acet.	Topsoil (0-5 cm) <2 mm total (INAA)	B horizon <2 mm aqua regia	B horizon <2 mm total (XRF)
Median	0.023	0.345	-	2.6	6	-
Min	<0.004	0.063	-	<0.2	<3	-
Max	1.14	15.4	-	24	51	-

Medium	C horizon <2 mm aqua regia	C horizon <2 mm total (XRF)	C horizon <2 mm total (INAA)	Lake water unfiltered (mg/l)
Median	6	-	5.8	<0.000,02
Min	<3	-	1	<0.000,02
Max	66	-	54	0.000,06

THORIUM   Th

Concentrations in Soils and Sediments (mg/kg)

Medium	Soil	Agricultural soil - Ap horizon	Agricultural soil - Top (0-25 cm)	Agricultural Soil - Bottom (50-75 cm)	Topsoil (0-15 cm)	Urban soil (0-2 cm)
	World	Canada	Finland	Finland	England & Wales	Trondheim
	<2 mm	<2 mm	<2 mm	<2 mm	<2 mm	<2 mm
	total	total (INAA)	aqua regia	aqua regia	aqua regia	aqua regia
Median	9.4*	7.4	-	-	-	<5
Min	-	0.2	-	-	-	<5
Max	-	16	-	-	-	11

*Estimated mean

Medium	Forest soil - Humus	Forest soil - B horizon	Forest soil - C horizon	Till (C horizon)	Till (C horizon)	Laterite (25 ±15 cm)
	Norway	Norway	Norway	Finland	Finland	Australia
	<2 mm	<2 mm	<2 mm	<0.063 mm	<0.063 mm	0.45-2 mm
	7N HNO$_3$	7N HNO$_3$	7N HNO$_3$	aqua regia	total	total
Median	-	-	-	5.8	10.2	6.6
Min	-	-	-	-	-	0.3
Max	-	-	-	-	-	40

Medium	Stream sediment	Stream sediment	Stream sediment	Overbank sediment	Overbank sediment	Organic stream sediment
	Austria	Southern Scotland	Harz, Germany	Norway	Norway	Finland
	<0.18 mm	<0.10 mm	<0.063 mm	<0.063 mm	<0.063 mm	<2 mm
	total (XRF)	total	total (ICP-MS)	total (XRF)	7N HNO$_3$	Conc. HNO$_3$
Median	12.7	-	17	-	-	4.93
Min	-	-	2.7	-	-	0.95#
Max	1919	-	42	-	-	14.9^

#2nd percentile   ^98th percentile

Concentrations in Waters (mg/l)

Medium	Ocean water	Ocean water	Ocean water	Stream water	Stream water	Stream water
	World (1)	World (2)	North Pacific	World	Nova Scotia, Canada	Finland
					<0.45 µm	<0.45 µm
Median	0.000,001*	0.000,01*	0.000,000,02*	<0.000,1*	-	-
Min	-	-	-	-	-	-
Max	-	-	-	-	-	-

*Estimated mean   #2nd percentile   ^98th percentile

Medium	Stream water	Stream water	Stream water	Lake water	Ground water
	Romania	Eastern India	Harz, Germany	Norway	Southern Norway
	unfiltered	<0.2 µm	unfiltered	unfiltered	unfiltered
	ICP-MS	ICP-MS	ICP-MS	ICP-MS	ICP-MS
Median	0.000,048	0.000,58	0.000,03	<0.000,015	0.000,013
Min	0.000,001	0.000,12	<0.000,001	<0.000,015	<0.000,001
Max	0.005	0.013,89	0.000,7	0.000,131	0.021,4

Th THORIUM

Concentrations in Precipitation (mg/l)

Medium	Rain water	Rain water	Snow melt-water	Snow melt-water	Snow filter residue	Snow filter residue
	Kola, remote <0.45 µm ICP-MS	Kola, polluted <0.45 µm ICP-MS	Kola, remote <0.45 µm ICP-MS	Kola, polluted <0.45 µm ICP-MS	Kola, remote >0.45 µm ICP-AES	Kola, polluted >0.45 µm ICP-AES
Median	<0.000,02	<0.000,02	<0.000,01	<0.000,01	<0.000,5	<0.000,5
Min	<0.000,02	<0.000,02	<0.000,01	<0.000,01	<0.000,5	<0.000,5
Max	<0.000,02	<0.000,02	<0.000,01	<0.000,01	<0.000,5	0.000,8

Medium	Rain water	Rain water				
	Coastal, Norway unfiltered ICP-MS	Inland, Norway unfiltered ICP-MS				
Median	<0.000,01	<0.000,01				
Min	<0.000,01	<0.000,01				
Max	0.000,12	0.000,14				

Concentrations in Air (ng/m^3) and Yearly Deposition (kg/km^2/yr)

Medium	Air	Air	Bulk deposition	Throughfall deposition	Bulk deposition	Bulk deposition
	World, remote	World, polluted	West Germany	West Germany	Kola, remote	Kola, polluted
Median	0.03 / 0.016	0.16	-	-	-	-
Min	0.000,05	0.03	-	-	-	-
Max	0.15	1.3	-	-	-	-

*Estimated value

Concentrations in Plants (mg/kg)

Medium	Moss	Moss	Crustose lichen	Lichen	Dandelion	Spruce bark
	Norway conc. HNO$_3$ (1990)	Germany conc. HNO$_3$ + H$_2$O$_2$ (1995)	Germany conc. HNO$_3$	Northwest Territories oven-dried	Europe oven-dried	Canada ashed total (INAA)
Median	0.08	0.094	1.8	-	-	2.5
Min	<0.01	0.02	0.26	-	-	0.3
Max	1.8	2.5	17	-	-	7.5

*Mean #Geometric mean

Concentrations in Human Fluids (mg/l)

Medium	Human blood	Human serum	Human urine			
	Lombardy, Italy	Lombardy, Italy	Lombardy, Italy			
Mean	0.000,21	<0.000,1	0.000,085			
Min	0.000,03	<0.000,1	0.000,01			
Max	0.000,73	<0.000,1	0.000,7			

THORIUM   Th

Environmental Geochemistry

Biological impacts	Considered non-essential. Chemotoxic and radiotoxic. Carcinogenic
Uses	Nuclear industry, ignition gauze of portable gas lamp, coating of W wires, coating of optical lenses, Mg-Ni alloys
Environmental pathways	P fertilisers, U mining and processing, coal combustion, geogenic dust
Environmental mobility	Oxidising conditions: very low Acid conditions: very low Reducing conditions: very low Neutral to alkaline conditions: very low Comments: -
Geochemical barriers	Mechanical, minor adsorption (clays and Al hydroxides)
Natural association	K-Th-U (general association), REE-Li-Rb-Cs-Be-Nb-Ta-Zr-B-Th-U-F (pegmatites), REE-Th-P-Zr-Fe-Cu (monazite veins), REE-Th-Ba-Sr-P-F-N-C (carbonatites), REE-U-Th-P-F (phosphorites), REE-Au-Ti-Sn-Sr-Th (placers)
Action levels	-
Remarks	Chemical properties of Th are similar to those of REE. Causes liver and bone cancer
Suggested analytical method(s)	ICP-MS, INAA, fluorometry

World Yearly Production (t/y)	700 (in 1984)
Price of Purest Form, in Small Quantities ($/kg)	15,000
Market Price ($/kg)	-

Ti TITANIUM

Physico-Chemical Properties

Atomic number	22	Melting/boiling point (K)	1941 / 3560
Atomic mass	47.867(1)	Isotopes & isomers	5 stable + 15 unstable
Atomic radius (pm)	200	Acid/base of oxide	amphoteric
Main oxidation state(s)	+4 (+2,+3)	State (at 300 K, 1 atm.)	solid
Ionic radius (pm)	100 (+2), 81 (+3), 56-88 (+4)	Metallic character	pred. metal
		Element group(s)	heavy metal
Electronegativity (Pauling)	1.54	Affinity	lithophile
Density (g/cm^3)	4.54		

Naturally occurring isotopes	Natural abundance (%)	Atomic mass	Half-life
^{46}Ti	8.25	45.952,63	stable
^{47}Ti	7.44	46.951,76	stable
^{48}Ti	73.72	47.947,947	stable
^{49}Ti	5.41	48.947,871	stable
^{50}Ti	5.18	49.944,79	stable

Unstable isotopes	Longest half-life
^{39}Ti to ^{58}Ti	67 y

Concentrations in Crust/Rocks (mg/kg)

Bulk continental crust	4010 / 5650 / 5400
Upper continental crust	3117 / 3000
Ultramafic rock	3000
Ocean ridge basalt	9000
Gabbro, basalt	10,000
Granite, granodiorite	3000
Sandstone	1500
Greywacke	4316
Shale, schist	6000
Limestone	400
Coal	1000

Typical Minerals
ilmenite (FeTiO$_3$), rutile (TiO$_2$), sphene (CaTiSiO$_5$)

Possible Host Minerals
pyroxenes, amphiboles, micas, garnets

Mass (kg) in

Continental crust	9.46x10^{+19}	Oceans	1.32x10^{+12}	Plants	9.20x10^{+9}

Concentrations in Media Sampled for the Kola Project (mg/kg)

Medium	Moss conc. HNO$_3$	Humus <2 mm	Humus <2 mm amm. acet.	Topsoil (0-5 cm) <2 mm	B horizon <2 mm aqua regia	B horizon <2 mm total (XRF)
Median	-	-	0.15	-	920	3477
Min	-	-	<0.05	-	51.5	659
Max	-	-	1.59	-	4940	11,870

Medium	C horizon <2 mm aqua regia	C horizon <2 mm total (XRF)	C horizon <2 mm total (INAA)		Lake water unfiltered (mg/l)
Median	806	3470	-		-
Min	48.8	530	-		-
Max	5730	19,000	-		-

TITANIUM — Ti

Concentrations in Soils and Sediments (mg/kg)

Medium	Soil	Agricultural soil - Ap horizon	Agricultural soil - Top (0-25 cm)	Agricultural Soil - Bottom (50-75 cm)	Topsoil (0-15 cm)	Urban soil (0-2 cm)
	World	Canada	Finland	Finland	England & Wales	Trondheim
	<2 mm	<2 mm	<2 mm	<2 mm	<2 mm	<2 mm
	total		aqua regia	aqua regia	aqua regia	aqua regia
Median	4000*	-	402	718	-	1110
Min	-	-	6.78	<5	-	84
Max	-	-	2190	3010	-	3170

*Estimated mean

Medium	Forest soil - Humus	Forest soil - B horizon	Forest soil - C horizon	Till (C horizon)	Till (C horizon)	Laterite (25 ±15 cm)
	Norway	Norway	Norway	Finland	Finland	Australia
	<2 mm	<2 mm	<2 mm	<0.063 mm	<0.063 mm	0.45-2 mm
	7N HNO$_3$	7N HNO$_3$	7N HNO$_3$	aqua regia	total	total
Median	110	1300	1300	1100	3900	1858
Min	10	20	40	-	-	300
Max	1900	5700	5000	-	-	16,067

Medium	Stream sediment	Stream sediment	Stream sediment	Overbank sediment	Overbank sediment	Organic stream sediment
	Austria	Southern Scotland	Harz, Germany	Norway	Norway	Finland
	<0.18 mm	<0.10 mm	<0.063 mm	<0.063 mm	<0.063 mm	<2 mm
	"total" (ICP-AES)	total (DCES)	total (XRF)	total (XRF)	7N HNO$_3$	Conc. HNO$_3$
Median	5100	5395	4616	5800	-	1270
Min	-	600	588	1500	-	270#
Max	75,800	43,164	8873	114,600	-	2570^

#2nd percentile   ^98th percentile

Concentrations in Waters (mg/l)

Medium	Ocean water	Ocean water	Ocean water	Stream water	Stream water	Stream water
	World (1)	World (2)	North Pacific	World	Nova Scotia, Canada	Finland
					<0.45 µm	<0.45 µm
					ICP-MS	
Median	0.001*	0.001*	0.000,006,5*	0.003*	0.001,2	-
Min	-	-	-	-	<0.000,5	-
Max	-	-	-	-	0.007,7	-

*Estimated mean   #2nd percentile   ^98th percentile

Medium	Stream water	Stream water	Stream water	Lake water	Ground water
	Romania	Eastern India	Harz, Germany	Norway	Southern Norway
	unfiltered	<0.2 µm	unfiltered	unfiltered	unfiltered
	ICP-MS	ICP-MS	ICP-MS	ICP-MS	ICP-MS
Median	0.008,5	0.092,6	0.003,7	0.004,86	0.000,635
Min	0.000,4	0.020,8	<0.000,1	<0.000,4	0.000,14
Max	0.06	2.282,5	0.019	1.22	0.355

Ti — TITANIUM

Concentrations in Precipitation (mg/l)

Medium	Rain water	Rain water	Snow melt-water	Snow melt-water	Snow filter residue	Snow filter residue
	Kola, remote <0.45 μm	Kola, polluted <0.45 μm	Kola, remote <0.45 μm	Kola, polluted <0.45 μm	Kola, remote >0.45 μm ICP-AES	Kola, polluted >0.45 μm ICP-AES
Median	-	-	-	-	0.000,43	0.031,1
Min	-	-	-	-	0.000,34	0.015,1
Max	-	-	-	-	0.003,41	0.139

Medium	Rain water	Rain water
	Coastal, Norway unfiltered ICP-MS	Inland, Norway unfiltered ICP-MS
Median	-	-
Min	-	-
Max	-	-

Concentrations in Air (ng/m^3) and Yearly Deposition (kg/km^2/yr)

Medium	Air	Air	Bulk deposition	Throughfall deposition	Bulk deposition	Bulk deposition
	World, remote	World, polluted	West Germany	West Germany	Kola, remote	Kola, polluted
Median	7 / 6.4	85	-	-	-	-
Min	0.1	<10	-	-	-	-
Max	100	230	-	-	-	-

*Estimated value

Concentrations in Plants (mg/kg)

Medium	Moss	Moss	Crustose lichen	Lichen	Dandelion	Spruce bark
	Norway conc. HNO$_3$ (1990)	Germany conc. HNO$_3$ + H$_2$O$_2$ (1995)	Germany conc. HNO$_3$	Northwest Territories oven-dried total (INAA)	Europe oven-dried	Canada ashed aqua regia
Median	-	21.8	-	62.7*	-	305
Min	-	4.7	-	-	-	26
Max	-	235	-	-	-	1562

*Mean   #Geometric mean

Concentrations in Human Fluids (mg/l)

Medium	Human blood	Human serum	Human urine
	Lombardy, Italy	Lombardy, Italy	Lombardy, Italy
Mean	-	-	0.002,1
Min	-	-	0.000,6
Max	-	-	0.003,7

TITANIUM Ti

Environmental Geochemistry

Biological impacts	Considered non-essential. Non-toxic
Uses	Pigments, alloys, aeronautical industry, tubing, catalyst in polyethylene production, toothpaste, food additive
Environmental pathways	Geogenic dust is the main source of Ti in the environment
Environmental mobility	Oxidising conditions: very low Acid conditions: very low Reducing conditions: very low Neutral to alkaline conditions: very low Comments: -
Geochemical barriers	Mechanical
Natural association	Fe-Mg-Mn-V-Ti-Sc-S (many silicates, sulphides), REE-Au-Ti-Sn-Sr-Th (placers), Nb-Ta-Sn-W-Li-Be-Ti-Rb-Cs-U-Th-B-Zr-Hf-P-F-REE (granites and syenitic pegmatites), Nb-Ta-Na-K-Ba-Sr-Ti-Zr-U-Th-Cu-Zn-P-S-F-REE (carbonatites), Nb-Ta-Ti-Ga-Be-Al (bauxite developed on alkaline rocks)
Action levels	-
Remarks	Ti possibly has a positive influence on growth rate of grain. Severe environmental problems can be associated with the production of Ti (acid disposal). $TiCl_4$ is used to create artificial fog. In many practical applications Ti is considered a "light metal" (e.g., bicycle frames)
Suggested analytical method(s)	XRF, ICP-AES, ICP-MS (water)

World Yearly Production (t/y)	2,378,000 (in 1995)
Price of Purest Form, in Small Quantities ($/kg)	4563 (99.99%)
Market Price ($/kg)	0.65 (rutile, 95% TiO_2)

Tl THALLIUM

Physico-Chemical Properties

Atomic number	81	Melting/boiling point (K)	577 / 1746
Atomic mass	204.383,3(2)	Isotopes & isomers	2 stable + 45 unstable
Atomic radius (pm)	208	Acid/base of oxide	base
Main oxidation state(s)	+1 (+3)	State (at 300 K, 1 atm.)	solid
Ionic radius (pm)	164-184 (+1), 89-112 (+3)	Metallic character	metal
Electronegativity (Pauling)	2.04	Element group(s)	heavy metal
Density (g/cm^3)	11.84	Affinity	chalcophile

Naturally occurring isotopes	Natural abundance (%)	Atomic mass	Half-life	Unstable isotopes	Longest half-life
^{203}Tl	29.524	202.972,33	stable	^{179}Tl to ^{210}Tl	3.78 y
^{205}Tl	70.476	204.974,41	stable		

Concentrations in Crust/Rocks (mg/kg)

Bulk continental crust	0.52 / 0.85 / 0.36
Upper continental crust	0.75 / 0.75
Ultramafic rock	0.05
Ocean ridge basalt	0.01
Gabbro, basalt	0.18
Granite, granodiorite	1.1
Sandstone	0.4
Greywacke	-
Shale, schist	1
Limestone	0.05
Coal	3

Typical Minerals
crookesite ((Cu,Tl,Ag)$_2$Se), lorandite (TlAsS$_2$)

Possible Host Minerals
K-minerals (e.g., micas, feldspars), many sulphides

Mass (kg) in

Continental crust	1.23x10^{+16}	Oceans	1.32x10^{+10}	Plants	9.20x10^{+7}

Concentrations in Media Sampled for the Kola Project (mg/kg)

Medium	Moss	Humus <2 mm	Humus <2 mm	Topsoil (0-5 cm) <2 mm	B horizon <2 mm	B horizon <2 mm
	conc. HNO$_3$	conc. HNO$_3$	amm. acet.		aqua regia	total (XRF)
Median	0.023	0.092	-	-	-	-
Min	<0.004	0.02	-	-	-	-
Max	0.35	0.56	-	-	-	-

Medium	C horizon <2 mm	C horizon <2 mm	C horizon <2 mm	Lake water unfiltered
	aqua regia	total (XRF)	total (INAA)	(mg/l)
Median	-	-	-	<0.000,01
Min	-	-	-	<0.000,01
Max	-	-	-	0.000,01

THALLIUM — TL

Concentrations in Soils and Sediments (mg/kg)

Medium	Soil	Agricultural soil - Ap horizon	Agricultural soil - Top (0-25 cm)	Agricultural Soil - Bottom (50-75 cm)	Topsoil (0-15 cm)	Urban soil (0-2 cm)
	World	Canada	Finland	Finland	England & Wales	Trondheim
	<2 mm	<2 mm	<2 mm	<2 mm	<2 mm	<2 mm
	total		aqua regia	aqua regia	aqua regia	aqua regia
Median	0.5*	-	-	-	-	-
Min	-	-	-	-	-	-
Max	-	-	-	-	-	-

*Estimated mean

Medium	Forest soil - Humus	Forest soil - B horizon	Forest soil - C horizon	Till (C horizon)	Till (C horizon)	Laterite (25 ±15 cm)
	Norway	Norway	Norway	Finland	Finland	Australia
	<2 mm	<2 mm	<2 mm	<0.063 mm	<0.063 mm	0.45-2 mm
	7N HNO₃	7N HNO₃	7N HNO₃	aqua regia	total	total
Median	-	-	-	-	-	-
Min	-	-	-	-	-	-
Max	-	-	-	-	-	-

Medium	Stream sediment	Stream sediment	Stream sediment	Overbank sediment	Overbank sediment	Organic stream sediment
	Austria	Southern Scotland	Harz, Germany	Norway	Norway	Finland
	<0.18 mm	<0.10 mm	<0.063 mm	<0.063 mm	<0.063 mm	<2 mm
		total	total (ICP-MS)	total (XRF)	7N HNO₃	Conc. HNO₃
Median	-	-	0.79	-	-	0.17
Min	-	-	0.2	-	-	0.04#
Max	-	-	2.5	-	-	0.58^

#2nd percentile   ^98th percentile

Concentrations in Waters (mg/l)

Medium	Ocean water	Ocean water	Ocean water	Stream water	Stream water	Stream water
	World (1)	World (2)	North Pacific	World	Nova Scotia, Canada	Finland
					<0.45 μm	<0.45 μm
					ICP-MS	ICP-MS
Median	0.000,019*	0.000,01*	0.000,013*	0.000,04*	<0.000,005	<0.000,015
Min	-	-	-	-	<0.000,005	<0.000,015#
Max	-	-	-	-	0.000,065	0.000,019^

*Estimated mean   #2nd percentile   ^98th percentile

Medium	Stream water	Stream water	Stream water	Lake water	Ground water
	Romania	Eastern India	Harz, Germany	Norway	Southern Norway
	unfiltered	<0.2 μm	unfiltered	unfiltered	unfiltered
	ICP-MS	ICP-MS	ICP-MS	ICP-MS	ICP-MS
Median	0.000,008	0.000,04	0.000,02	<0.000,006	0.000,003
Min	0.000,002	0.000,01	<0.000,005	<0.000,006	<0.000,002
Max	0.000,71	0.000,81	0.001,1	0.000,046	0.000,16

Tl THALLIUM

Concentrations in Precipitation (mg/l)

Medium	Rain water	Rain water	Snow melt-water	Snow melt-water	Snow filter residue	Snow filter residue
	Kola, remote <0.45 μm ICP-MS	Kola, polluted <0.45 μm ICP-MS	Kola, remote <0.45 μm ICP-MS	Kola, polluted <0.45 μm ICP-MS	Kola, remote >0.45 μm	Kola, polluted >0.45 μm
Median	<0.000,02	0.000,02	<0.000,01	0.000,02	-	-
Min	<0.000,02	<0.000,02	<0.000,01	0.000,01	-	-
Max	<0.000,02	0.000,14	<0.000,01	0.000,04	-	-

Medium	Rain water	Rain water
	Coastal, Norway unfiltered ICP-MS	Inland, Norway unfiltered ICP-MS
Median	0.000,02	<0.000,01
Min	<0.000,01	<0.000,01
Max	0.000,08	0.000,08

Concentrations in Air (ng/m^3) and Yearly Deposition (kg/km^2/yr)

Medium	Air	Air	Bulk deposition	Throughfall deposition	Bulk deposition	Bulk deposition
	World, remote	World, polluted	West Germany	West Germany	Kola, remote	Kola, polluted
Median	-	-	-	-	-	-
Min	-	-	-	-	-	-
Max	-	-	-	-	-	-

*Estimated value

Concentrations in Plants (mg/kg)

Medium	Moss	Moss	Crustose lichen	Lichen	Dandelion	Spruce bark
	Norway conc. HNO$_3$ (1990)	Germany conc. HNO$_3$ + H$_2$O$_2$ (1995)	Germany conc. HNO$_3$	Northwest Territories oven-dried	Europe oven-dried	Canada ashed
Median	0.08	0.039	0.2	-	-	-
Min	0.01	<0.001	0.08	-	-	-
Max	0.68	0.69	1	-	-	-

*Mean   #Geometric mean

Concentrations in Human Fluids (mg/l)

Medium	Human blood	Human serum	Human urine
	Lombardy, Italy	Lombardy, Italy	Lombardy, Italy
Mean	0.000,39	0.000,18	0.000,42
Min	0.000,1	<0.000,05	0.000,06
Max	0.001,1	0.000,4	0.000,82

THALLIUM   TL

Environmental Geochemistry

Biological impacts	Considered non-essential. Highly toxic
Uses	Alloys (Pb, Sb, Sn), low temperature thermometers, electronics, glass (with As, Se), rodenticide
Environmental pathways	Petroleum refining, smelting of Pb, Zn (and Cu) sulphides, waste incineration, coal combustion, cement factories (from FeS_2 and/or clay minerals)
Environmental mobility	Oxidising conditions: low Acid conditions: low Reducing conditions: very low Neutral to alkaline conditions: low Comments: -
Geochemical barriers	-
Natural association	K-Rb-Tl (phyllosilicates), Tl-Fe-Zn-Cu-Pb-Ag-As-Se (sulphide deposits)
Action levels	Drinking water, MAC: 0.05 mg/l (US EPA), 0.000,1 mg/l (Russia MoH). Agricultural soil, maximum tolerable concentration: 1 mg/kg (Germany)
Remarks	Tl_2SO_4 is used as rat poison (1 g lethal for humans). Some plants are very sensitive to Tl poisoning (e.g., tobacco); others are very resistant. Plants can easily take up Tl by the roots. Tl has a strong tendency to accumulate in aquatic life. The US EPA has set a Maximum Contaminant Level Goal for Tl in drinking water of 0.000,5 mg/l. The US EPA suggests that drinking water exceeding action levels can lead to gastrointestinal irritation and nerve damage in the short term, and to changes in blood chemistry, damage to liver, kidney, intestinal and testicular tissues, and hair loss in the long term
Suggested analytical method(s)	ICP-MS

World Yearly Production (t/y)	30 (in 1984)
Price of Purest Form, in Small Quantities ($/kg)	1255 (99.999%)
Market Price ($/kg)	-

Tm THULIUM

Physico-Chemical Properties

Atomic number	69	Melting/boiling point (K)	1818 / 2223
Atomic mass	168.934,21(2)	Isotopes & isomers	1 stable + 37 unstable
Atomic radius (pm)	242	Acid/base of oxide	weak base
Main oxidation state(s)	+3 (+2)	State (at 300 K, 1 atm.)	solid
Ionic radius (pm)	117-123 (+2), 102-119 (+3)	Metallic character	metal
		Element group(s)	heavy metal, REE
Electronegativity (Pauling)	1.25	Affinity	lithophile
Density (g/cm^3)	9.321		

Naturally occurring isotopes	Natural abundance (%)	Atomic mass	Half-life
^{169}Tm	100	168.934,21	stable

Unstable isotopes	Longest half-life
^{146}Tm to ^{176}Tm	1.92 y

Concentrations in Crust/Rocks (mg/kg)

Bulk continental crust	0.3 / 0.52 / 0.32
Upper continental crust	- / 0.33
Ultramafic rock	-
Ocean ridge basalt	-
Gabbro, basalt	-
Granite, granodiorite	-
Sandstone	-
Greywacke	-
Shale, schist	-
Limestone	-
Coal	<0.1

Typical Minerals

monazite ((Ce,La,Nd,Th,Tm)(PO$_4$,SiO$_4$)),
bastnaesite ((Ce,La,Tm)CO$_3$(F,OH)),
cerite ((Ce,La,Tm)$_9$(Mg,Fe)Si$_7$(O,OH,F)$_{28}$),
allanite ((Ca,Ce,La,Tm)$_2$FeAl$_2$OSi$_3$O$_{11}$(OH))

Possible Host Minerals
biotite, apatite, pyroxenes, feldspars, zircon

Mass (kg) in

Continental crust	7.08x10^{+15}	Oceans	2.64x10^{+8}	Plants	7.36x10^{+6}

Concentrations in Media Sampled for the Kola Project (mg/kg)

Medium	Moss conc. HNO$_3$	Humus <2 mm	Humus <2 mm amm. acet.	Topsoil (0-5 cm) <2 mm	B horizon <2 mm aqua regia	B horizon <2 mm total (XRF)
Median	-	-	-	-	-	-
Min	-	-	-	-	-	-
Max	-	-	-	-	-	-

Medium	C horizon <2 mm aqua regia	C horizon <2 mm total (XRF)	C horizon <2 mm total (INAA)			Lake water unfiltered (mg/l)
Median	-	-	-			-
Min	-	-	-			-
Max	-	-	-			-

THULIUM   Tm

Concentrations in Soils and Sediments (mg/kg)

Medium	Soil	Agricultural soil - Ap horizon	Agricultural soil - Top (0-25 cm)	Agricultural Soil - Bottom (50-75 cm)	Topsoil (0-15 cm)	Urban soil (0-2 cm)
	World	Canada	Finland	Finland	England & Wales	Trondheim
	<2 mm total	<2 mm	<2 mm aqua regia	<2 mm aqua regia	<2 mm aqua regia	<2 mm aqua regia
Median	-	-	-	-	-	-
Min	-	-	-	-	-	-
Max	-	-	-	-	-	-

*Estimated mean

Medium	Forest soil - Humus	Forest soil - B horizon	Forest soil - C horizon	Till (C horizon)	Till (C horizon)	Laterite (25 ±15 cm)
	Norway	Norway	Norway	Finland	Finland	Australia
	<2 mm 7N HNO$_3$	<2 mm 7N HNO$_3$	<2 mm 7N HNO$_3$	<0.063 mm aqua regia	<0.063 mm total	0.45-2 mm total
Median	-	-	-	-	-	-
Min	-	-	-	-	-	-
Max	-	-	-	-	-	-

Medium	Stream sediment	Stream sediment	Stream sediment	Overbank sediment	Overbank sediment	Organic stream sediment
	Austria	Southern Scotland	Harz, Germany	Norway	Norway	Finland
	<0.18 mm	<0.10 mm total	<0.063 mm total	<0.063 mm total (XRF)	<0.063 mm 7N HNO$_3$	<2 mm Conc. HNO$_3$
Median	-	-	-	-	-	-
Min	-	-	-	-	-	-
Max	-	-	-	-	-	-

#2nd percentile ^98th percentile

Concentrations in Waters (mg/l)

Medium	Ocean water World (1)	Ocean water World (2)	Ocean water North Pacific	Stream water World	Stream water Nova Scotia, Canada	Stream water Finland
					<0.45 µm ICP-MS	<0.45 µm
Median	0.000,000,17*	0.000,000,2*	0.000,000,2*	-	<0.000,005	-
Min	-	-	-	-	<0.000,005	-
Max	-	-	-	-	0.000,016	-

*Estimated mean   #2nd percentile   ^98th percentile

Medium	Stream water Romania	Stream water Eastern India	Stream water Harz, Germany	Lake water Norway	Ground water Southern Norway
	unfiltered ICP-MS	<0.2 µm ICP-MS	unfiltered ICP-MS	unfiltered ICP-MS	unfiltered ICP-MS
Median	-	0.000,03	0.000,01	<0.000,003	0.000,002
Min	-	0.000,01	<0.000,001	<0.000,003	<0.000,001
Max	-	0.000,44	0.000,12	0.000,53	0.000,23

Tm THULIUM

Concentrations in Precipitation (mg/l)

Medium	Rain water	Rain water	Snow melt-water	Snow melt-water	Snow filter residue	Snow filter residue
	Kola, remote <0.45 µm	Kola, polluted <0.45 µm	Kola, remote <0.45 µm	Kola, polluted <0.45 µm	Kola, remote >0.45 µm	Kola, polluted >0.45 µm
Median	-	-	-	-	-	-
Min	-	-	-	-	-	-
Max	-	-	-	-	-	-

Medium	Rain water	Rain water				
	Coastal, Norway unfiltered ICP-MS	Inland, Norway unfiltered ICP-MS				
Median	-	-				
Min	-	-				
Max	-	-				

Concentrations in Air (ng/m^3) and Yearly Deposition (kg/km^2/yr)

Medium	Air	Air	Bulk deposition	Throughfall deposition	Bulk deposition	Bulk deposition
	World, remote	World, polluted	West Germany	West Germany	Kola, remote	Kola, polluted
Median	-	-	-	-	-	-
Min	-	-	-	-	-	-
Max	-	-	-	-	-	-

*Estimated value

Concentrations in Plants (mg/kg)

Medium	Moss	Moss	Crustose lichen	Lichen	Dandelion	Spruce bark
	Norway conc. HNO$_3$ (1990)	Germany conc. HNO$_3$ + H$_2$O$_2$ (1995)	Germany conc. HNO$_3$	Northwest Territories oven-dried	Europe oven-dried	Canada ashed
Median	-	-	-	-	-	-
Min	-	-	-	-	-	-
Max	-	-	-	-	-	-

*Mean   #Geometric mean

Concentrations in Human Fluids (mg/l)

Medium	Human blood	Human serum	Human urine			
	Lombardy, Italy	Lombardy, Italy	Lombardy, Italy			
Mean	-	-	-			
Min	-	-	-			
Max	-	-	-			

THULIUM   Tm

Environmental Geochemistry

Biological impacts	Considered non-essential. Generally low toxicity, but data to assess the health relevance of REE are scarce
Uses	Fluorescent materials, lasers
Environmental pathways	Geogenic dust
Environmental mobility	Oxidising conditions: very low Acid conditions: very low Reducing conditions: very low Neutral to alkaline conditions: very low Comments: -
Geochemical barriers	Mechanical
Natural association	REE-Li-Rb-Cs-Be-Nb-Ta-Zr-B-Th-U-F (pegmatites), REE-Th-P-Zr-Fe-Cu (monazite veins), REE-Th-Ba-Sr-P-F-N-C (carbonatites), REE-U-P-F (phosphorites), REE-Au-Ti-Sn-Sr-Th (placers)
Action levels	-
Remarks	Toxicity of REE decreases as atomic number increases. Inhaled REE probably cause pneumoconiosis. REE taken up orally (e.g., via drinking water) accumulate in the skeleton, teeth, lungs, liver and kidneys. Tm is scarcely used
Suggested analytical method(s)	ICP-MS

World Yearly Production (t/y)	54,000 REE-minerals (in 1991)
Price of Purest Form, in Small Quantities ($/kg)	316,875 (99.99%)
Market Price ($/kg)	-

U   URANIUM

Physico-Chemical Properties

Atomic number	92	Melting/boiling point (K)	1408 / 4404
Atomic mass	238.028,9(1)	Isotopes & isomers	0 stable + 23 unstable
Atomic radius (pm)	-	Acid/base of oxide	amphoteric
Main oxidation state(s)	+6 (+2,+3,+4,+5)	State (at 300 K, 1 atm.)	solid
Ionic radius (pm)	116.5 (+3), 103-131 (+4), 90-98 (+5), 59-100 (+6)	Metallic character	metal
		Element group(s)	actinide, heavy metal
Electronegativity (Pauling)	1.38	Affinity	lithophile
Density (g/cm^3)	11.95		

Naturally occurring isotopes	Natural abundance (%)	Atomic mass	Half-life
^{234}U	0.005,5	234.040,95	2.45x10^{+5} y
^{235}U	0.72	235.043,92	7.04x10^{+8} y
^{238}U	99.274,5	238.050,78	4.46x10^{+9} y

Unstable isotopes	Longest half-life
^{218}U to ^{242}U	4.46x10^{+9} y

Concentrations in Crust/Rocks (mg/kg)

Bulk continental crust	1.7 / 2.7 / 0.91
Upper continental crust	2.5 / 2.8
Ultramafic rock	0.02
Ocean ridge basalt	0.1
Gabbro, basalt	0.5
Granite, granodiorite	4
Sandstone	1.3
Greywacke	2
Shale, schist	3.2
Limestone	1
Coal	2

Typical Minerals
uraninite/pitchblende (UO_2), brannerite ($(U,Ca,Ce)(Ti,Fe)_2O_6$), carnotite ($K_2(UO_2)_2(VO_4)_2 \cdot 3H_2O$)

Possible Host Minerals
zircon, apatite, allanite, monazite, Nb-Ta minerals

Mass (kg) in

Continental crust	4.01x10^{+16}	Oceans	4.23x10^{+12}	Plants	1.84x10^{+7}

Concentrations in Media Sampled for the Kola Project (mg/kg)

Medium	Moss	Humus <2 mm	Humus <2 mm	Topsoil (0-5 cm) <2 mm	B horizon <2 mm	B horizon <2 mm
	conc. HNO_3	conc. HNO_3	amm. acet.	total (INAA)	aqua regia	total (XRF)
Median	0.011	0.099	-	<0.5	-	-
Min	<0.004	0.008	-	<0.5	-	-
Max	0.451	14.3	-	30	-	-

Medium	C horizon <2 mm aqua regia	C horizon <2 mm total (XRF)	C horizon <2 mm total (INAA)			Lake water unfiltered (mg/l)
Median	-	-	0.85			0.000,02
Min	-	-	<0.5			<0.000,01
Max	-	-	16			0.000,68

URANIUM U

Concentrations in Soils and Sediments (mg/kg)

Medium	Soil	Agricultural soil - Ap horizon	Agricultural soil - Top (0-25 cm)	Agricultural Soil - Bottom (50-75 cm)	Topsoil (0-15 cm)	Urban soil (0-2 cm)
	World	Canada	Finland	Finland	England & Wales	Trondheim
	<2 mm	<2 mm	<2 mm	<2 mm	<2 mm	<2 mm
	total	total (INAA)	aqua regia	aqua regia	aqua regia	aqua regia
Median	2.7*	2.2	-	-	-	-
Min	-	0.2	-	-	-	-
Max	-	18	-	-	-	-

*Estimated mean

Medium	Forest soil - Humus	Forest soil - B horizon	Forest soil - C horizon	Till (C horizon)	Till (C horizon)	Laterite (25 ±15 cm)
	Norway	Norway	Norway	Finland	Finland	Australia
	<2 mm	<2 mm	<2 mm	<0.063 mm	<0.063 mm	0.45-2 mm
	7N HNO$_3$	7N HNO$_3$	7N HNO$_3$	aqua regia	total	total
Median	-	-	-	-	3.1	1.75
Min	-	-	-	-	-	<0.06
Max	-	-	-	-	-	30

Medium	Stream sediment	Stream sediment	Stream sediment	Overbank sediment	Overbank sediment	Organic stream sediment
	Austria	Southern Scotland	Harz, Germany	Norway	Norway	Finland
	<0.18 mm	<0.10 mm	<0.063 mm	<0.063 mm	<0.063 mm	<2 mm
	total (XRF)	total (INAA)	total (ICP-MS)	total (XRF)	7N HNO$_3$	Conc. HNO$_3$
Median	5	2.9	6.5	-	-	2.01
Min	-	0.2	2	-	-	0.39#
Max	172	827	71	-	-	20.3^

#2nd percentile   ^98th percentile

Concentrations in Waters (mg/l)

Medium	Ocean water World (1)	Ocean water World (2)	Ocean water North Pacific	Stream water World	Stream water Nova Scotia, Canada	Stream water Finland
					<0.45 μm	<0.45 μm
					ICP-MS	ICP-MS
Median	0.003,2*	0.003,2*	0.003,2*	0.000,04*	0.000,036	0.000,073
Min	-	-	-	-	<0.000,005	<0.000,01#
Max	-	-	-	-	0.001,026	0.001,45^

*Estimated mean   #2nd percentile   ^98th percentile

Medium	Stream water	Stream water	Stream water	Lake water	Ground water	
	Romania	Eastern India	Harz, Germany	Norway	Southern Norway	
	unfiltered	<0.2 μm	unfiltered	unfiltered	unfiltered	
	ICP-MS	ICP-MS	ICP-MS	ICP-MS	ICP-MS	
Median	0.001,17	0.000,39	0.000,05	0.000,041	0.003,5	
Min	0.000,007	0.000,08	<0.000,001	<0.000,004	0.000,008	
Max	0.033	0.005,85	0.006,3	0.002,22	2.017,86	

355

U   URANIUM

Concentrations in Precipitation (mg/l)

Medium	Rain water	Rain water	Snow melt-water	Snow melt-water	Snow filter residue	Snow filter residue
	Kola, remote <0.45 μm ICP-MS	Kola, polluted <0.45 μm ICP-MS	Kola, remote <0.45 μm ICP-MS	Kola, polluted <0.45 μm ICP-MS	Kola, remote >0.45 μm	Kola, polluted >0.45 μm
Median	<0.000,01	<0.000,01	<0.000,01	<0.000,01	-	-
Min	<0.000,01	<0.000,01	<0.000,01	<0.000,01	-	-
Max	<0.000,01	0.000,01	<0.000,01	<0.000,01	-	-

Medium	Rain water	Rain water
	Coastal, Norway unfiltered ICP-MS	Inland, Norway unfiltered ICP-MS
Median	<0.000,01	<0.000,01
Min	<0.000,01	<0.000,01
Max	0.000,05	0.000,43

Concentrations in Air (ng/m^3) and Yearly Deposition (kg/km^2/yr)

Medium	Air	Air	Bulk deposition	Throughfall deposition	Bulk deposition	Bulk deposition
	World, remote	World, polluted	West Germany	West Germany	Kola, remote	Kola, polluted
Median	-	-	-	-	-	-
Min	-	0.02	-	-	-	-
Max	-	<0.5	-	-	-	-

*Estimated value

Concentrations in Plants (mg/kg)

Medium	Moss	Moss	Crustose lichen	Lichen	Dandelion	Spruce bark
	Norway conc. HNO$_3$ (1990)	Germany conc. HNO$_3$ + H$_2$O$_2$ (1995)	Germany conc. HNO$_3$	Northwest Territories oven-dried	Europe oven-dried	Canada ashed total (INAA)
Median	0.04	0.029	0.7	-	-	1.1
Min	<0.01	0.006	0.1	-	-	<0.1
Max	0.89	1.1	11	-	-	4.2

*Mean   #Geometric mean

Concentrations in Human Fluids (mg/l)

Medium	Human blood	Human serum	Human urine			
	Lombardy, Italy	Lombardy, Italy	Lombardy, Italy			
Mean	<0.000,1	-	<0.000,1			
Min	-	-	-			
Max	-	-	-			

URANIUM U

Environmental Geochemistry

Biological impacts	Considered non-essential. Chemotoxic and radiotoxic. Carcinogenic
Uses	Nuclear industry, nuclear bombs, glass, ship ballast, counterweight for planes
Environmental pathways	U mining and milling, phosphate fertilisers, coal combustion
Environmental mobility	Oxidising conditions: high Acid conditions: high Reducing conditions: very low Neutral to alkaline conditions: very high Comments: -
Geochemical barriers	Reduction, adsorption, special ion precipitates (e.g., vanadate)
Natural association	U-Mo-Se-V-Cu-C (sandstone or roll-front type U deposits), U-Cu-Ag-Co-V-Ni-As-Au-Mo-Se-Bi (vein type U deposits), U-Cu-Ag-Co-Ni-As-V-Se-Mo-Au (unconformity vein type U deposits), U-Th-Mo-Nb-Ti-REE (pegmatite U deposits), U-Nb-Th-Cu-F-P-Ti-Zr (carbonatite hosted U deposits), U-Th-Ti-Au-Zr-REE (placer U deposits)
Action levels	Drinking water, no adequate data to permit recommendation of heath-based guideline value (WHO); MAC: 0.02 mg/l (Canada CWQG), 1.7 mg/l (Russia MoH); see also Remarks. Agricultural soil, maximum tolerable concentration: 5 mg/kg (Germany)
Remarks	Radiotoxicity of U in nuclear fallout is marginal compared to other radionuclides. Chemical properties resemble those of REE. Leaching of Ra from U mine tailings is of more concern to the environment than the U itself. Decay of U releases Rn, a radiotoxic and highly mobile (gas) element. Drinking water, discussed MAC: 0.1 mg/l (USA)
Suggested analytical method(s)	ICP-MS, INAA, fluorometry

World Yearly Production (t/y)	32,700 (in 1995)
Price of Purest Form, in Small Quantities ($/kg)	-
Market Price ($/kg)	35 (U_3O_8)

V   VANADIUM

Physico-Chemical Properties

Atomic number	23	Melting/boiling point (K)	2183 / 3680
Atomic mass	50.941,5(1)	Isotopes & isomers	1 stable + 18 unstable
Atomic radius (pm)	192	Acid/base of oxide	amphoteric
Main oxidation state(s)	+5 (+2,+3,+4)	State (at 300 K, 1 atm.)	solid
Ionic radius (pm)	93 (+2), 78 (+3), 67-86 (+4), 49.5-68 (+5)	Metallic character	pred. non-metal
		Element group(s)	heavy non-metal
Electronegativity (Pauling)	1.63	Affinity	lithophile
Density (g/cm^3)	6.11		

Naturally occurring isotopes	Natural abundance (%)	Atomic mass	Half-life
^{50}V	0.25	49.947,16	>1.4x10^{+17} y
^{51}V	99.75	50.943,96	stable

Unstable isotopes	Longest half-life
^{43}V to ^{61}V	>1.4x10^{+17} y

Concentrations in Crust/Rocks (mg/kg)

Bulk continental crust	98 / 120 / 230
Upper continental crust	53 / 60
Ultramafic rock	80
Ocean ridge basalt	250
Gabbro, basalt	260
Granite, granodiorite	70
Sandstone	20
Greywacke	98
Shale, schist	130
Limestone	15
Coal	40

Typical Minerals
carnotite (K$_2$(UO$_2$)$_2$(VO$_4$)$_2$.3H$_2$O), roscoelite (K(V,Al,Mg)$_2$(AlSi$_3$)O$_{10}$(OH)$_2$), vanadinite (Pb$_5$(VO$_4$)$_3$Cl), patronite (VS$_4$), V-magnetite ((Fe,V)$_3$O$_4$), descloizite (Pb(Zn,Cu)(VO$_4$)(OH))

Possible Host Minerals
pyroxenes, amphiboles, micas, apatite, magnetite, sphene, rutile

Mass (kg) in

Continental crust	2.31x10^{+18}	Oceans	3.31x10^{+12}	Plants	9.20x10^{+6}

Concentrations in Media Sampled for the Kola Project (mg/kg)

Medium	Moss	Humus <2 mm	Humus <2 mm	Topsoil (0-5 cm) <2 mm	B horizon <2 mm	B horizon <2 mm
	conc. HNO$_3$	conc. HNO$_3$	amm. acet.		aqua regia	total (XRF)
Median	1.6	4.86	0.1	-	42.1	-
Min	0.28	1.08	<0.02	-	7.8	-
Max	83.8	48.9	1.49	-	146	-

Medium	C horizon <2 mm	C horizon <2 mm	C horizon <2 mm		Lake water unfiltered
	aqua regia	total (XRF)	total (INAA)		(mg/l)
Median	30.9	-	-		0.000,19
Min	4.5	-	-		0.000,06
Max	183	-	-		0.003,02

VANADIUM V

Concentrations in Soils and Sediments (mg/kg)

Medium	Soil	Agricultural soil -	Agricultural soil - Top	Agricultural Soil - Bottom	Topsoil (0-15 cm)	Urban soil (0-2 cm)
	World	Ap horizon Canada	(0-25 cm) Finland	(50-75 cm) Finland	England & Wales	Trondheim
	<2 mm total	<2 mm total (AAS)	<2 mm aqua regia	<2 mm aqua regia	<2 mm aqua regia	<2 mm aqua regia
Median	90*	94	25.4	26.1	-	55
Min	-	5	<1	1.55	-	6.7
Max	-	304	176	122	-	144

*Estimated mean

Medium	Forest soil - Humus	Forest soil - B horizon	Forest soil - C horizon	Till (C horizon)	Till (C horizon)	Laterite (25 ±15 cm)
	Norway	Norway	Norway	Finland	Finland	Australia
	<2 mm	<2 mm	<2 mm	<0.063 mm	<0.063 mm	0.45-2 mm
	7N HNO$_3$	7N HNO$_3$	7N HNO$_3$	aqua regia	total	total
Median	6.1	43.7	38.3	35	79	48
Min	0.5	6.2	4.3	-	-	10
Max	99.3	312	167.1	-	-	820

Medium	Stream sediment	Stream sediment	Stream sediment	Overbank sediment	Overbank sediment	Organic stream sediment
	Austria	Southern Scotland	Harz, Germany	Norway	Norway	Finland
	<0.18 mm	<0.10 mm	<0.063 mm	<0.063 mm	<0.063 mm	<2 mm
	"total" (ICP-AES)	total (DCES)	total (XRF)	total (XRF)	7N HNO$_3$ ICP-AES	Conc. HNO$_3$
Median	98	98	58	118	21	44.1
Min	-	<9	17	40	3.3	15.6#
Max	927	380	151	1029	256	100^

#2nd percentile ^98th percentile

Concentrations in Waters (mg/l)

Medium	Ocean water	Ocean water	Ocean water	Stream water	Stream water	Stream water
	World (1)	World (2)	North Pacific	World	Nova Scotia, Canada	Finland
					<0.45 µm	<0.45 µm
					ICP-AES	ICP-MS
Median	0.002,5*	0.002,5*	0.002*	0.000,9*	<0.005	0.000,53
Min	-	-	-	-	<0.005	0.000,08#
Max	-	-	-	-	0.006	0.002,1^

*Estimated mean   #2nd percentile   ^98th percentile

Medium	Stream water	Stream water	Stream water	Lake water	Ground water
	Romania	Eastern India	Harz, Germany	Norway	Southern Norway
	unfiltered	<0.2 µm	unfiltered	unfiltered	unfiltered
	ICP-MS	ICP-MS	ICP-MS	ICP-MS	ICP-MS
Median	0.002,7	0.009,7	0.000,52	<0.000,3	0.000,5
Min	0.000,03	0.002,6	<0.000,1	<0.000,3	<0.000,01
Max	0.139	0.086,6	0.013	0.002,43	0.016,3

V   VANADIUM

Concentrations in Precipitation (mg/l)

Medium	Rain water	Rain water	Snow melt-water	Snow melt-water	Snow filter residue	Snow filter residue
	Kola, remote	Kola, polluted	Kola, remote	Kola, polluted	Kola, remote	Kola, polluted
	<0.45 µm	<0.45 µm	<0.45 µm	<0.45 µm	>0.45 µm	>0.45 µm
	ICP-MS	ICP-MS	ICP-MS	ICP-MS	ICP-AES	ICP-AES
Median	0.000,13	0.003,51	0.000,32	0.015,7	<0.000,05	0.018,4
Min	0.000,07	0.001,36	0.000,25	0.004	<0.000,05	0.006,46
Max	0.000,23	0.015,1	0.000,46	0.024,2	0.000,09	0.056,7

Medium	Rain water	Rain water
	Coastal, Norway unfiltered ICP-MS	Inland, Norway unfiltered ICP-MS
Median	0.000,32	0.000,44
Min	<0.000,01	0.000,05
Max	0.004	0.004,8

Concentrations in Air (ng/m^3) and Yearly Deposition (kg/km^2/yr)

Medium	Air	Air	Bulk deposition	Throughfall deposition	Bulk deposition	Bulk deposition
	World, remote	World, polluted	West Germany	West Germany	Kola, remote	Kola, polluted
Median	0.8 / 0.53	200	-	-	0.24*	19.3*
Min	0.000,9	1	-	-	-	-
Max	3 (20)	2000	-	-	-	-

*Estimated value

Concentrations in Plants (mg/kg)

Medium	Moss	Moss	Crustose lichen	Lichen	Dandelion	Spruce bark
	Norway conc. HNO$_3$ (1990)	Germany conc. HNO$_3$ + H$_2$O$_2$ (1995)	Germany conc. HNO$_3$	Northwest Territories oven-dried total (INAA)	Europe oven-dried	Canada ashed aqua regia
Median	2.4	1.71	29	1.78*	-	61
Min	0.63	0.34	7.2	-	-	19
Max	35	40	87	-	-	478

*Mean   #Geometric mean

Concentrations in Human Fluids (mg/l)

Medium	Human blood	Human serum	Human urine
	Lombardy, Italy	Lombardy, Italy	Lombardy, Italy
Mean	0.000,35	0.000,62	0.000,8
Min	0.000,09	0.000,07	0.000,05
Max	0.001,1	0.001,8	0.001,44

VANADIUM   V

Environmental Geochemistry

Biological impacts	Essential for some organisms. Toxic
Uses	Steel production, Ti alloys, catalysts (e.g., in production of sulphuric acid), polymerisation catalyst for propylene and ethylene
Environmental pathways	Oil and coal combustion, steel production, traffic, geogenic dust, weathering
Environmental mobility	Oxidising conditions: high Acid conditions: high Reducing conditions: very low Neutral to alkaline conditions: very high Comments: -
Geochemical barriers	Reduction, precipitation
Natural association	V-Ti-Fe-P (vanadiferous magnetite), V-Cu-Pb-Zn-Mo-Ag-Au-As (polymetallic sulphide deposits), U-V-Se-Mo-Cu-K-Ca-C (sandstone type U deposits), P-V-U-F-Se-As-etc. (phosphorites and black shales), V-Fe-Mn-P (some sedimentary Fe ores), V-S-C-Ni-Fe-Ca (tar and heavy oil deposits)
Action levels	Drinking water: listed in Norwegian regulations, but no value given; MAC: 0.1 mg/l (Russia MoH). Agricultural soil, maximum tolerable concentration: 50 mg/kg (Germany)
Remarks	Essential nutrient for many animals; some compounds can significantly increase the growth rate (up to 40%). V affects biomass production. V in drinking water prevents caries (as does F). Radish, parsley and potatoes can accumulate V. Certain mushrooms accumulate V to very high concentrations. Toxicity of V is greatly dependent on speciation and oxidation state. Oil and derivatives can, but must not, contain high V concentrations. Coal ash can have high V content
Suggested analytical method(s)	XRF, ICP-AES, ICP-MS (water)

World Yearly Production (t/y)	32,300 (in 1995)
Price of Purest Form, in Small Quantities ($/kg)	22,050 (99.9+%)
Market Price ($/kg)	7.3 (98% V_2O_5)

W TUNGSTEN

Physico-Chemical Properties

Atomic number	74	Melting/boiling point (K)	3695 / 5828
Atomic mass	183.84(1)	Isotopes & isomers	5 stable + 32 unstable
Atomic radius (pm)	202	Acid/base of oxide	weak acid
Main oxidation state(s)	+6 (+2,+3,+4,+5)	State (at 300 K, 1 atm.)	solid
Ionic radius (pm)	80 (+4), 76 (+5), 56-74 (+6)	Metallic character	non-metal
Electronegativity (Pauling)	2.36	Element group(s)	heavy non-metal
Density (g/cm^3)	19.3	Affinity	lithophile, siderophile

Naturally occurring isotopes	Natural abundance (%)	Atomic mass	Half-life
^{180}W	0.12	179.946,71	stable
^{182}W	26.498	181.948,21	stable
^{183}W	14.314	182.950,22	stable
^{184}W	30.642	183.950,93	stable
^{186}W	28.426	185.954,36	stable

Unstable isotopes	Longest half-life
^{158}W to ^{190}W	121.2 d

Concentrations in Crust/Rocks (mg/kg)

Bulk continental crust	1 / 1.25 / 1
Upper continental crust	1.4 / 2
Ultramafic rock	0.3
Ocean ridge basalt	0.5
Gabbro, basalt	0.6
Granite, granodiorite	1.5
Sandstone	1
Greywacke	-
Shale, schist	1.8
Limestone	0.5
Coal	1

Typical Minerals
scheelite (CaWO$_4$), wolframite ((Fe,Mn)WO$_4$)

Possible Host Minerals
muscovite, Nb-Ta minerals, powellite, Mn-oxides

Mass (kg) in

Continental crust	2.36x10^{+16}	Oceans	1.32x10^{+11}	Plants	3.68x10^{+8}

Concentrations in Media Sampled for the Kola Project (mg/kg)

Medium	Moss <2 mm conc. HNO$_3$	Humus <2 mm	Humus <2 mm amm. acet.	Topsoil (0-5 cm) <2 mm total (INAA)	B horizon <2 mm aqua regia	B horizon <2 mm total (XRF)
Median	-	-	-	<1	-	-
Min	-	-	-	<1	-	-
Max	-	-	-	7	-	-

Medium	C horizon <2 mm aqua regia	C horizon <2 mm total (XRF)	C horizon <2 mm total (INAA)		Lake water unfiltered (mg/l)
Median	-	-	<1		-
Min	-	-	<1		-
Max	-	-	10		-

TUNGSTEN W

Concentrations in Soils and Sediments (mg/kg)

Medium	Soil	Agricultural soil - Ap horizon	Agricultural soil - Top (0-25 cm)	Agricultural Soil - Bottom (50-75 cm)	Topsoil (0-15 cm)	Urban soil (0-2 cm)
	World	Canada	Finland	Finland	England & Wales	Trondheim
	<2 mm	<2 mm	<2 mm	<2 mm	<2 mm	<2 mm
	total	aqua regia	aqua regia	aqua regia	aqua regia	aqua regia
Median	1.5*	-	-	-	-	-
Min	-	-	-	-	-	-
Max	-	-	-	-	-	-

*Estimated mean

Medium	Forest soil - Humus	Forest soil - B horizon	Forest soil - C horizon	Till (C horizon)	Till (C horizon)	Laterite (25 ±15 cm)
	Norway	Norway	Norway	Finland	Finland	Australia
	<2 mm	<2 mm	<2 mm	<0.063 mm	<0.063 mm	0.45-2 mm
	7N HNO$_3$	7N HNO$_3$	7N HNO$_3$	aqua regia	total	total
Median	-	-	-	0.2	1.2	0.9
Min	-	-	-	-	-	<0.1
Max	-	-	-	-	-	12

Medium	Stream sediment	Stream sediment	Stream sediment	Overbank sediment	Overbank sediment	Organic stream sediment
	Austria	Southern Scotland	Harz, Germany	Norway	Norway	Finland
	<0.18 mm	<0.10 mm	<0.063 mm	<0.063 mm	<0.063 mm	<2 mm
	total (XRF)	total	total (ICP-MS)	total (XRF)	7N HNO$_3$	Conc. HNO$_3$
Median	1	-	4.4	17	-	-
Min	-	-	0.25	<1	-	-
Max	2693	-	70	55	-	-

#2nd percentile ^98th percentile

Concentrations in Waters (mg/l)

Medium	Ocean water	Ocean water	Ocean water	Stream water	Stream water	Stream water
	World (1)	World (2)	North Pacific	World	Nova Scotia, Canada	Finland
					<0.45 µm	<0.45 µm
Median	0.000,1*	0.000,1*	0.000,01*	0.000,03*	-	-
Min	-	-	-	-	-	-
Max	-	-	-	-	-	-

*Estimated mean #2nd percentile ^98th percentile

Medium	Stream water	Stream water	Stream water	Lake water	Ground water
	Romania	Eastern India	Harz, Germany	Norway	Southern Norway
	unfiltered	<0.2 µm	unfiltered	unfiltered	unfiltered
	ICP-MS	ICP-MS	ICP-MS	ICP-MS	ICP-MS
Median	0.000,011	0.000,05	0.000,03	<0.000,02	0.000,046,5
Min	0.000,002	0.000,02	<0.000,01	<0.000,02	<0.000,002
Max	0.000,38	0.000,89	0.000,72	0.000,653	0.060,8

W — TUNGSTEN

Concentrations in Precipitation (mg/l)

Medium	Rain water	Rain water	Snow melt-water	Snow melt-water	Snow filter residue	Snow filter residue
	Kola, remote <0.45 μm	*Kola, polluted <0.45 μm*	*Kola, remote <0.45 μm*	*Kola, polluted <0.45 μm*	*Kola, remote >0.45 μm*	*Kola, polluted >0.45 μm*
Median	-	-	-	-	-	-
Min	-	-	-	-	-	-
Max	-	-	-	-	-	-

Medium	Rain water	Rain water				
	Coastal, Norway unfiltered ICP-MS	*Inland, Norway unfiltered ICP-MS*				
Median	-	-				
Min	-	-				
Max	-	-				

Concentrations in Air (ng/m^3) and Yearly Deposition (kg/km^2/yr)

Medium	Air	Air	Bulk deposition	Throughfall deposition	Bulk deposition	Bulk deposition
	World, remote	*World, polluted*	*West Germany*	*West Germany*	*Kola, remote*	*Kola, polluted*
Median	0.02 / 0.005	2.5	-	-	-	-
Min	0.001,5	0.03	-	-	-	-
Max	<0.5	6	-	-	-	-

*Estimated value

Concentrations in Plants (mg/kg)

Medium	Moss	Moss	Crustose lichen	Lichen	Dandelion	Spruce bark
	Norway conc. HNO$_3$ (1990)	*Germany conc. HNO$_3$ + H$_2$O$_2$ (1995)*	*Germany conc. HNO$_3$*	*Northwest Territories oven-dried*	*Europe oven-dried*	*Canada ashed total (INAA)*
Median	-	0.057	-	-	-	<1
Min	-	0.01	-	-	-	<1
Max	-	2.6	-	-	-	11

*Mean   #Geometric mean

Concentrations in Human Fluids (mg/l)

Medium	Human blood	Human serum	Human urine			
	Lombardy, Italy	*Lombardy, Italy*	*Lombardy, Italy*			
Mean	0.000,39	0.000,045	0.000,32			
Min	0.000,05	0.000,004	0.000,07			
Max	0.000,75	0.000,5	0.000,9			

TUNGSTEN — W

Environmental Geochemistry

Biological impacts	Considered non-essential. Little is known about toxicity, but not considered to be a serious health hazard
Uses	Alloys (steel), electrodes, abrasives and cutting/drilling tools (W-carbide), light bulbs, X-ray tubes, catalysts
Environmental pathways	Geogenic sources thought to be more important than anthropogenic sources, W mining and smelting, steel industry
Environmental mobility	Oxidising conditions: very low Acid conditions: very low Reducing conditions: very low Neutral to alkaline conditions: low Comments: -
Geochemical barriers	pH, adsorption by Mn, mechanical
Natural association	W-Mo-Sn-Cu-As-Nb-Ta-Bi-Li-B-F-REE (pegmatites and aplites), W-Mo-Bi-Cu-Pb-Zn-S-As-Au-Ag-B-F (skarns)
Action levels	Drinking water, MAC: 0.05 mg/l (Russia MoH)
Remarks	Can replace Mo in minerals. Na tungstate solutions increase growth, yield, sugar content and N-fixation of grapes and alfalfa
Suggested analytical method(s)	ICP-MS, INAA

World Yearly Production (t/y)	22,800 (in 1995)
Price of Purest Form, in Small Quantities ($/kg)	7040 (99.999%)
Market Price ($/kg)	0.06 (wolframite, 65%)

Xe   XENON

Physico-Chemical Properties

Atomic number	54	Melting/boiling point (K)	161.4 / 165.11
Atomic mass	131.29(2)	Isotopes & isomers	9 stable + 35 unstable
Atomic radius (pm)	124	Acid/base of oxide	-
Main oxidation state(s)	0	State (at 300 K, 1 atm.)	gas
Ionic radius (pm)	-	Metallic character	-
Electronegativity (Pauling)	-	Element group(s)	noble gas
Density (g/cm^3)	0.005,887	Affinity	atmophile

Naturally occurring isotopes	Natural abundance (%)	Atomic mass	Half-life
^{124}Xe	0.1	123.905,895	stable
^{126}Xe	0.09	125.904,27	stable
^{128}Xe	1.91	127.903,53	stable
^{129}Xe	26.4	128.904,78	stable
^{130}Xe	4.1	129.903,51	stable
^{131}Xe	21.2	130.905,08	stable
^{132}Xe	26.9	131.904,155	stable
^{134}Xe	10.4	133.905,395	stable
^{136}Xe	8.9	135.907,22	stable

Unstable isotopes	Longest half-life
^{110}Xe to ^{145}Xe	36.4 d

Concentrations in Crust/Rocks (mg/kg)

Bulk continental crust	- / 0.000,03 / -
Upper continental crust	- / -
Ultramafic rock	-
Ocean ridge basalt	-
Gabbro, basalt	-
Granite, granodiorite	-
Sandstone	-
Greywacke	-
Shale, schist	-
Limestone	-
Coal	-

Typical Minerals
-

Possible Host Minerals
-

Mass (kg) in

Continental crust	7.08x10^{+11}	Oceans	6.61x10^{+10}	Plants	-

Concentrations in Media Sampled for the Kola Project (mg/kg)

Medium	Moss	Humus <2 mm amm. acet.	Humus <2 mm amm. acet.	Topsoil (0-5 cm) <2 mm	B horizon <2 mm aqua regia	B horizon <2 mm total (XRF)
	conc. HNO$_3$					
Median	-	-	-	-	-	-
Min	-	-	-	-	-	-
Max	-	-	-	-	-	-

Medium	C horizon <2 mm aqua regia	C horizon <2 mm total (XRF)	C horizon <2 mm total (INAA)	Lake water unfiltered (mg/l)
Median	-	-	-	-
Min	-	-	-	-
Max	-	-	-	-

XENON   Xe

Concentrations in Soils and Sediments (mg/kg)

Medium	Soil	Agricultural soil - Ap horizon	Agricultural soil - Top (0-25 cm)	Agricultural Soil - Bottom (50-75 cm)	Topsoil (0-15 cm)	Urban soil (0-2 cm)
	World	Canada	Finland	Finland	England & Wales	Trondheim
	<2 mm total	<2 mm	<2 mm aqua regia	<2 mm aqua regia	<2 mm aqua regia	<2 mm aqua regia
Median	-	-	-	-	-	-
Min	-	-	-	-	-	-
Max	-	-	-	-	-	-

*Estimated mean

Medium	Forest soil - Humus	Forest soil - B horizon	Forest soil - C horizon	Till (C horizon)	Till (C horizon)	Laterite (25 ±15 cm)
	Norway	Norway	Norway	Finland	Finland	Australia
	<2 mm	<2 mm	<2 mm	<0.063 mm	<0.063 mm	0.45-2 mm
	7N HNO$_3$	7N HNO$_3$	7N HNO$_3$	aqua regia	total	total
Median	-	-	-	-	-	-
Min	-	-	-	-	-	-
Max	-	-	-	-	-	-

Medium	Stream sediment	Stream sediment	Stream sediment	Overbank sediment	Overbank sediment	Organic stream sediment
	Austria	Southern Scotland	Harz, Germany	Norway	Norway	Finland
	<0.18 mm	<0.10 mm	<0.063 mm	<0.063 mm	<0.063 mm	<2 mm
		total	total	total (XRF)	7N HNO$_3$	Conc. HNO$_3$
Median	-	-	-	-	-	-
Min	-	-	-	-	-	-
Max	-	-	-	-	-	-

#2nd percentile   ^98th percentile

Concentrations in Waters (mg/l)

Medium	Ocean water	Ocean water	Ocean water	Stream water	Stream water	Stream water
	World (1)	World (2)	North Pacific	World	Nova Scotia, Canada	Finland
					<0.45 µm	<0.45 µm
Median	0.000,05*	0.000,05*	0.000,066*	-	-	-
Min	-	-	-	-	-	-
Max	-	-	-	-	-	-

*Estimated mean   #2nd percentile   ^98th percentile

Medium	Stream water	Stream water	Stream water	Lake water	Ground water
	Romania	Eastern India	Harz, Germany	Norway	Southern Norway
	unfiltered	<0.2 µm	unfiltered	unfiltered	unfiltered
	ICP-MS	ICP-MS		ICP-MS	
Median	-	-	-	-	-
Min	-	-	-	-	-
Max	-	-	-	-	-

Xe　　XENON

Concentrations in Precipitation (mg/l)

Medium	Rain water Kola, remote <0.45 µm	Rain water Kola, polluted <0.45 µm	Snow melt- water Kola, remote <0.45 µm	Snow melt- water Kola, polluted <0.45 µm	Snow filter residue Kola, remote >0.45 µm	Snow filter residue Kola, polluted >0.45 µm
Median	-	-	-	-	-	-
Min	-	-	-	-	-	-
Max	-	-	-	-	-	-

Medium	Rain water Coastal, Norway unfiltered ICP-MS	Rain water Inland, Norway unfiltered ICP-MS				
Median	-	-				
Min	-	-				
Max	-	-				

Concentrations in Air (ng/m^3) and Yearly Deposition (kg/km^2/yr)

Medium	Air World, remote	Air World, polluted	Bulk deposition West Germany	Throughfall deposition West Germany	Bulk deposition Kola, remote	Bulk deposition Kola, polluted
Median	-	-	-	-	-	-
Min	-	-	-	-	-	-
Max	-	-	-	-	-	-

*Estimated value

Concentrations in Plants (mg/kg)

Medium	Moss Norway conc. HNO_3 (1990)	Moss Germany conc. HNO_3 + H_2O_2 (1995)	Crustose lichen Germany conc. HNO_3	Lichen Northwest Territories oven-dried	Dandelion Europe oven-dried	Spruce bark Canada ashed
Median	-	-	-	-	-	-
Min	-	-	-	-	-	-
Max	-	-	-	-	-	-

*Mean　#Geometric mean

Concentrations in Human Fluids (mg/l)

Medium	Human blood Lombardy, Italy	Human serum Lombardy, Italy	Human urine Lombardy, Italy
Mean	-	-	-
Min	-	-	-
Max	-	-	-

XENON   Xe

Environmental Geochemistry

Biological impacts	Considered non-essential. Harmless to plants. Anaesthetic to mammals
Uses	Filling gas for light bulbs
Environmental pathways	Natural trace component of atmosphere
Environmental mobility	Oxidising conditions: very high — Acid conditions: very high Reducing conditions: very high — Neutral to alkaline conditions: very high Comments: Inert gas
Geochemical barriers	-
Natural association	-
Action levels	-
Remarks	Occurs in atmosphere at low concentrations (0.5 mg/m^3), together with the other noble gasses
Suggested analytical method(s)	-

World Yearly Production (t/y)	-
Price of Purest Form, in Small Quantities ($/kg)	-
Market Price ($/kg)	-

Y — YTTRIUM

Physico-Chemical Properties

Atomic number	39
Atomic mass	88.905,85(2)
Atomic radius (pm)	227
Main oxidation state(s)	+3
Ionic radius (pm)	104-121.5
Electronegativity (Pauling)	1.22
Density (g/cm^3)	4.469
Melting/boiling point (K)	1795 / 3618
Isotopes & isomers	1 stable + 37 unstable
Acid/base of oxide	weak base
State (at 300 K, 1 atm.)	solid
Metallic character	metal
Element group(s)	light metal
Affinity	lithophile

Naturally occurring isotopes	Natural abundance (%)	Atomic mass	Half-life
^{89}Y	100	88.905,85	stable

Unstable isotopes	Longest half-life
^{78}Y to ^{102}Y	106.6 d

Concentrations in Crust/Rocks (mg/kg)

Bulk continental crust	24 / 33 / 20
Upper continental crust	20.7 / 22
Ultramafic rock	2
Ocean ridge basalt	32
Gabbro, basalt	20
Granite, granodiorite	35
Sandstone	15
Greywacke	26
Shale, schist	30
Limestone	4
Coal	3

Typical Minerals

monazite ((Ce,La,Nd,Th,Y)(PO$_4$,SiO$_4$)), bastnaesite ((Ce,La,Y)CO$_3$(F,OH)), xenotime (YPO$_4$), yttrialite ((Y,Th)$_2$Si$_2$O$_7$), euxinite ((Y,Ca,Ce,U,Th)(Nb,Ta,Ti)$_2$O$_6$)

Possible Host Minerals

biotite, feldspars, pyroxenes, apatite

Mass (kg) in

Continental crust	5.66x10^{+17}	Oceans	1.73x10^{+9}	Plants	3.68x10^{+8}

Concentrations in Media Sampled for the Kola Project (mg/kg)

Medium	Moss conc. HNO$_3$	Humus <2 mm conc. HNO$_3$	Humus <2 mm amm. acet.	Topsoil (0-5 cm) <2 mm	B horizon <2 mm aqua regia	B horizon <2 mm total (XRF)
Median	<0.1	0.9	-	-	3.4	-
Min	<0.1	0.2	-	-	0.7	-
Max	5.9	68.8	-	-	122	-

Medium	C horizon <2 mm aqua regia	C horizon <2 mm total (XRF)	C horizon <2 mm total (INAA)			Lake water unfiltered (mg/l)
Median	4.4	-	-			-
Min	0.9	-	-			-
Max	169	-	-			-

YTTRIUM Y

Concentrations in Soils and Sediments (mg/kg)

Medium	Soil	Agricultural soil - Ap horizon	Agricultural soil - Top (0-25 cm)	Agricultural Soil - Bottom (50-75 cm)	Topsoil (0-15 cm)	Urban soil (0-2 cm)
	World	Canada	Finland	Finland	England & Wales	Trondheim
	<2 mm total	<2 mm	<2 mm aqua regia	<2 mm aqua regia	<2 mm aqua regia	<2 mm aqua regia
Median	20*	-	-	-	-	8
Min	-	-	-	-	-	0.6
Max	-	-	-	-	-	17.5

*Estimated mean

Medium	Forest soil - Humus	Forest soil - B horizon	Forest soil - C horizon	Till (C horizon)	Till (C horizon)	Laterite (25 ±15 cm)
	Norway	Norway	Norway	Finland	Finland	Australia
	<2 mm	<2 mm	<2 mm	<0.063 mm	<0.063 mm	0.45-2 mm
	7N HNO$_3$	7N HNO$_3$	7N HNO$_3$	aqua regia	total	total
Median	-	-	-	8.3	21	4
Min	-	-	-	-	-	<1
Max	-	-	-	-	-	17

Medium	Stream sediment	Stream sediment	Stream sediment	Overbank sediment	Overbank sediment	Organic stream sediment
		Southern Scotland	Harz, Germany	Norway	Norway	Finland
	Austria					
	<0.18 mm	<0.10 mm	<0.063 mm	<0.063 mm	<0.063 mm	<2 mm
	total (XRF)	total (DCES)	total (XRF)	total (XRF)	7N HNO$_3$	Conc. HNO$_3$
Median	37	29	47	-	-	11.5
Min	-	<3	16	-	-	4.1#
Max	2055	731	485	-	-	39.7^

#2nd percentile ^98th percentile

Concentrations in Waters (mg/l)

Medium	Ocean water	Ocean water	Ocean water	Stream water	Stream water	Stream water
	World (1)	World (2)	North Pacific	World	Nova Scotia, Canada	Finland
					<0.45 µm ICP-MS	<0.45 µm
Median	0.000,013*	0.000,001,3*	0.000,017*	0.000,04*	0.000,16	-
Min	-	-	-	-	<0.000,01	-
Max	-	-	-	-	0.001,02	-

*Estimated mean   #2nd percentile   ^98th percentile

Medium	Stream water	Stream water	Stream water	Lake water	Ground water
	Romania	Eastern India	Harz, Germany	Norway	Southern Norway
	unfiltered ICP-MS	<0.2 µm ICP-MS	unfiltered ICP-MS	unfiltered ICP-MS	unfiltered ICP-MS
Median	0.000,4	0.003	0.000,74	0.000,093	0.000,140,5
Min	0.000,03	0.000,36	<0.000,1	<0.000,003	<0.000,01
Max	0.098	0.044,98	0.014	0.002,7	0.018,454

Y   YTTRIUM

Concentrations in Precipitation (mg/l)

Medium	Rain water	Rain water	Snow melt-water	Snow melt-water	Snow filter residue	Snow filter residue
	Kola, remote <0.45 µm	Kola, polluted <0.45 µm	Kola, remote <0.45 µm	Kola, polluted <0.45 µm	Kola, remote >0.45 µm ICP-AES	Kola, polluted >0.45 µm ICP-AES
Median	-	-	-	-	<0.000,01	0.000,12
Min	-	-	-	-	<0.000,01	0.000,06
Max	-	-	-	-	0.000,01	0.000,7

Medium	Rain water	Rain water
	Coastal, Norway unfiltered ICP-MS	Inland, Norway unfiltered ICP-MS
Median	<0.000,01	0.000,02
Min	<0.000,01	<0.000,01
Max	0.000,32	0.002,2

Concentrations in Air (ng/m^3) and Yearly Deposition (kg/km^2/yr)

Medium	Air	Air	Bulk deposition	Throughfall deposition	Bulk deposition	Bulk deposition
	World, remote	World, polluted	West Germany	West Germany	Kola, remote	Kola, polluted
Median	-	-	-	-	-	-
Min	-	0.22	-	-	-	-
Max	-	6	-	-	-	-

*Estimated value

Concentrations in Plants (mg/kg)

Medium	Moss	Moss	Crustose lichen	Lichen	Dandelion	Spruce bark
	Norway conc. HNO$_3$ (1990)	Germany conc. HNO$_3$ + H$_2$O$_2$ (1995)	Germany conc. HNO$_3$	Northwest Territories oven-dried	Europe oven-dried	Canada ashed
Median	0.22	0.23	5	-	-	-
Min	0.05	0.04	1.1	-	-	-
Max	2	6.5	30	-	-	-

*Mean   #Geometric mean

Concentrations in Human Fluids (mg/l)

Medium	Human blood	Human serum	Human urine			
	Lombardy, Italy	Lombardy, Italy	Lombardy, Italy			
Mean	-	-	-			
Min	-	-	-			
Max	-	-	-			

YTTRIUM — Y

Environmental Geochemistry

Biological impacts	Considered non-essential. Toxicity considered low, but seems to have a higher acute toxicity than REE. Contradictory reports regarding carcinogenic effects
Uses	Ceramic industry (zirconia stabiliser, sintering additive, Si-nitride, sialon), alloys (Cu-Ni steel, or with Zr), high-temperature superconductors, lasers, fluorescent materials, catalyst in ethylene polymerisation, glass (with Th)
Environmental pathways	Geogenic sources thought to be more important than anthropogenic sources, REE mining and processing
Environmental mobility	Oxidising conditions: very low Acid conditions: very low Reducing conditions: very low Neutral to alkaline conditions: very low Comments: -
Geochemical barriers	Mechanical
Natural association	REE-Y-Li-Rb-Cs-Be-Nb-Ta-Zr-B-Th-U-F (pegmatites), REE-Y-Th-P-Zr-Fe-Cu (monazite veins), REE-Y-Th-Ba-Sr-P-F-N-C (carbonatites), REE-Y-U-P-F (phosphorites), REE-Y-Au-Ti-Sn-Sr-Th (placers)
Action levels	-
Remarks	Occurs generally together with the REE. Some plants (e.g., hickory (*Carya*)) accumulate Y. May replace Ca in its biological function. In 1991, 25 t Y-oxide were used in the glass industry, and 200 t in the ceramics industry
Suggested analytical method(s)	ICP-AES, ICP-MS (water)

World Yearly Production (t/y)	1150 (from 1460 Y_2O_3) (in 1995)
Price of Purest Form, in Small Quantities ($/kg)	39,650 (99.98%)
Market Price ($/kg)	ca. 25 (yttria, 99.99% Y_2O_3)

Yb YTTERBIUM

Physico-Chemical Properties

Atomic number	70	Melting/boiling point (K)	1092 / 1469
Atomic mass	173.04(3)	Isotopes & isomers	7 stable + 26 unstable
Atomic radius (pm)	247	Acid/base of oxide	weak base
Main oxidation state(s)	+3 (+2)	State (at 300 K, 1 atm.)	solid
Ionic radius (pm)	116-128 (+2), 101-118 (+3)	Metallic character	metal
		Element group(s)	heavy metal, REE
Electronegativity (Pauling)	1.1	Affinity	lithophile
Density (g/cm^3)	6.966 (beta)		

Naturally occurring isotopes	Natural abundance (%)	Atomic mass	Half-life
^{168}Yb	0.13	167.933,895	stable
^{170}Yb	3.05	169.934,76	stable
^{171}Yb	14.3	170.936,32	stable
^{172}Yb	21.9	171.936,38	stable
^{173}Yb	16.12	172.938,21	stable
^{174}Yb	31.8	173.938,86	stable
^{176}Yb	12.7	175.942,57	stable

Unstable isotopes	Longest half-life
^{151}Yb to ^{180}Yb	32.03 d

Concentrations in Crust/Rocks (mg/kg)

Bulk continental crust	2 / 3.2 / 2.2
Upper continental crust	1.5 / 2.2
Ultramafic rock	0.3
Ocean ridge basalt	4
Gabbro, basalt	3.2
Granite, granodiorite	4.8
Sandstone	1.3
Greywacke	2.1
Shale, schist	3.2
Limestone	0.5
Coal	1

Typical Minerals
monazite ((Ce,La,Nd,Th,Yb)(PO$_4$,SiO$_4$)),
bastnaesite ((Ce,La,Yb)CO$_3$(F,OH)),
cerite ((Ce,La,Yb)$_9$(Mg,Fe)Si$_7$(O,OH,F)$_{28}$),
allanite ((Ca,Ce,La,Yb)$_2$FeAl$_2$OSi$_3$O$_{11}$(OH))

Possible Host Minerals
biotite, apatite, pyroxenes, feldspars, zircon

Mass (kg) in

Continental crust	4.72x10^{+16}	Oceans	1.06x10^{+9}	Plants	3.68x10^{+7}

Concentrations in Media Sampled for the Kola Project (mg/kg)

| Medium | Moss | Humus | Humus | Topsoil (0-5 cm) | B horizon | B horizon |
| | <2 mm | <2 mm | <2 mm | <2 mm | <2 mm | <2 mm |
	conc. HNO$_3$		amm. acet.	total (INAA)	aqua regia	total (XRF)
Median	-	-	-	0.9	-	-
Min	-	-	-	<0.2	-	-
Max	-	-	-	9.3	-	-

| Medium | C horizon | C horizon | C horizon | | Lake water |
| | <2 mm | <2 mm | <2 mm | | unfiltered |
	aqua regia	total (XRF)	total (INAA)		(mg/l)
Median	-	-	1.9		-
Min	-	-	0.3		-
Max	-	-	19.9		-

YTTERBIUM   Yb

Concentrations in Soils and Sediments (mg/kg)

Medium	Soil World <2 mm total	Agricultural soil - Ap horizon Canada <2 mm	Agricultural soil - Top (0-25 cm) Finland <2 mm aqua regia	Agricultural Soil - Bottom (50-75 cm) Finland <2 mm aqua regia	Topsoil (0-15 cm) England & Wales <2 mm aqua regia	Urban soil (0-2 cm) Trondheim <2 mm aqua regia
Median	3.3*	-	-	-	-	-
Min	-	-	-	-	-	-
Max	-	-	-	-	-	-

*Estimated mean

Medium	Forest soil - Humus Norway <2 mm 7N HNO$_3$	Forest soil - B horizon Norway <2 mm 7N HNO$_3$	Forest soil - C horizon Norway <2 mm 7N HNO$_3$	Till (C horizon) Finland <0.063 mm aqua regia	Till (C horizon) Finland <0.063 mm total	Laterite (25 ±15 cm) Australia 0.45-2 mm total
Median	-	-	-	0.3	2.3	-
Min	-	-	-	-	-	-
Max	-	-	-	-	-	-

Medium	Stream sediment Austria <0.18 mm	Stream sediment Southern Scotland <0.10 mm total	Stream sediment Harz, Germany <0.063 mm total	Overbank sediment Norway <0.063 mm total (XRF)	Overbank sediment Norway <0.063 mm 7N HNO$_3$	Organic stream sediment Finland <2 mm Conc. HNO$_3$
Median	-	-	-	-	-	-
Min	-	-	-	-	-	-
Max	-	-	-	-	-	-

#2nd percentile ^98th percentile

Concentrations in Waters (mg/l)

Medium	Ocean water World (1)	Ocean water World (2)	Ocean water North Pacific	Stream water World	Stream water Nova Scotia, Canada <0.45 µm ICP-MS	Stream water Finland <0.45 µm
Median	0.000,000,82*	0.000,000,8*	0.000,001,2*	0.000,004*	0.000,014	-
Min	-	-	-	-	<0.000,005	-
Max	-	-	-	-	0.000,109	-

*Estimated mean   #2nd percentile   ^98th percentile

Medium	Stream water Romania unfiltered ICP-MS	Stream water Eastern India <0.2 µm ICP-MS	Stream water Harz, Germany unfiltered ICP-MS	Lake water Norway unfiltered ICP-MS	Ground water Southern Norway unfiltered ICP-MS
Median	-	0.000,26	0.000,06	0.000,009	0.000,014
Min	-	0.000,03	<0.000,002	<0.000,006	<0.000,001
Max	-	0.003,73	0.000,45	0.000,271	0.001,75

Yb   YTTERBIUM

Concentrations in Precipitation (mg/l)

Medium	Rain water	Rain water	Snow melt-water	Snow melt-water	Snow filter residue	Snow filter residue
	Kola, remote <0.45 µm	*Kola, polluted <0.45 µm*	*Kola, remote <0.45 µm*	*Kola, polluted <0.45 µm*	*Kola, remote >0.45 µm*	*Kola, polluted >0.45 µm*
Median	-	-	-	-	-	-
Min	-	-	-	-	-	-
Max	-	-	-	-	-	-

Medium	Rain water	Rain water
	Coastal, Norway unfiltered ICP-MS	*Inland, Norway unfiltered ICP-MS*
Median	-	-
Min	-	-
Max	-	-

Concentrations in Air (ng/m^3) and Yearly Deposition (kg/km^2/yr)

Medium	Air	Air	Bulk deposition	Throughfall deposition	Bulk deposition	Bulk deposition
	World, remote	*World, polluted*	*West Germany*	*West Germany*	*Kola, remote*	*Kola, polluted*
Median	-	-	-	-	-	-
Min	-	-	-	-	-	-
Max	-	-	-	-	-	-

*Estimated value

Concentrations in Plants (mg/kg)

Medium	Moss	Moss	Crustose lichen	Lichen	Dandelion	Spruce bark
	Norway conc. HNO$_3$ (1990)	*Germany conc. HNO$_3$ + H$_2$O$_2$ (1995)*	*Germany conc. HNO$_3$*	*Northwest Territories oven-dried*	*Europe oven-dried*	*Canada ashed total (INAA)*
Median	-	-	-	-	-	1.42
Min	-	-	-	-	-	<0.05
Max	-	-	-	-	-	4.97

*Mean   #Geometric mean

Concentrations in Human Fluids (mg/l)

Medium	Human blood	Human serum	Human urine
	Lombardy, Italy	*Lombardy, Italy*	*Lombardy, Italy*
Mean	0.000,15	-	0.000,028
Min	0.000,05	-	0.000,001,5
Max	0.000,3	-	0.000,09

YTTERBIUM   Yb

Environmental Geochemistry

Biological impacts	Considered non-essential. Generally low toxicity, but data to assess the health relevance of REE are scarce
Uses	Condensers
Environmental pathways	Geogenic dust, mining and processing of alkaline rocks. Generally, geogenic sources more important than anthropogenic ones
Environmental mobility	Oxidising conditions: very low Acid conditions: very low Reducing conditions: very low Neutral to alkaline conditions: very low Comments: -
Geochemical barriers	Mechanical
Natural association	REE-Li-Rb-Cs-Be-Nb-Ta-Zr-B-Th-U-F (pegmatites), REE-Th-P-Zr-Fe-Cu (monazite veins), REE-Th-Ba-Sr-P-F-N-C (carbonatites), REE-U-P-F (phosphorites), REE-Au-Ti-Sn-Sr-Th (placers)
Action levels	-
Remarks	Toxicity of REE decreases as atomic number increases. Inhaled REE probably cause pneumoconiosis. REE taken up orally (e.g., via drinking water) accumulate in the skeleton, teeth, lungs, liver and kidneys. Occurs generally together with the other REE. Some plants (e.g., hickory (*Carya*)) accumulate Yb. Yb is scarcely used
Suggested analytical method(s)	ICP-MS

World Yearly Production (t/y)	54,000 REE-minerals (in 1991)
Price of Purest Form, in Small Quantities ($/kg)	72,000 (99.99%)
Market Price ($/kg)	-

Zn ZINC

Physico-Chemical Properties

Atomic number	30	Melting/boiling point (K)	692.68 / 1180
Atomic mass	65.39(2)	Isotopes & isomers	5 stable + 23 unstable
Atomic radius (pm)	153	Acid/base of oxide	amphoteric
Main oxidation state(s)	+2	State (at 300 K, 1 atm.)	solid
Ionic radius (pm)	74-104	Metallic character	pred. metal
Electronegativity (Pauling)	1.65	Element group(s)	heavy metal
Density (g/cm^3)	7.133	Affinity	chalcophile

Naturally occurring isotopes	Natural abundance (%)	Atomic mass	Half-life
^{64}Zn	48.6	63.929,15	stable
^{66}Zn	27.9	65.926,04	stable
^{67}Zn	4.1	66.927,13	stable
^{68}Zn	18.8	67.924,85	stable
^{70}Zn	0.6	69.925,33	stable

Unstable isotopes	Longest half-life
^{57}Zn to ^{81}Zn	243.8 d

Concentrations in Crust/Rocks (mg/kg)

Bulk continental crust	65 / 70 / 80
Upper continental crust	52 / 71
Ultramafic rock	60
Ocean ridge basalt	70
Gabbro, basalt	100
Granite, granodiorite	50
Sandstone	20
Greywacke	76
Shale, schist	100
Limestone	40
Coal	50

Typical Minerals
sphalerite (ZnS, cubic), wurtzite (ZnS, hexagonal), smithonite (ZnSO$_4$), zincite (ZnO)

Possible Host Minerals
pyroxenes, amphiboles, micas, garnets, magnetite

Mass (kg) in

Continental crust	1.53x10^{+18}	Oceans	6.61x10^{+11}	Plants	9.20x10^{+10}

Concentrations in Media Sampled for the Kola Project (mg/kg)

Medium	Moss	Humus	Humus	Topsoil (0-5 cm)	B horizon	B horizon
		<2 mm	<2 mm	<2 mm	<2 mm	<2 mm
	conc. HNO$_3$	conc. HNO$_3$	amm. acet.	total (INAA)	aqua regia	total (XRF)
Median	32.2	46	19.3	<50	25.5	-
Min	11.7	12	3.82	<50	3.7	-
Max	81.9	198	108	223	209	-

Medium	C horizon	C horizon	C horizon		Lake water
	<2 mm	<2 mm	<2 mm		unfiltered
	aqua regia	total (XRF)	total (INAA)		(mg/l)
Median	20.9	-	66		0.000,99
Min	3.7	-	<50		0.000,07
Max	348	-	356		0.020,1

ZINC   Zn

Concentrations in Soils and Sediments (mg/kg)

Medium	Soil	Agricultural soil - Ap horizon	Agricultural soil - Top (0-25 cm)	Agricultural Soil - Bottom (50-75 cm)	Topsoil (0-15 cm)	Urban soil (0-2 cm)
	World	Canada	Finland	Finland	England & Wales	Trondheim
	<2 mm	<2 mm	<2 mm	<2 mm	<2 mm	<2 mm
	total	total (AAS)	aqua regia	aqua regia	aqua regia	aqua regia
Median	70*	72	22.3	16.3	82	99
Min	-	20	<1	<1	5	7.4
Max	-	835	121	140	3648	3420

*Estimated mean

Medium	Forest soil - Humus	Forest soil - B horizon	Forest soil - C horizon	Till (C horizon)	Till (C horizon)	Laterite (25 ±15 cm)
	Norway	Norway	Norway	Finland	Finland	Australia
	<2 mm	<2 mm	<2 mm	<0.063 mm	<0.063 mm	0.45-2 mm
	7N HNO_3	7N HNO_3	7N HNO_3	aqua regia	total	total
Median	42	24.7	40	32	57	24
Min	4.2	0.7	4.7	-	-	<2
Max	1600	237	644.5	-	-	110

Medium	Stream sediment	Stream sediment	Stream sediment	Overbank sediment	Overbank sediment	Organic stream sediment
	Austria	Southern Scotland	Harz, Germany	Norway	Norway	Finland
	<0.18 mm	<0.10 mm	<0.063 mm	<0.063 mm	<0.063 mm	<2 mm
	"total" (ICP-AES)	total (DCES)	total (XRF)	total (XRF)	7N HNO_3 ICP-AES	Conc. HNO_3
Median	80	191	209	75	44	45.9
Min	-	<12	20	1	1.7	14.2#
Max	4229	874	4554	1158	1000	165^

#2nd percentile   ^98th percentile

Concentrations in Waters (mg/l)

Medium	Ocean water	Ocean water	Ocean water	Stream water	Stream water	Stream water
	World (1)	World (2)	North Pacific	World	Nova Scotia, Canada	Finland
					<0.45 µm	<0.45 µm
					ICP-AES	ICP-MS
Median	0.004,9*	0.000,5*	0.000,35*	0.015*	<0.005	0.003,6
Min	-	-	-	-	<0.005	0.001,1#
Max	-	-	-	-	0.011	0.022,7^

*Estimated mean   #2nd percentile   ^98th percentile

Medium	Stream water	Stream water	Stream water	Lake water	Ground water
	Romania	Eastern India	Harz, Germany	Norway	Southern Norway
	unfiltered	<0.2 µm	unfiltered	unfiltered	unfiltered
	ICP-MS	ICP-MS	ICP-MS	ICP-MS	ICP-MS
Median	0.003,3	0.010,3	0.026	0.001,7	0.023,4
Min	0.000,1	0.004,5	<0.000,1	<0.000,3	<0.000,5
Max	0.324	0.073,5	0.643	0.139	1.324,5

Zn ZINC

Concentrations in Precipitation (mg/l)

Medium	Rain water	Rain water	Snow melt-water	Snow melt-water	Snow filter residue	Snow filter residue
	Kola, remote <0.45 μm ICP-MS	Kola, polluted <0.45 μm ICP-MS	Kola, remote <0.45 μm ICP-MS	Kola, polluted <0.45 μm ICP-MS	Kola, remote >0.45 μm ICP-AES	Kola, polluted >0.45 μm ICP-AES
Median	0.005,7	0.056	0.003,29	0.011,9	0.000,23	0.005,38
Min	0.002,54	0.033	0.002,61	0.007,51	0.000,02	0.003,04
Max	0.014,1	0.197	0.006,51	0.019,8	0.000,98	0.014,1

Medium	Rain water	Rain water
	Coastal, Norway unfiltered ICP-MS	Inland, Norway unfiltered ICP-MS
Median	0.001,2	0.006,6
Min	<0.000,02	0.000,08
Max	0.039	0.063

Concentrations in Air (ng/m^3) and Yearly Deposition (kg/km^2/yr)

Medium	Air	Air	Bulk deposition	Throughfall deposition	Bulk deposition	Bulk deposition
	World, remote	World, polluted	West Germany	West Germany	Kola, remote	Kola, polluted
Median	7 / 6.1	900	-	10	-	-
Min	0.03	<10,000	11	300	-	-
Max	15 (110)	16,000	190	-	-	-

*Estimated value

Concentrations in Plants (mg/kg)

Medium	Moss	Moss	Crustose lichen	Lichen	Dandelion	Spruce bark
	Norway conc. HNO$_3$ (1990)	Germany conc. HNO$_3$ + H$_2$O$_2$ (1995)	Germany conc. HNO$_3$	Northwest Territories oven-dried total (INAA)	Europe oven-dried conc. HNO$_3$ (AAS)	Canada ashed total (INAA)
Median	36	54	130	24.1*	43#	1500
Min	9	14	63	-	-	290
Max	576	245	1540	-	-	5800

*Mean   #Geometric mean

Concentrations in Human Fluids (mg/l)

Medium	Human blood	Human serum	Human urine			
	Lombardy, Italy	Lombardy, Italy	Lombardy, Italy			
Mean	6.34	0.922	0.456			
Min	3.5	0.54	0.302			
Max	8.8	1.51	1.3			

ZINC Zn

Environmental Geochemistry

Biological impacts	Essential for all organisms. Toxicity low; deficiency more important. Not thought to be carcinogenic, but growth rate of tumours is affected by dietary Zn intake
Uses	Galvanising, alloys, rubber industry, pigments, chemicals, paint, glass, plastics, lubricants, batteries, pesticide (wood preservative), pharmaceutical industry, fungicide, building industry (e.g., gutters and drains)
Environmental pathways	Zn smelters, combustion, traffic (e.g., wear of tyres), waste water, sewage sludge, geogenic dust
Environmental mobility	Oxidising conditions: high　　　　　　Acid conditions: high Reducing conditions: very low　　　　Neutral to alkaline conditions: very low Comments: -
Geochemical barriers	pH, adsorption (clays, Fe-Mn oxides, organic matter), co-precipitation (with Fe, Mn)
Natural association	Zn-Cd (essentially ubiquitous), Zn-Cd-Pb-Ba-F (Mississippi Valley type deposits), Zn-Pb-Mn-Ba-Fe (stratiform volcanogenic deposits), Zn-Pb-Fe-Cu-Ag-Ba-Te-etc. (veins and massive sulphide deposits), Zn-Pb-Cu-Ag-B-Mo-W-Be (skarns), Mn-Ni-Cu-Co-Zn (deep sea nodules), Cu-Mo-Re-Fe-Au-Ag-Zn (some porphyry Cu deposits), Cu-Pb-Zn (Cu shales)
Action levels	Drinking water, recommended value: 0.3 mg/l (Norway Shd); MAC: 5 mg/l (Canada CWQG); recommended: 5 mg/l (WHO); MAC: 1 mg/l (Russia MoH). Ground water, background: 0.065 mg/l; remediate: 0.8 mg/l (Netherlands VROM). Soil, background: 140 mg/kg; remediate: 720 mg/kg (Netherlands VROM). Agricultural soil, maximum tolerable concentration: 300 mg/kg (Germany)
Remarks	The main Zn ore mineral, sphalerite, generally contains many trace elements such as Cd, Se, Mn, Ag, Cu, Ga, Hg, In, Co, Ge, Sn, Ni, As, Tl, Sb, Bi. Zn-compounds are used in pig and chicken farming as food additives. Zn toxicity has been observed in cattle feeding on very Zn-rich soils. May lead to growth depression in plants at levels >300 mg/kg in soil. May be found at elevated concentration in some P fertilisers. Used as fertiliser (e.g., $ZnSO_4 \cdot 7H_2O$, ZnO, Zn-EDTA). Preferentially taken up by carrot, salad and spinach, low uptake in potato. Concentrations of 3 mg/l in drinking water may give rise to consumer complaints due to appearance and taste
Suggested analytical method(s)	ICP-AES, XRF

World Yearly Production (t/y)	6,791,000 (in 1995)
Price of Purest Form, in Small Quantities ($/kg)	4409 (99.999,9%)
Market Price ($/kg)	1

Zr ZIRCONIUM

Physico-Chemical Properties

Atomic number	40	Melting/boiling point (K)	2128 / 4682
Atomic mass	91.224(2)	Isotopes & isomers	5 stable + 26 unstable
Atomic radius (pm)	216	Acid/base of oxide	amphoteric
Main oxidation state(s)	+4 (+2,+3)	State (at 300 K, 1 atm.)	solid
Ionic radius (pm)	73-103 (+4)	Metallic character	pred. metal
Electronegativity (Pauling)	1.33	Element group(s)	heavy metal
Density (g/cm^3)	6.506	Affinity	lithophile

Naturally occurring isotopes	Natural abundance (%)	Atomic mass	Half-life
^{90}Zr	51.45	89.904,7	stable
^{91}Zr	11.22	90.905,64	stable
^{92}Zr	17.15	91.905,04	stable
^{94}Zr	17.38	93.906,31	stable
^{96}Zr	2.8	95.908,275	stable

Unstable isotopes	Longest half-life
^{80}Zr to ^{105}Zr	1,500,000 y

Concentrations in Crust/Rocks (mg/kg)

Bulk continental crust	203 / 165 / 100
Upper continental crust	237 / 190
Ultramafic rock	30
Ocean ridge basalt	100
Gabbro, basalt	120
Granite, granodiorite	200
Sandstone	250
Greywacke	302
Shale, schist	160
Limestone	20
Coal	20

Typical Minerals
zircon (ZrSiO$_4$), baddeleyite (ZrO$_2$)

Possible Host Minerals
pyroxenes, amphiboles, micas, garnets, ilmenite, rutile

Mass (kg) in

Continental crust	4.79x10^{+18}	Oceans	3.90x10^{+10}	Plants	1.84x10^{+8}

Concentrations in Media Sampled for the Kola Project (mg/kg)

Medium	Moss	Humus	Humus	Topsoil (0-5 cm)	B horizon	B horizon
	<2 mm	<2 mm	<2 mm	<2 mm	<2 mm	<2 mm
	conc. HNO$_3$		amm. acet.		aqua regia	total (XRF)
Median	-	-	-	-	-	-
Min	-	-	-	-	-	-
Max	-	-	-	-	-	-

Medium	C horizon	C horizon	C horizon			Lake water
	<2 mm	<2 mm	<2 mm			unfiltered
	aqua regia	total (XRF)	total (INAA)			(mg/l)
Median	-	-	-			-
Min	-	-	-			-
Max	-	-	-			-

ZIRCONIUM — Zr

Concentrations in Soils and Sediments (mg/kg)

Medium	Soil	Agricultural soil - Ap horizon	Agricultural soil - Top (0-25 cm)	Agricultural Soil - Bottom (50-75 cm)	Topsoil (0-15 cm)	Urban soil (0-2 cm)
	World	Canada	Finland	Finland	England & Wales	Trondheim
	<2 mm total	<2 mm	<2 mm aqua regia	<2 mm aqua regia	<2 mm aqua regia	<2 mm aqua regia
Median	230*	-	-	-	-	-
Min	-	-	-	-	-	-
Max	-	-	-	-	-	-

*Estimated mean

Medium	Forest soil - Humus	Forest soil - B horizon	Forest soil - C horizon	Till (C horizon)	Till (C horizon)	Laterite (25 ±15 cm)
	Norway	Norway	Norway	Finland	Finland	Australia
	<2 mm 7N HNO$_3$	<2 mm 7N HNO$_3$	<2 mm 7N HNO$_3$	<0.063 mm aqua regia	<0.063 mm total	0.45-2 mm total
Median	0.8	7.9	12.3	1.2	200	70
Min	<0.2	0.3	0.6	-	-	7
Max	9.4	43.2	89.7	-	-	142

Medium	Stream sediment	Stream sediment	Stream sediment	Overbank sediment	Overbank sediment	Organic stream sediment
	Austria	Southern Scotland	Harz, Germany	Norway	Norway	Finland
	<0.18 mm total (XRF)	<0.10 mm total (DCES)	<0.063 mm total (XRF)	<0.063 mm total (XRF)	<0.063 mm 7N HNO$_3$	<2 mm Conc. HNO$_3$
Median	256	680	818	612	-	-
Min	-	<30	120	47	-	-
Max	17,307	149	2434	2636	-	-

#2nd percentile   ^98th percentile

Concentrations in Waters (mg/l)

Medium	Ocean water	Ocean water	Ocean water	Stream water	Stream water	Stream water
	World (1)	World (2)	North Pacific	World	Nova Scotia, Canada	Finland
					<0.45 µm	<0.45 µm
Median	0.000,03*	0.000,03*	0.000,015*	0.001*	-	-
Min	-	-	-	-	-	-
Max	-	-	-	-	-	-

*Estimated mean   #2nd percentile   ^98th percentile

Medium	Stream water	Stream water	Stream water	Lake water	Ground water
	Romania	Eastern India	Harz, Germany	Norway	Southern Norway
	unfiltered ICP-MS	<0.2 µm ICP-MS	unfiltered ICP-MS	unfiltered ICP-MS	unfiltered ICP-MS
Median	0.000,089	0.000,89	0.000,2	<0.000,015	0.000,047
Min	0.000,002	0.000,5	<0.000,01	<0.000,015	<0.000,002
Max	0.003	0.022,09	0.005,4	0.000,464	-

Zr ZIRCONIUM

Concentrations in Precipitation (mg/l)

Medium	Rain water	Rain water	Snow melt-water	Snow melt-water	Snow filter residue	Snow filter residue
	Kola, remote <0.45 µm	Kola, polluted <0.45 µm	Kola, remote <0.45 µm	Kola, polluted <0.45 µm	Kola, remote >0.45 µm	Kola, polluted >0.45 µm
Median	-	-	-	-	-	-
Min	-	-	-	-	-	-
Max	-	-	-	-	-	-

Medium	Rain water	Rain water				
	Coastal, Norway unfiltered ICP-MS	Inland, Norway unfiltered ICP-MS				
Median	-	-				
Min	-	-				
Max	-	-				

Concentrations in Air (ng/m^3) and Yearly Deposition (kg/km^2/yr)

Medium	Air	Air	Bulk deposition	Throughfall deposition	Bulk deposition	Bulk deposition
	World, remote	World, polluted	West Germany	West Germany	Kola, remote	Kola, polluted
Median	-	3	-	-	-	-
Min	-	0.7	-	-	-	-
Max	-	26	-	-	-	-

*Estimated value

Concentrations in Plants (mg/kg)

Medium	Moss	Moss	Crustose lichen	Lichen	Dandelion	Spruce bark
	Norway conc. HNO$_3$ (1990)	Germany conc. HNO$_3$ + H$_2$O$_2$ (1995)	Germany conc. HNO$_3$	Northwest Territories oven-dried	Europe oven-dried	Canada ashed
Median	-	0.478	-	-	-	-
Min	-	0.11	-	-	-	-
Max	-	19	-	-	-	-

*Mean   #Geometric mean

Concentrations in Human Fluids (mg/l)

Medium	Human blood	Human serum	Human urine			
	Lombardy, Italy	Lombardy, Italy	Lombardy, Italy			
Mean	-	-	<0.002			
Min	-	-	-			
Max	-	-	-			

ZIRCONIUM   Zr

Environmental Geochemistry

Biological impacts	Considered non-essential. Scarce data on toxicity to humans, but not considered very toxic. Carcinogenic effects not known
Uses	Glass and ceramic industry, abrasives, nuclear industry, textile impregnation, alloys, tanning
Environmental pathways	Geogenic sources probably more important than anthropogenic ones
Environmental mobility	Oxidising conditions: very low Acid conditions: very low Reducing conditions: very low Neutral to alkaline conditions: very low Comments: -
Geochemical barriers	Mechanical
Natural association	REE-Y-Li-Rb-Cs-Be-Nb-Ta-Zr-B-Th-U-F (pegmatites), REE-Y-Th-P-Zr-Fe-Cu (monazite veins), U-Nb-Th-Cu-F-P-Ti-Zr (carbonatite hosted U deposits), U-Th-Ti-Au-Zr-REE (placer U deposits), Nb-Ta-Sn-W-Li-Be-Ti-Rb-Cs-U-Th-B-Zr-Hf-P-F-REE (granites and syenitic pegmatites), Nb-Ta-Na-K-Ba-Sr-Ti-Zr-U-Th-Cu-Zn-P-S-F-REE (carbonatites)
Action levels	-
Remarks	U and Th content of Zr sands can lead to environmental concerns. Quality of Zr sand for industrial applications is determined by low Fe and Ti content. Traded grade of Zr sands: ca. 65% (ZrO_2 + HfO_2). Zircon is a typical placer mineral
Suggested analytical method(s)	XRF, ICP-AES, ICP-MS (water)

World Yearly Production (t/y)	950,000 Zr sand (in 1995)
Price of Purest Form, in Small Quantities ($/kg)	104,750 (99.99+%)
Market Price ($/kg)	0.5 (Zr-sand with 67% ZrO_2)

APPENDIX

Table A1. Conversion between element and oxide masses

From	To	(1) Multiply by	From	To	(2) Multiply by
Ac	Ac_2O_3	1.1057	Ac_2O_3	Ac	0.9044
Ag	Ag_2O	1.0742	Ag_2O	Ag	0.9310
Al	Al_2O_3	1.8895	Al_2O_3	Al	0.5293
Ar	-	-	-	-	-
As	As_2O_3	1.3203	As_2O_3	As	0.7574
At	AtO	1.0762	AtO	At	0.9292
Au	Au_2O_3	1.1218	Au_2O_3	Au	0.8914
B	B_2O_3	3.2197	B_2O_3	B	0.3106
Ba	BaO	1.1165	BaO	Ba	0.8957
Be	BeO	2.7753	BeO	Be	0.3603
Bi	Bi_2O_3	1.1148	Bi_2O_3	Bi	0.8970
Br	BrO_3	1.6007	BrO_3	Br	0.6247
C	CO	2.3321	CO	C	0.4288
C	CO_2	3.6642	CO_2	C	0.2729
C	CO_3	4.9963	CO_3	C	0.2001
Ca	CaO	1.3992	CaO	Ca	0.7147
Cd	CdO	1.1423	CdO	Cd	0.8754
Ce	Ce_2O_3	1.1713	Ce_2O_3	Ce	0.8538
Cl	ClO_2	1.9026	ClO_2	Cl	0.5256
Co	CoO	1.2715	CoO	Co	0.7865
Co	Co_3O_4	1.3620	Co_3O_4	Co	0.7342
Cr	Cr_2O_3	1.4616	Cr_2O_3	Cr	0.6842
Cs	Cs_2O	1.0602	Cs_2O	Cs	0.9432
Cu	CuO	1.2518	CuO	Cu	0.7989
Dy	Dy_2O_3	1.1477	Dy_2O_3	Dy	0.8713
Er	Er_2O_3	1.1435	Er_2O_3	Er	0.8745
Eu	Eu_2O_3	1.1579	Eu_2O_3	Eu	0.8636
F	F_2O	1.4211	F_2O	F	0.7037
Fe	FeO	1.2865	FeO	Fe	0.7773
Fe	Fe_2O_3	1.4297	Fe_2O_3	Fe	0.6994
Fe	Fe_3O_4	1.3820	Fe_3O_4	Fe	0.7236
Fr	Fr_2O	1.0359	Fr_2O	Fr	0.9654
Ga	Ga_2O_3	1.3442	Ga_2O_3	Ga	0.7439
Gd	Gd_2O_3	1.1526	Gd_2O_3	Gd	0.8676
Ge	GeO_2	1.4407	GeO_2	Ge	0.6941

Table A1. Conversion between element and oxide masses (continued)

From	To	(1) Multiply by	From	To	(2) Multiply by
H	H_2O	8.9366	H_2O	H	0.1119
He	-	-	-	-	-
Hf	HfO_2	1.1793	HfO_2	Hf	0.8480
Hg	HgO	1.0798	HgO	Hg	0.9261
Hg	Hg_2O	1.0399	Hg_2O	Hg	0.9616
Ho	Ho_2O_3	1.1455	Ho_2O_3	Ho	0.8730
I	IO_3	1.3782	IO_3	I	0.7256
In	In_2O_3	1.2090	In_2O_3	In	0.8271
Ir	IrO_2	1.1665	IrO_2	Ir	0.8573
K	K_2O	1.2046	K_2O	K	0.8301
Kr	-	-	-	-	-
La	La_2O_3	1.1728	La_2O_3	La	0.8527
Li	Li_2O	2.1525	Li_2O	Li	0.4646
Lu	Lu_2O_3	1.1372	Lu_2O_3	Lu	0.8794
Mg	MgO	1.6583	MgO	Mg	0.6030
Mn	MnO	1.2912	MnO	Mn	0.7745
Mn	MnO_2	1.5825	MnO_2	Mn	0.6319
Mn	Mn_2O_3	1.4368	Mn_2O_3	Mn	0.6960
Mn	Mn_3O_4	1.3883	Mn_3O_4	Mn	0.7203
Mo	MoO_3	1.5003	MoO_3	Mo	0.6665
Mo	MoO_4	1.6671	MoO_4	Mo	0.5999
N	NO_2	3.2845	NO_2	N	0.3045
Na	Na_2O	1.3480	Na_2O	Na	0.7419
Nb	Nb_2O_5	1.4305	Nb_2O_5	Nb	0.6990
Nd	Nd_2O_3	1.1664	Nd_2O_3	Nd	0.8574
Ne	-	-	-	-	-
Ni	NiO	1.2726	NiO	Ni	0.7858
Np	Np_2O_5	1.1688	Np_2O_5	Np	0.8556
O	-	-	-	-	-
Os	OsO_2	1.1682	OsO_2	Os	0.8560
P	PO_4	3.0662	PO_4	P	0.3261
P	P_2O_5	2.2914	P_2O_5	P	0.4364
Pa	Pa_2O_5	1.1731	Pa_2O_5	Pa	0.8524

Table A1. Conversion between element and oxide masses (continued)

From	To	(1) Multiply by	From	To	(2) Multiply by
Pb	PbO	1.0772	PbO	Pb	0.9283
Pb	PbO_2	1.1544	PbO_2	Pb	0.8662
Pd	PdO	1.1503	PdO	Pd	0.8693
Po	PoO_2	1.1531	PoO_2	Po	0.8672
Pr	Pr_2O_3	1.1703	Pr_2O_3	Pr	0.8545
Pt	PtO	1.0820	PtO	Pt	0.9242
Pt	PtO_2	1.1640	PtO_2	Pt	0.8591
Pu	PuO_2	1.1311	PuO_2	Pu	0.8841
Ra	RaO	1.0708	RaO	Ra	0.9339
Rb	Rb_2O	1.0936	Rb_2O	Rb	0.9144
Re	Re_2O_7	1.3007	Re_2O_7	Re	0.7688
Rh	Rh_2O_3	1.2332	Rh_2O_3	Rh	0.8109
Rn	-	-	-	-	-
Ru	RuO_2	1.3166	RuO_2	Ru	0.7595
S	SO_2	1.9979	SO_2	S	0.5005
S	SO_3	2.4968	SO_3	S	0.4005
S	SO_4	2.9958	SO_4	S	0.3338
Sb	Sb_2O_3	1.1971	Sb_2O_3	Sb	0.8354
Sb	Sb_2O_5	1.3285	Sb_2O_5	Sb	0.7527
Sc	Sc_2O_3	1.5338	Sc_2O_3	Sc	0.6520
Se	SeO_2	1.4052	SeO_2	Se	0.7116
Se	SeO_3	1.6079	SeO_3	Se	0.6219
Si	SiO_2	2.1393	SiO_2	Si	0.4674
Sm	Sm_2O_3	1.1596	Sm_2O_3	Sm	0.8624
Sn	SnO	1.1348	SnO	Sn	0.8812
Sn	SnO_2	1.2696	SnO_2	Sn	0.7877
Sr	SrO	1.1826	SrO	Sr	0.8456
Ta	Ta_2O_5	1.2211	Ta_2O_5	Ta	0.8190
Tb	Tb_2O_3	1.1510	Tb_2O_3	Tb	0.8688
Te	TeO_2	1.2508	TeO_2	Te	0.7995
Te	TeO_3	1.3762	TeO_3	Te	0.7267
Th	ThO_2	1.1379	ThO_2	Th	0.8788
Ti	TiO_2	1.6685	TiO_2	Ti	0.5993
Tl	Tl_2O_3	1.1174	Tl_2O_3	Tl	0.8949
Tm	Tm_2O_3	1.1421	Tm_2O_3	Tm	0.8756

Table A1. Conversion between element and oxide masses (continued)

From	To	(1) Multiply by	From	To	(2) Multiply by
U	UO_2	1.1344	UO_2	U	0.8815
U	UO_3	1.2016	UO_3	U	0.8322
U	U_2O_7	1.2353	U_2O_7	U	0.8095
U	U_3O_8	1.1792	U_3O_8	U	0.8480
V	VO_3	1.9422	VO_3	V	0.5149
V	V_2O_3	1.4711	V_2O_3	V	0.6798
V	V_2O_5	1.7852	V_2O_5	V	0.5602
W	WO_3	1.2611	WO_3	W	0.7930
Xe	-	-	-	-	-
Y	Y_2O_3	1.2699	Y_2O_3	Y	0.7874
Yb	Yb_2O_3	1.1387	Yb_2O_3	Yb	0.8782
Zn	ZnO	1.2447	ZnO	Zn	0.8034
Zr	ZrO_2	1.3508	ZrO_2	Zr	0.7403

Formulas:

First multiplication factor = $(MW)_{oxide} / \{(AW)_{element} \times (stoichiometry)_{element\ in\ oxide}\}$

Second multiplication factor = 1 / (First multiplication factor)

AW: atomic weight; MW: molecular weight

Table A1 is based on the most recent IUPAC atomic weight figures (IUPAC, 1996)

Table A2. Conversion between mg, mmol and meq

Species	To obtain *mmol* multiply *mg* by	Assumed charge (absolute value)	To obtain *meq* multiply *mg* by
Ac	0.0044053	-	-
Ag	0.0092706	1	0.0092706
Ag	0.0092706	2	0.0185411
Al	0.0370624	3	0.1111871
Ar	0.0250325	-	-
As	0.0133473	3	0.0400419
At	0.0047619	-	-
Au	0.005077	1	0.005077
Au	0.005077	3	0.015231
B	0.0924924	3	0.2774772
Ba	0.0072819	2	0.0145637
Be	0.1109609	2	0.2219218
Be	0.1109609	3	0.3328827
Bi	0.0047851	3	0.0143554
Br	0.012515	1	0.012515
C	0.0832585	4	0.3330342
CO_3^{2-}	0.0166641	2	0.0333283
HCO_3^-	0.0163889	1	0.0163889
Ca	0.0249511	2	0.0499022
Cd	0.0088959	2	0.0177917
Ce	0.0071369	3	0.0214108
Cl	0.0282065	1	0.0282065
Co	0.0169684	2	0.0339367
Cr	0.0192322	2	0.0384644
Cr	0.0192322	3	0.0576966
Cr	0.0192322	6	0.1153931
Cs	0.0075241	1	0.0075241
Cu	0.0157366	2	0.0314731
Dy	0.0061537	2	0.0123075
Dy	0.0061537	3	0.0184612
Dy	0.0061537	4	0.0246149
Er	0.0059786	3	0.0179358
Eu	0.0065805	3	0.0197415

Table A2. Conversion between mg, mmol and meq (continued)

Species	To obtain *mmol* multiply *mg* by	Assumed charge (absolute value)	To obtain *meq* multiply *mg* by
F	0.052636	1	0.052636
Fe	0.0179066	2	0.0358133
Fe	0.0179066	3	0.0537199
Fr	0.0044843	-	-
Ga	0.0143424	3	0.0430273
Gd	0.0063592	3	0.0190775
Ge	0.0137718	4	0.0550873
H	0.9921157	1	0.9921157
He	0.2498375	-	-
Hf	0.0056025	4	0.02241
Hg	0.0049852	2	0.0099705
Ho	0.0060632	3	0.0181895
I	0.0078799	1	0.0078799
In	0.0087094	3	0.0261282
Ir	0.0052024	4	0.0208098
K	0.0255766	1	0.0255766
Kr	0.011933	-	-
La	0.0071991	3	0.0215974
Li	0.1440673	1	0.1440673
Lu	0.0057154	3	0.0171461
Mg	0.0411437	2	0.0822874
Mn	0.0182023	2	0.0364046
Mo	0.0104231	2	0.0208461
Mo	0.0104231	6	0.0625384
N	0.0713942	3	0.2141825
NO_2^-	0.0217365	1	0.0217365
NO_3^-	0.0161277	1	0.0161277
NH_4^+	0.0554369	1	0.0554369
Na	0.0434976	1	0.0434976
Nb	0.0107635	5	0.0538176
Nd	0.0069327	2	0.0138655
Nd	0.0069327	3	0.0207982
Nd	0.0069327	5	0.0346637
Ne	0.0495546	-	-

Table A2. Conversion between mg, mmol and meq (continued)

Species	To obtain *mmol* multiply *mg* by	Assumed charge (absolute value)	To obtain *meq* multiply *mg* by
Ni	0.0170377	1	0.0170377
Ni	0.0170377	2	0.0340754
Np	0.0042194	-	-
O	0.0625022	2	0.1250045
OH^-	0.058798	1	0.058798
Os	0.0052567	3	0.0157701
Os	0.0052567	4	0.0210268
Os	0.0052567	6	0.0315403
Os	0.0052567	8	0.0420537
P	0.0322854	5	0.161427
PO_4^{3-}	0.0105295	3	0.0315884
HPO_4^{2-}	0.0104189	2	0.0208378
$H_2PO_4^-$	0.0103106	1	0.0103106
Pa	0.0043283	5	0.0216417
Pb	0.004826	2	0.009652
Pd	0.0093966	2	0.0187933
Po	0.0047847	-	-
Pr	0.0070968	3	0.0212905
Pt	0.0051261	4	0.0205046
Pu	0.0040984	4	0.0163934
Ra	0.0044248	2	0.0088496
Rb	0.0117003	1	0.0117003
Re	0.0053704	7	0.0375926
Rh	0.0097177	3	0.029153
Rn	0.0045045	-	-
Ru	0.0098939	3	0.0296818
Ru	0.0098939	4	0.0395757
S	0.0311851	2	0.0623702
S	0.0311851	6	0.1871106
SO_4^{2-}	0.0104097	2	0.0208194
HS^-	0.0302347	1	0.0302347
Sb	0.0082129	3	0.0246386
Sc	0.022244	3	0.066732
Se	0.0126642	2	0.0253283
Se	0.0126642	4	0.0506566

Table A2. Conversion between mg, mmol and meq (continued)

Species	To obtain *mmol* multiply *mg* by	Assumed charge (absolute value)	To obtain *meq* multiply *mg* by
Si	0.0356055	4	0.1424221
SiO_2	0.0166433	-	-
H_4SiO_4	0.0104042	-	-
Sm	0.0066506	3	0.0199517
Sn	0.0084238	4	0.0336954
Sr	0.0114128	2	0.0228256
Ta	0.0055265	5	0.0276323
Tb	0.0062923	3	0.0188768
Te	0.0078368	4	0.0313472
Th	0.0043096	4	0.0172385
Ti	0.0208912	4	0.0835647
Tl	0.0048928	1	0.0048928
Tm	0.0059195	3	0.0177584
U	0.0042012	6	0.025207
V	0.0196304	5	0.0981518
W	0.0054395	6	0.0326369
Xe	0.0076166	-	-
Y	0.0112479	3	0.0337436
Yb	0.0057789	3	0.0173367
Zn	0.0152924	2	0.0305848
Zr	0.010962	4	0.043848

Formulas:
mmol = mg / MW
meq = mmol × charge = (mg × charge) / MW
MW: molecular weight

Table A2 is based on the most recent IUPAC atomic weight figures (IUPAC, 1996)

Table A3. Conversion between selected units

Weights

1 t = 1000 kg = 1,000,000 g
1 kg = 1000 g = 1,000,000 mg
1 mg = 10^{-3} g
1 μg = 10^{-6} g
1 ng = 10^{-9} g

1 t = 1000 kg = 0,98421 long tons = 1.1023 short tons
1 kg = 2.2046 avoirdupois pounds
1 g = 15.432 grain = 0.03251 troy ounce = 0.035274 avoirdupois pounds
1 mg = 0.005 carat

1 long ton = 1016 kg = 1.016 t (metric ton)
1 short ton = 907.18 kg = 0.98718 t (metric ton)
1 avoirdupois pound = 0.45359 kg
1 troy pound = 0.37324 kg
1 troy ounce = 31.103 g
1 carat = 0.002 g = 200 mg
1 grain = 0.064799 g = 64.799 mg

Areas

1 km^2 = 0.38608 square miles
1 m^2 = 1.1960 square yards = 10.7639 square feet
1 cm^2 = 0.15500 square inch

1 square mile = 640 acres = 2.5900 km^2
1 square yard = 9 square feet = 0.83613 m^2
1 square foot = 0.092903 m^2 = 929.03 cm^2
1 square inch = 6.4516 cm^2

1 ha = 10,000 m^2 = 2.471 acres
1 are = 100 m^2 = 119.6 square yards = 1/100 ha
1 acre = 43,560 square feet = 1/640 square mile = 0.4047 ha = 4047 m^2

Table A3. Conversion between selected units (continued)

Volumes

$1\ m^3 = 1000\ dm^3 = 1{,}000{,}000\ cm^3$
$1\ dm^3 = 1\ l = 1000\ cm^3 = 0.001\ m^3$
$1\ cm^3 = 0.000{,}001\ m^3$

1 cubic inch = 16.387 cm^3 = 0.016387 l
1 cubic foot = 0.028317 m^3 = 1728 cubic inches
1 cubic yard = 0.76455 m^3 = 27 cubic feet
1 cubic mile = 4.61818 km^3

1 l = 1.0567 quarts = 0.24617 gallon
1 quart = 0.94635 l
1 pint = 0.4732 l
1 U.S. gallon = 4 quarts = 231 cubic inches = 3.7854 l
1 British Imperial gallon = 277.4 cubic inches = 4.546 l
1 barrel = 31.5 gallons = 119.24 l
1 fluid ounce = 29.573 ml

Deposition

1 kg/m^2/year = 10,000 kg/ha/year
1 kg/km^2/year = 1 mg/m^2/year
1 kg/km^2/year = 0.01 kg/ha/year

Table A4. Conversion between common concentration units

wt.%	mg/kg (ppm, g/t)	µg/kg (ppb)	ng/kg (ppt)
10	100,000	100,000,000	100,000,000,000
1	10,000	10,000,000	10,000,000,000
0.1	1,000	1,000,000	1,000,000,000
0.01	100	100,000	100,000,000
0.001	10	10,000	10,000,000
0.000,1	1	1,000	1,000,000
0.000,01	0.1	100	100,000
0.000,001	0.01	10	10,000
0.000,000,1	0.001	1	1,000
0.000,000,01	0.000,1	0.1	100
0.000,000,001	0.000,01	0.01	10
0.000,000,000,1	0.000,001	0.001	1

pure water of density 1 g cm^{-3}: 1 mg/kg = 1 mg/l
(density of ocean water = 1.024763 g cm^{-3} at 293 K and normal atmospheric pressure)